Vegetation Cover and Environment of the "Mammoth Epoch" in Siberia

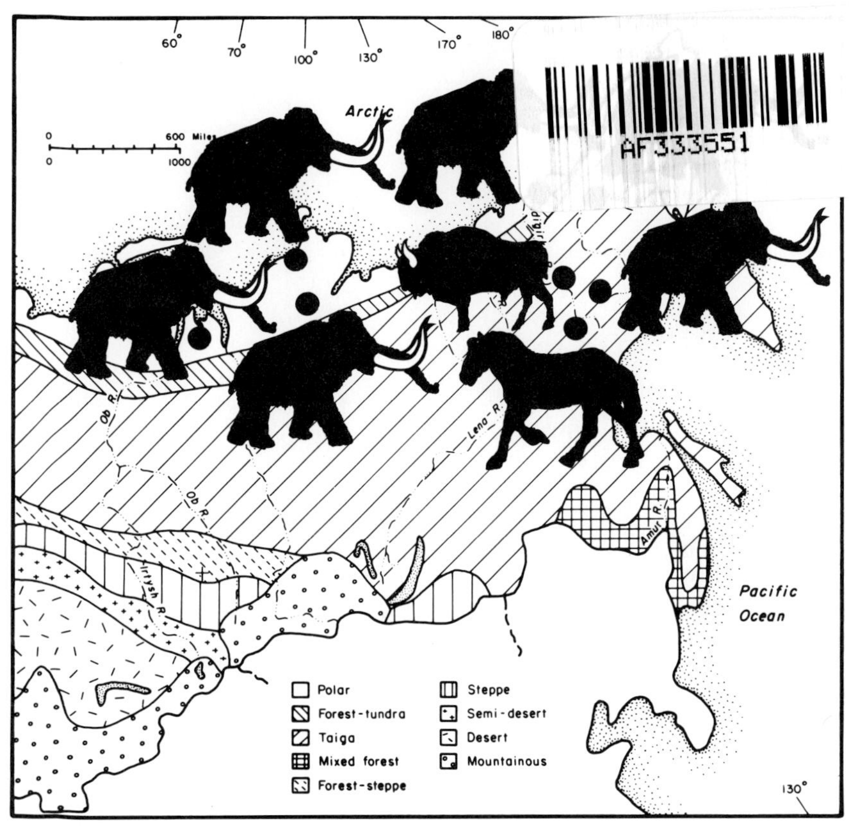

by
Valentina V. Ukraintseva

edited by
Larry D. Agenbroad, Jim I. Mead, Richard H. Hevly

VEGETATION COVER AND ENVIRONMENT OF THE "MAMMOTH EPOCH" IN SIBERIA produced by The Mammoth Site of Hot Springs, South Dakota, Inc., 1800 Highway 18-Truck Route, Hot Springs, South Dakota 57747-0606. (605) 745-6017 or FAX (605) 745-3038.

Written by Valentina V. Ukraintseva and edited by Larry D. Agenbroad, Jim I. Mead, and Richard H. Hevly.

Printed in Rapid City, South Dakota by Fenske Printing Inc.

ISBN 0-9624750-3-3

Paleoenvironments, Mammoths, and Siberia: An Introduction

by

Jim I. Mead, Larry D. Agenbroad, Richard H. Hevly

When one thinks of the Pleistocene Epoch, the Ice Age, (Anthropogene of Russia) it is easy to conjure up thoughts about glaciers covering vast stretches of land (such as the majority of Canada or Scandinavia), or extremely cold and stark landscapes, or maybe strange extinct, giant mammals. Probably the universal symbol representing the most recent ice age (the Wisconsin Glacial in North America and the Zyryanka or Valdai Glacial in Siberia; approximately 80,000 to 11,000 years ago) is the mammoth (an extinct elephant in the genus *Mammuthus*.) Mammoths roamed most, if not all, of North America, northern Asia, and Europe – the Holarctic. Readers, professional and avocational alike, seem to continually seek out information about this gentle beast, now long extinct. Probably nothing is more interesting and awe-inspiring than the realization that there actually exist remains of the mammoth carcasses with hide, hair, muscle, and internal organs, "forever" encased in the permanently frozen grounds of the high Arctic of Siberia and Alaska.

Much has been written about the mammoth and the various habitats in which it lived – all intriguing bits of information. Books have been published that discuss the reconstructed Pleistocene environments of the Arctic region of North America, the haunts of the woolly mammoth, and others. *Late Pleistocene Environments of the United States, The Pleistocene* by H.E. Wright (1983, editor) provides an important overview of the reconstructed late Pleistocene environments of North America. Similarly, A.A. Velichko's (1984, editor) book *Late Quaternary Environments of the Soviet Union* is an invaluable reference to the overviews concerning northern Eurasia. *Paleoecology of Beringia* by Hopkins, Matthews, Schweger and Young (1982) provides a thorough review with data for the reconstructed environments of Beringia, that area that is now submerged underneath the Bering and Chukchi seas that once connected Siberia with Alaska – the Bering Land Bridge. We even have a description of the 'Mammoth Steppe'– a biotic zone that no longer exists – a paradigm by Guthrie (1982, 1990). Overviews of archaeology (West, 1981) and various reasons for extinctions (numerous articles in Martin and Klein, 1984) are easily found. Details about mammoths and their relatives can be discovered in Haynes' (1991) book *Mammoths, Mastodons, and Elephants,* an especially important resource about their biology and behavior.

A great deal is known about environments, climate, floras, and faunas of the North American side of the Holarctic. However, not all authors agree with the various results and conclusions; much is still to be discovered and understood. What is missing in the quest to understand the extinct mammoth and its (possibly extinct) habitat, is the wealth of knowledge from northern Asia-Siberia. This data is well published, in the Russian language — material often "lost" to the English readers of the world. Conclusions and summations abound in the literature and are revered by the English-language audience (Markova, published in English in 1984; Tolmachev, 1970, published in English in 1982). What is needed, even required, is the detailed data from the frozen carcasses (of mammoths, bison, and horses) from Siberia. Bits and pieces of these unique data have been made available in English (Ukraintseva, 1981, 1985). Never has there occurred, in a single volume, written in English, a discussion of all the complete data and the various conclusions regarding the frozen fauna and their associated plant communities.

It is precisely this English version that is produced here — Ukraintseva's entire data set about "vegetation cover and environment of the mammoth epoch." We have edited the translation of Ukraintseva's original Russian manuscript, but we felt that it was important to "leave the flavor" of the English as produced by the author. We feel that this English version of her data is important and useful, and will be of tremendous interest to the professional, student, and layperson. A lot of this information has only been viewed in detail in Russia. Most of the overall editing has been provided by Agenbroad, with Mead working on the copy-editing and biology, and Hevly editing for botany. Below we provide a short overview of the physiographic setting of Siberia, and the most recent ice age of northern Asia — a backdrop for which to place Ukraintseva's work. We have also provided an abbreviated index to make it easier to locate selected subjects and taxa.

Siberia

Siberia is part of Russia, the former USSR. It covers a vast region of north central to northeastern Asia, and therefore, encompasses varied environments and physiographic zones. Siberia is said to be bound on the west by the Ural Mountains, to the southwest by Kazakhstan, to the south by the mountains of the Altai and Transbaikalian that abutt China and Mongolia, and the Arctic and north Pacific oceans to the north and east.

Figure I-1 provides a generalized map of greater Siberia locating major elevational regions, including plains, plateaus, rivers, and mountains. It is easy to see how the low West Siberian Plain and the Central Siberian Plateau act as sinks for the cold Arctic air masses that become entrapped north of the barrier mountains (Tien Shan, Altai, Sayan, Transbaikalian and Stanovoy) to the south.

Figure I-1. Map of Siberia locating major physiographic provinces and elevations.

Quaternary - age (=Anthropogene of Ukraintseva) sediments are often located in the lowest and typically coldest regions.

It is important to understand the present situation of Siberia's environments and climates to fully appreciate the differences during and since the most recent glacial age. Siberia is easily divided climatically into halves, east and west. The coldest January temperatures are typically found in northeastern Siberia. More importantly to the preservation of the frozen fauna is the degree of warmth achieved during the summer months. Again, much of the north and northeast, areas often with Quaternary deposits, show mean July isotherms of less than 16°C (60° F). Western Siberia has a continental climate, influenced predominantly by air masses from the Arctic. In summer the Arctic air interacts with the warm continental air to create clouds and rain. From November to March, temperatures usually remain below -30°C (-22°F). The summer lasts for three months, winters are severe. Eastern Siberia has the most continental climate of Russia. Throughout the year, cold Arctic air is dominant, although, summer temperatures can be warm. The coldest place on earth outside of part of Antarctica is the region of Verkhoyansk, the center of eastern Siberia (January temperatures reach -68°C [-98°F]) (Knystautas, 1987).

Knystautas (1987) has provided brief descriptions (below), including soil type, physiographic zones and vegetation, and other natural conditions, of the physical zones observed in the greater Siberia region.

West Siberian Plain—This vast land lies on the West Siberian Plain where its surface is flat and poorly drained; much of its is marshland. The northern portion of this region has a thick layer of permafrost. Throughout this area today there is tundra, wooded tundra, forest and forest-steppe vegetation.

Central Siberia—Most of this zone is situated in the Cengral Siberian Tableland, which has a relatively flat surface and deeply etched river valleys. The climate is markedly continental; permafrost is predominant.

Altai-Sayan Mountains—This zone includes the Altai, Sayan, Kuznetskii Alatau, and Tuva mountain ranges. All of these dramatic mountain masses have been extensively carved by Pleistocene glacial erosion. To the west of this zone, in the wetter climates, dense conifer forest grow. Up-slope these forests are replaced by open stands of pine (*Pinus sibirica*) and alpine meadows. In the valleys are steppes and semi-deserts. Still higher in elevations are mountain tundra, permafrost, and alpine glaciers (remnants from the icc age past).

Lake Baikal Zone—This is a region of mountains, with both flat-topped and conical peaks, separated by long, sinuous valleys. The climate is markedly continental, dry in the valleys, but moister and colder in the mountains; permafrost is widespread. Steppes occur in the drier valleys while forests live on the mountain slopes.

North-Eastern Zone—This zone is a mountainous region with a harsh, continental climate, especially in particular valleys. It is here that temperatures drop to their lowest. Open stands of larch (*Larix*) forest predominate, with mountains covered with tundra.

North Pacific Zone—This zone includes the Chukot Peninsula, the Anadyr Tableland, the Anadyr Lowland, the Koryak Highlands, and the Kamchatka Peninsula. In the northern and middle portions of the zone permafrost is widespread. Tundra stretches far to the south in the Chukot Peninsula. Taiga is found only in the south of this zone.

The present vegetation of Siberia can be grouped into different zones: High Arctic, Tundra, North Pacific Mountain, Taiga, and Eurasian Steppe. Figure I-2 provides a generalized distribution of the present major vegetation zones (Knystautas, 1987).

Vegetation Types of the Arctic Polar Lands

The polar lands are roughly those lying north of arborescent growth ('treeline'). This tree limit approximately coincides with a mean temperature of 10°C(50°F) for the warmest month (normally July in the Arctic). This relatively treeless zone occurs northward of the stunted 'elfin wood' and twisted 'krumholz' limits of the 'taiga' or spruce-fir forest. The Arctic is also characterized by winters which are dark and cold, a growing season less than 50 days, with the subsoil permanently frozen, and annual precipitation below 50 cm (often, 25 cm), mostly in the form of snow. Several different plant associations can be recognized; but these usually occur in a mosaic or an interdigitating pattern depending on local microenvironmental conditions or the relative youth (successional status) of the community resulting from its repeated readjustment to almost perpetual frost related disturbance.

Forest-Tundra or Taiga occurring toward the northern limit of arborescent growth is really only the product of depauperation of the various facies of the Northern Coniferous Forest. Precipitation ranges from 25-100 cm per annum and the mean temperature of the warmest month is 10-15°C with a growing season of 50-70 days. In Eurasia and North America this forest is composed of various assortments (or occasionally a single species) of such evergreen gymnosperms as spruce (*Picea* spp.), fir (*Abies* spp.), pine (*Pinus* spp.) and larch (*Larix* spp.) with deciduous angiosperm trees including the birches (*Betula* spp.), alders (*Alnus* spp.), willows (*Salix* spp.), and poplars (*Populus* spp.). Longitudinal replacement of trees occurs in both Eurasia and North America

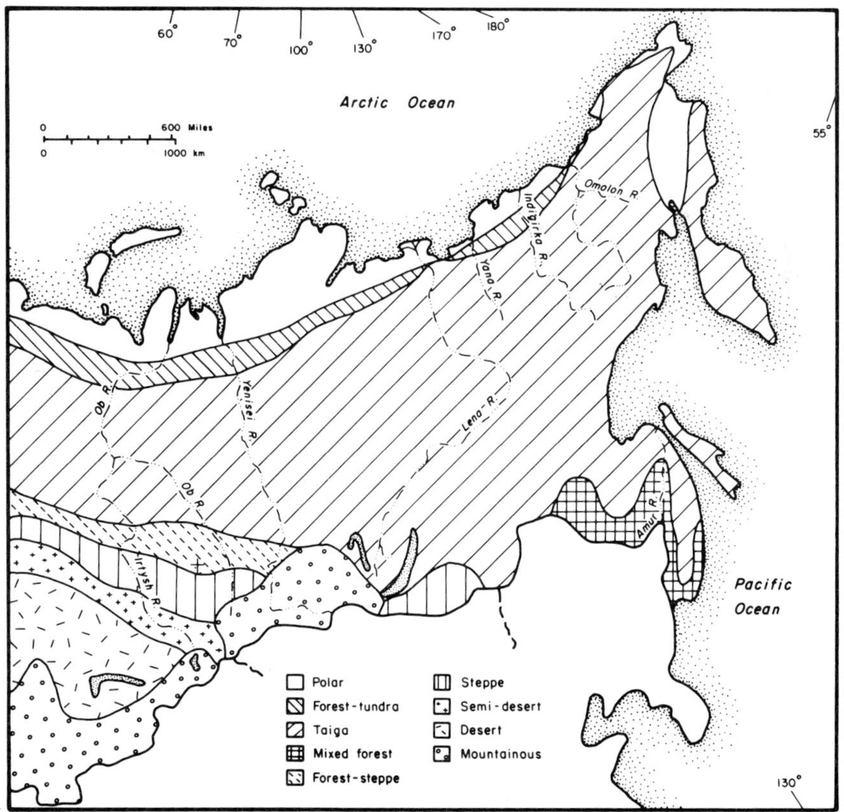

Figure I-2. Map of Siberia locating major vegetation zones.

in response to climatic gradients. In Siberia the dominant gymnosperms include: *Pinus siberica, P. pumila* and *P. sylvestris; Abies siberica; Larix siberica* and *L. dahurica;* and *Picea obovata.* Other woody plants include representatives of the heath family (Ericaceae) such as crowberry (*Empetrum nigrum*), laurel (*Kalmia* spp.), blueberry (*Vaccinium* spp.) and Labrador tea (*Ledum* spp.). Lichens such as the reindeer-mosses (*Cladonia* spp.) and Iceland-mosses (*Cetraria* spp.) occur in drier habitats with true mosses more common in the more moist habitats.

Arctic scrub and heathlands are developed on the most favorable slopes, in damp depressions and especially along water courses and the margins of lakes. Except for the generally more moist conditions, the climate is that of the tundra rather than the taiga. The Arctic Scrub is composed of low, shrubby willows (*Salix* spp.) alders (*Alnus* spp.) and/or birches (*Betula* spp.) with scattered, very depauperate specimens of larch or more rarely spruce. The plants may occur so closely as to exclude other woody taxa or in drier habitats be sufficiently spaced as to permit the occurrence of an understory of various 'heaths' or 'ericads' such as crowberry, blueberry and cranberry, arctic bell-heather (*Cassiope tetragona*), Labrador-tea, *Andromeda* and *Pyrola*. Graminoids or 'grassy' components include: the true grasses (*Calamagrostis, Poa* spp. and *Arctogrostis*); the sedges (*Carex* spp., *Eriophorum*, and *Luzula*); and the rushes (*Juncus* spp.). Lichens (*Cladonia* spp. and *Sterocaulon* spp.) and various coarse, true mosses (e.g. *Dicranium, Polytrichum,* and *Rhacomitrium*) occur on drier sites with bog-mosses such as *Sphagnum* spp. occurring in more moist habitats.

Arctic tundras are cold, wind swept and drier than the preceding environments, occurring on valley bottoms and lower mountain slopes or even around lakes. The vegetation of arctic tundra is gramminoid dominated, including true grasses such as *Calamagrostis, Poa, Arctophila* and *Arctogrostis,* and a variety of sedges (*Carex* spp. *Eriophorum* spp., and *Kobresia* spp.) and rushes. Woody plants, when present, are of very low stature, few in number and diversity, but include alder, birch and a number of willows. Although often widely scattered, a number of forbs belonging to the following families may be observed: Asteraceae (*Armeria, Artemisia, Aster, Erigeron, Tanacetum* and *Taraxacum*); Brassicaceae (*Draba*); Campanulaceae (*Campanula*); Caryophyllaceae (*Arenaria, Silene, Stellaria*); Fabaceae (*Astragalus, Oxytropus, Vicia*); Gentianaceae (*Gentiana, Menyanthes*); Lamiaceae (*Thymus*); Papaveraceae (*Papaver*); Polemoniaceae (*Polmonium*); Polygonaceae (*Polygonum, Rumex*); Ranunculaceae (*Aconitum, Aquilegia, Ranunculus, Thallictrum*); Rosaceae (*Comarium, Dasiphora, Dryas, Potentilla, Rosa, Rubus*); Saxifragaceae (*Saxifraga, Ribes*); Scrophulariaceae (*Pedicularis*); and Valerianaceae (*Valeriana*).

Table I-1 provides a listing of the composition of Siberian plant communities for the forested regions and the tundra.

Table I-1. Composition of Siberian Plant communities. (Ukraintseva and Kozhevnika 1978, 1981)

Scientific name	Common name	Taiga, Krumholz, Elvin Forest	Tund Shrub-Heath
Graminoids			
Cyperaceae			
Juncaceae			
Poaceae			
Ericoids			
Andromedia spp.	Prince's Pine		+
Arctostopylos	bear berry		+
Cassiope tetragona	bell heather		+
Colluna spp.	heather		+
Empetrum nigrum	crowberry		+
Erica spp.	heath		+
Kalmia spp.	laurel		+
Ledum palustre	marsh tea		+
Ledum decunber	Labrador tea		+
Pyrola grandiflora	schinlead		+
Vaccinium spp.	blueberry		+
	cowberry		+
	bilberry		+
	cranberry		+
Rose & Saxifrages			
Ribes dikuscha			+
Ribes triste			+
Rosa aciculans			+
Rubus chamaeomorus			+
Herbs			
Dryos punctata			+
Trees			
Abies siberica	Siberian fir	+	
Alnus fruticosa	green alder	+	
Alnus hirsuta-incana	alder	+	
Alnus kamtschatica	alder	+	
Betula platyhylla m.	tree birch	+	
Betula costata	tree birch	+	
Betula exilis	dwarf, shrub birch	±	+
Betula nana	dwarf, shrub birch	±	+
Corylus cornuta	hazelnut	±	
Juniperus communis	Common juniper	+	
Juniperus Siberica	Siberian juniper	+	
Larix dahurica	Dahurian larch	+	
Larix siberica	Siberian larch	+	
Picea abies	Norway spruce	+	
Picea obovata	Siberian spruce	+	
Picea ojanensis	spruce	+	
Pinus pumila	Siberian dwarf pine	+	
Pinus siberica	Siberian stone pine	+	
Pinus sylvestris	Scots pine	+	
Populus suaveolans	poplar	+	
Salix exilis, lanata	shrub willow	±	+
Salix fuscescens s.	tree willow	+	
Salix myrtilloides	shrub willow		
Salix repans	shrub willow		
Ulmus pumila	Siberian elm	±	

The Late Pleistocene

If the reader is familiar with the most recent glacial episode on North America, the Wisconsin Glacial, then one understands that most of Alaska and basically all of Canada, and the northern stretch of United States was covered by a thick cap of continental ice. Actually two glaciers occurred, the Cordilleran in the western mountainous regions and the Laurentide in the east, and that they occasionally coalesced in the region of Edmonton, Alberta. Because of this bias, one might think that all of the northern Holarctic was covered by a veneer of glacial ice during the height, or maximum, of the most recent glacial stage. This is actually far from true.

Glaciers exist where it is cold and relatively moist. Although Scandinavia and the nearby region of Europe was covered by a continental ice sheet, which even reached into the northern part of the Russian Plain during the last glaciation, north central and northeastern Siberia was predominantly too cold for ice sheets, except for localized alpine glaciers (Figure I-3).

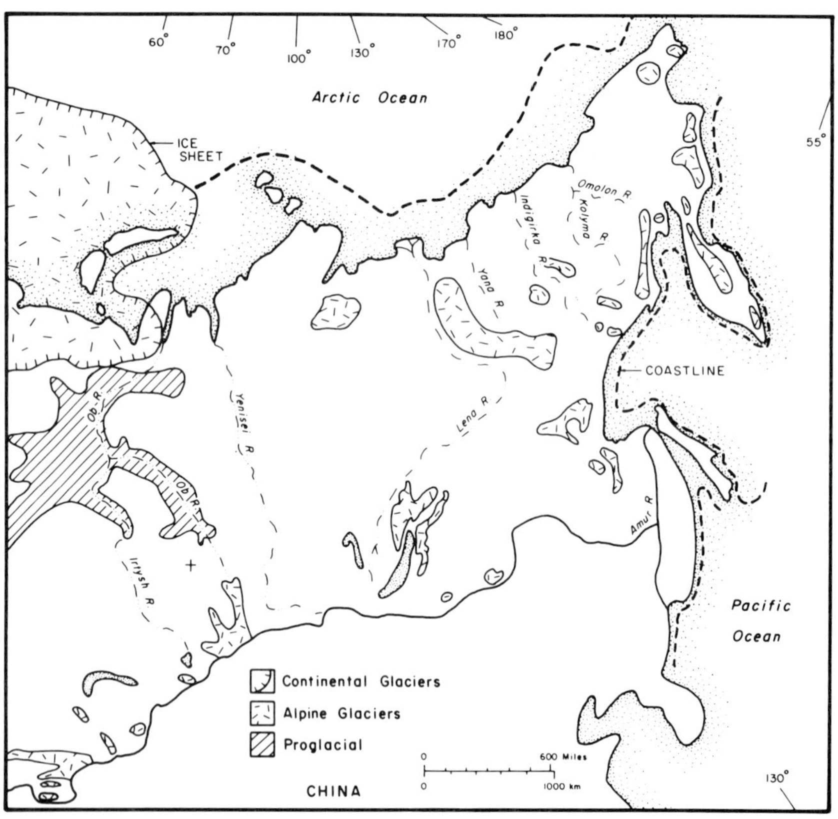

Figure I-3. Map of Siberia locating major ice fields and proglacial features.

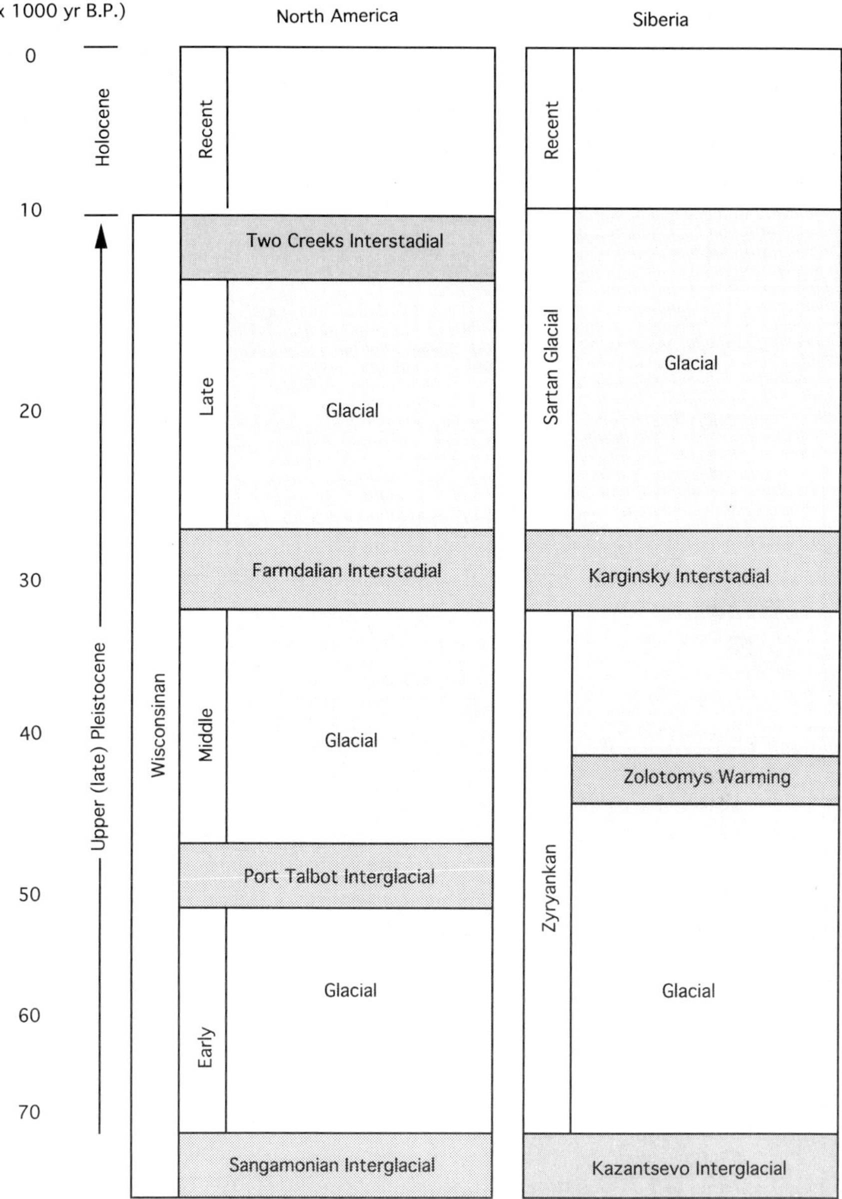

Figure I-4. Approximate correlation of upper Pleistocene glacial events: North America and Siberia.

Figure I-4 provides a synopsis of late Quaternary glacial and interglacial events in European Russia and Siberia. A detailed description of the various glacial events can be found in Velichko (1984). Wright and Barnosky (1984:xiv) provide a good summary:

"The Scandinavian ice sheet reached into the northern part of the Russian Plain during the last glaciation, subdivided into the early Valdai and Late Valdai separated by the Middle Valdai nonglacial interval, which includes several climatic fluctuations recorded by pollen evidence for alternating tundra and birch-conifer forest...The ice sheet near the margin was differentiated into a number of ice lobes fronted by proglacial lakes, and the eastern portion was confluent with ice from Novaya Zemlya. As the ice sheet began its retreat 16,000 to 15,000 years ago, distinct fluctuations of the margins occurred on the southwestern sector in Western Europe, with push moraines and other indications of active ice, but farther east on the more continental Russian Plain the marginal deposits were those of stagnant ice. Retreat was more rapid after 15,000 years ago, and especially after 12,000 years ago. The late glacial Salpausselka moraines of Finland extend eastward into Karelia and the Kola Peninsula, where ice persisted until about 9500 years ago. The Novaya Zemlya glacier also broke up about this time. In the West Siberian Plain in the lower Ob' and Yenisey River valleys, the last glaciation is similarly subdivided into two glacial...The taiga of today was converted to tundra or tundra-steppe.

Farther east in north-central Siberia the Middle Zryanka nonglacial interval, dated between 50,000 and 24,000 years ago, is represented by several phases in which trees invaded areas that are now tundra, implying warmer conditions than today...The Late Zyryanka Glaciation (Sartan, Late Valdai) that followed was centered on different mountain groups and plateaus in northeastern Siberia as well...Periglacial loess and syngenetic ice-wedge features in alluvium indicate a cold and windy climate. Tundra or forest-tundra extended 1000 km from the ice margin.

Interpretation of the paleoclimatic sequence in the mountains and the nearby loess areas differ, however, for some authors think that the glacial episodes were cold and wet, with conifer forests descending on the mountains, whereas others believe they were cold and dry."

Geochronologic Framework

Dr. Ukraintseva devotes her expertise to the analyses and description of modern environments in the regions of the spectacular frozen specimens described in this book. Her analyses of the microflora and macroflora associated with these animals, as gastrointestinal material and materials from the enclosing sediments provides a reconstruction of the paleoenvironment, for that place, at that point in time.

These data sets, to be truly significant, must be tied to an absolute chronology. Table I-2 provides a concise listing of the radiocarbon dates (provided in the author's text descriptions) of the mammoth fauna of Siberia. Of these fifteen dates six (40%) are from the interval 9750-13,700 yr B.P.; three (20%) are from the 29,500-39,500 yr B.P. interval; five (33.33%) are from the 40,350-44,540 yr B.P. interval; and one date (6.66%) is greater than 50,000 yr B.P.

Table I-2. Radiometric chronology of the mammoth fauna from Siberia, some of which provided associated gastrointestinal remains.

<u>Indigirka River Basin</u> (Eastern Siberia)

Mammoth	Shandrin River, tributary to lower Indigirka River	$40{,}350 \pm 880$
Mammoth	Berelekh River, left tributary of the Indigirka River	$39{,}590 \pm 870$ $13{,}700 \pm 400$ $12{,}240 \pm 160$ $11{,}830 \pm 110$
Horse	Selerikhan gold mine, Upper Indigirka River (on Balkhan Creek)	$38{,}590 \pm 120$
Bison	Shandrin River, tributary to lower Indigirka River	$29{,}500 \pm 1000$

<u>Kolyma River Basin</u> (Eastern Siberia)

Mammoth	Kirgilyakh Creek, left tributary of the Kolyma River	$41{,}000 \pm 1100$
Mammoth	Beresovka River	$44{,}000 \pm 3500$

<u>Khatanga River Basin</u> (Western Siberia)

Mammoth	Bolshayn Lesnaya Rassokha right tributary of the Novaya River	$53{,}170$
* Mammoth	Terekhtijh River	$44{,}540 \pm 1870$ $44{,}170 \pm 1870$
* Mammoth	Northern Taimyr	$11{,}450 \pm 450$
* Mammoth	Yuribei River, Gydan Peninsula	$10{,}000 \pm 70$ 9750 ± 100

* = No plant remains within the gastrointestinal tract.

Paleoenvironmental Reconstructions

Typically the geologist is provided in the field situation with stratified (layered) deposits of varying-sized sediments from which to make his or her analysis of the environment of deposition. The vertebrate paleontologist is usually left with teeth, bones, and rarely, organics, to provide a reconstruction of the various animals and the possible environmental scenario. Paleoecologists usually extract pollen from sediment cores taken from lakes and bogs from which they hope to illustrate how the vegetation has changed through time.

Under extreme xeric conditions in the arid southwestern portion of North America researchers have found packrat middens containing pollen and plant macrofragments, and caves containing dung of extinct animals where pollen, plant microhistological remains, and biochemical data are preserved. These provide unique situations for accurately reconstructing past environments and precise diets of now extinct mammals. However unfortunately, moisture provides the avenue for organic decay. The unique remains found in the North American Southwest are typically decomposed in other taphonomic (after death) situations.

Siberia in its permafrost region is different—it is unique in a most unusual way. The ground that froze some 30,000 to 40,000 years ago, after a warming trend, has not become unfrozen since. Whatever became part of the ground, has been preserved in the Siberian deepfreeze. Basically everything is preserved, right down to the hair and hide covering muscle tissue, internal organs, and the "last supper." What better way to reconstruct the diet of an animal than to look into its stomach or intestinal tract. And what better way to reconstruct the local plant community than to peer into the gut contents to see what was eaten and what was flowering at the time and not eaten. It is this unique paleoenvironmental data that Dr. Ukraintseva is presenting in the pages that follow.

Acknowledgements

The original manuscript was produced and translated in English by Dr. Ukraintseva. Manuscript pages were scanned onto computer discs and editing of the text was done by computer at Northern Arizona University. The intermediate and final versions of the text and tables were produced via desktop printing with PageMaker, by Finn T. Agenbroad and Maxine W. Campbell. Figures I-1 to I-3 were provided by Emilee M. Mead. The Mammoth Site of Hot Springs, South Dakota provided editorial services.

References Cited

Guthrie, R.D.
    1982   Mammals of the Mammoth Steppe as paleoenvironmental indicators in Hopkins, Matthews, Schweger, and Young (eds.) Paleoecology of Beringia. p. 307-26. Academic Press, New York.
    1990   Frozen Fauna of the Mammoth Steppe. University of Chicago Press, Chicago.

Haynes, G.
    1991   Mammoths, Mastodonts, and Elephants: Biology, Behavior, and the Fossil Record. Cambridge University Press.

Hopkins, D.M., J.V. Matthews, C. Schweger, and S. Young (eds.)
    1982   Paleoecology of Beringia. Academic Press, New York.

Kynstautas, A.
    1987   The natural history of the USSR. McGraw-Hill Book Co., New York.

Markova, A.K.
    1984   Late Pleistocene Mammalian Fauna of the Russian Plain. In A.A. Velicho (ed.) Late Quaternary environments of the Soviet Union. University of Minnesota Press, Minneapolis. pp. 209-222.

Martin, P.S. and R.G. Klein (eds.)
    1984   Quaternary Extinctions: A Prehistoric revolution. University of Arizona Press. Tucson.

Torlmachev, A.I.
1974 (1982) Endemic mountain plants of Central Asia. Nauka Press, Novosibirsk.

Tomachev, A.I.
    1974   Introduction to plant geography. Leningrad State University, Leningrad.

Ukraintseva, V.V.
    1981   Vegetation of warm late Pleistocene intervals and the extinction of some large herbivorous mammals. *Polar Geography and Geology.* Y.H. Winston and Sons. 4:189-203.
    1985   Forage of the large herbivorous mammals of the epoch of the mammoth. *Acta Zoologica Fennica* 170:215-220.

Velichko, A.A. (ed.)
    1984   Late Quaternary Environments of the Soviet Union. University of Minnesota Press, Minneapolis.

West, J.H.
    1981   The Archaeology of Beringia. Columbia University Press, New York.

Wright, H.E. Jr. (ed.)
    1983   Late-Quaternary Environments of the United States. University of Minnesota Press, Minneapolis.

Wright, H.E., Hr. and C.W. Barnowsky (eds. English edition)
    1984   Late Quaternary environments of the Soviet Union. (English translation of Velichko (ed.) University of Minnesota Press, Minneapolis.

Vegetation Cover and Environment of the ''Mammoth Epoch'' in Siberia

by
Valentina V. Ukraintseva

edited by
Larry D. Agenbroad, Jim I. Mead, Richard H. Hevly

Annotation

Results of the integrated paleobotanical study of (i) gastrointestinal contents of some fossil herbivorous mammals, representatives of the "mammoth" faunal complex (horse, bison, four mammoths) which died in various areas of Siberia during the past 53,000 yr B.P.; (ii) enclosing deposits; and (iii) sediments of synchronous age are first generalized in the book. Food remains of the fossil herbivores are shown to be a source of valuable information on ecosystems of the past, especially their main elements such as flora, vegetation, climate. Composition of floras, reliably dated as related to the Karginsky Interglacial (50,000-25,000 yr B.P.) and the Holocene, are given. The character of vegetational and climatic variations in the late Anthropogene (Quaternary) is traced over three marker areas (Indigirka-Kolyma Basin, Taimyr, Gydan Peninsula) where the above fossil animals were found and throughout the territory of Siberia as a whole. Paleoclimatic reconstructions, performed in the three above areas, are presented. New proposed here paleobotanical and paleogeographical data substantiate a new hypothesis concerning the causes of relatively fast extinction of some species of animals, especially, mammoth. Of special value is the methodical part of the work which can be regarded as a textbook to study unique paleontological finds of this kind.

The book is provided with numerous original photos, figures, histograms, spore and pollen diagrams, climagraphs. Of particular interest are microphotographs of pollen and spores of plants identified in analysis of food of the fossil animals. The book will be of interest to paleopalynologists, paleobotanists, and botanists dealing with the history of vegetational cover, paleontologists, geologists, paleogeographers, biologists with wide scientific interests in the history of the earth and its organic world.

Contents

List of Figures

List of Tables

Introduction

A better understanding of the present status of ecosystems of the Far North and a number of poorly studied areas of Siberia can be gained through going a few thousand years back to see what the environment was in that remote time when numerous herds of mammoths and their "associates", namely, horses, bison, woolly-rhinoceros, musk oxen, yaks and other mammals wandered about Siberia and adjacent areas of Mongolia, North China, and North America.

Food remains of fossil animals, preserved in permafrost in some areas of Siberia for tens of thousands of years, were those unique objects which allowed us to return in the past and to gain a more penetrating insight into habitats of concrete individuals and environment of representatives of the mammoth fauna as a whole, and especially to decipher the last dramatic pages of the "mammoth-epoch", recorded 55,000-60,000 yr B.P. (yrs before present; A.D. 1950).

The present work is the first in the world which shows the great potential of paleobotanical methods of studies and advantages of a complex approach to study unique paleontological objects.

Acknowledgments. - The author gratefully acknowledges the constant help and support of Professor B.A. Tikhomirov, Honored Science Worker, chief of studies on the problem, and Professor N.V. Lovelius, chief of the Polar Expedition, Botanical Institute, Academy of Sciences of the USSR; they provided support in studies and organization of field work in the areas of interesting finds. Multi-year contacts and discussions with Professor-in Paleontology K.K. Flerov were beneficial for the development of the scientific outlook of the author; the author is also indebted to him for advice and suggestions.

Academicians A.L. Takhtadzhyan, V.E. Sokolov, K.M. Sytnik, Professors A.T. Artyushenko, N.A. Sirenko, V.S. Volkova, A.A. Velichko, B.V. Zaverukha, G.I. Lazukov, N.A. Khotinsky, A.L. Chepalyga played important parts in different times in the fate of the book. The author expresses her sincere thanks to all of them.

Thanks are also due to Nikita Kultin, the author's son. Since the age of 12 Nikita has been her trustworthy support. He is a systems engineer today.

Dr. Larry D. Agenbroad, Dr. Richard Hevly, and Dr. Jim I. Mead have showed interest in the book and taken the trouble of editing. The author expresses her heartfelt thanks to them.

Chapter 1

History of Investigation

Remains of ancient animals which once lived in one or another area are being washed out from the earth's interiors along banks of rivers, lakes, streams, and on gully slopes. Discovery of their bones and even complete skeletons are well known from Anthropogene (Quaternary) deposits of Siberia and North America (Chersky, 1891; Pavlova, 1910; Popov, 1948; Guthrie, 1968, 1982; Sher, 1971; Vereshchagin, 1981; Lazarev, 1982; Agenbroad, 1984, and others). More than 50 almost wholly preserved skeletons of fossil animals have been found in various areas of Siberia. Nevertheless, frozen animal carcasses with well preserved entrails, including gastrointestinal tracts filled with plant remains, are extremely rare because they may have been preserved well only under favorable burial conditions at the site of animal's death. Some of the animals, such as mammoth, woolly rhinoceros, cave bear and others have died out quite recently, whereas others, for instance, bison, yak, saiga and others have only reduced their ranges, and others, for example, *Ovibos*, are on the verge of extinction.

Scientists have been aware for a long time that finds of fossil herbivorous animals with the preserved contents of their gastrointestinal tracts are very important for science. Attempts to examine food remains of fossil animals have been known since the second half of the 19th century. Academician F.F. Brandt was the first who as early as 1849 examined plant remains from the oral cavity of the mummified head of a woolly rhinoceros,and identified remains of coniferous species. This animal was discovered on the bank of the Vilyui River, near the town of Vilyuisk, Yakutia, K.A. Meyer and K.E. Marklin identified fruits of *Ephedra* sp. and twigs of *Salix* sp. in the food remains of this animal, respectively. Shmalgausen (1868) examined plant remains stuck among teeth of the rhinoceros. He noted that fragments of stems and leaves of monocotyledonous plants, probably of grasses, were the most common in the food remains studied. In addition to fragments of monocotyledonous and dicotyledonous plants, he identified twigs of the spruce *(Picea)*, larch *(Larix)*, willow *(Salix)* and *Ephedra*.

The opportunity to examine the composition of plants from the frozen gastrointestinal tract arose for the first time in history in 1900 when on the Berezovka River, right tributary of the Kolyma River, about 320 km northeast of the town of Srednekolymsk (Fig. 1) an almost wholly preserved frozen mammoth carcass (Fig. 2a) was found (Gerts, 1902). The Russian Academy of Sciences immediately organized an expedition to dig up the mammoth. The well known specimen has been taxidermically mounted and is on display at the Leningrad (St. Petersburg), Zoological Museum (Fig. 2b).

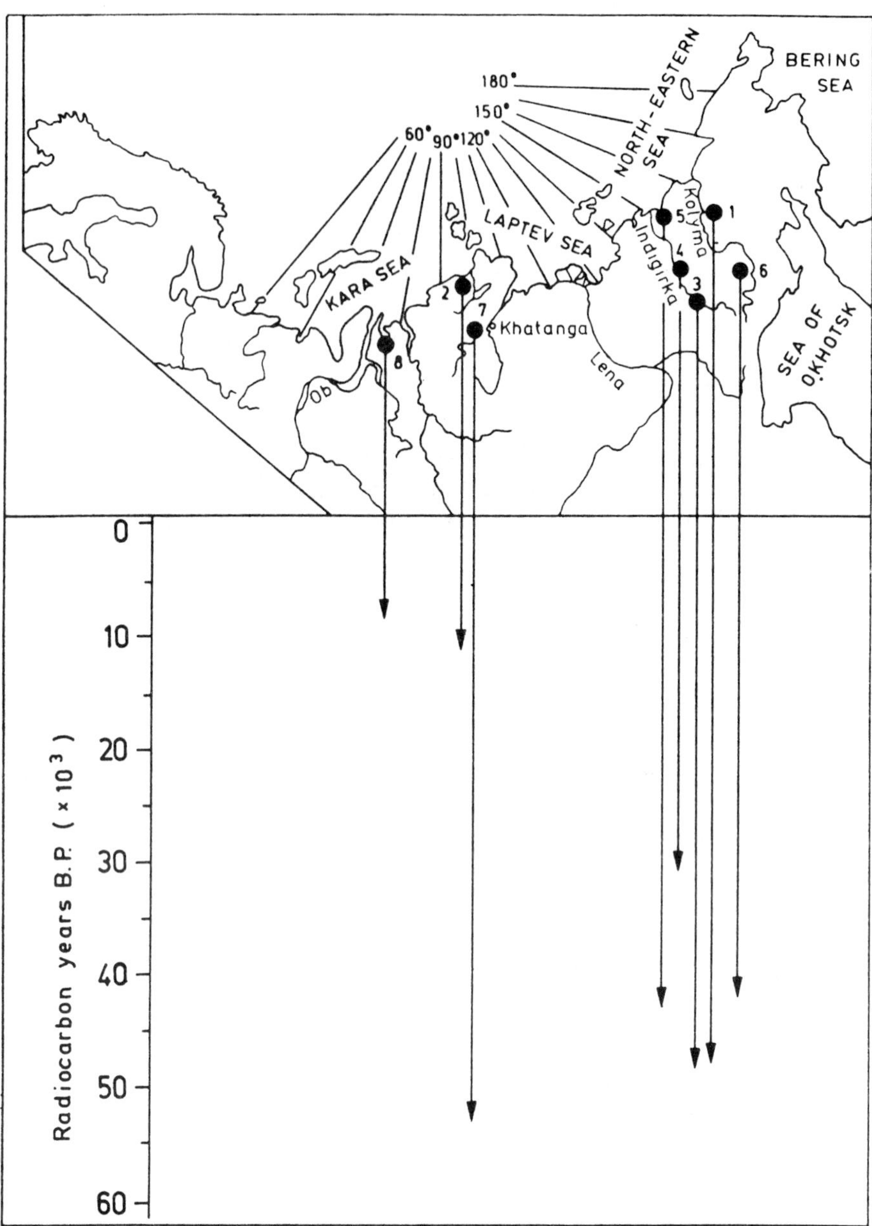

Figure 1. Sites of finds of frozen carcasses and skeletons (or their fragments) of animals which perished in different periods of the Pleistocene and Holocene. 1 mammoth, Berezovka River, 1900; 2 mammoth, Mamontovaya River, 1948; 3 horse, Elga River (Selerikhan Creek), 1968; 4 bison, middle Indigirka River, 1971; 5 mammoth, Shandrin River, 1971; 6 Kirgilyakh mammoth (baby mammoth "Dima"), 1977; 7 Vereshchagin's mammoth, 1978; 8 Yuribei mammoth, 1979.

Figure 2a.

Figure 2b.

V.N. Sukachev dealt with food remains of the Berezovka mammoth for years. As a result, he reliably identified 13 species of herbaceous plants and mosses in the food. According to Sukachev (1914), the food remains of the Berezovka mammoth were dominated by tissue fragments of grasses and sedges. Only three well preserved seeds of *Ranunculus acris* and some rather large and well preserved beans of *Oxytropis sordida* were found in the stomach content. Plant residues, stuck between the teeth, contained eight beans of such excellent preservation that they even had small hairs preserved in places. Some seeds resemble poppy-seeds of *Papaver nudicaule,* and three small fruits were similar to nuts of the thyme *(Thymus seryllum).* Three small twigs (3-5 mm long) of woody plants were also found, but it is impossible to identify them reliably. Some of the material brought by O.F. Gerz contained some pieces of wood and a piece of bark which had occurred beneath the mammoth in the enclosing deposits. According to Sukachev (1914), two pieces of wood belonged to twigs and two to roots of larch. The anatomic structure of the only piece of bark enabled Sukachev to regard it as the alder (*Alnus* sp.) Material available was not sufficient to determine the species; it was not ruled out that the bark could belong to the alder (*Alnus fruticosa*).

Deposits enclosing a fossil animal corpse were first paleobotanically studied by W. Szafer (1946). He identified 14 species of plant remains in strata encompassing the woolly rhinoceros corpse in the vicinity of Starun, West Ukraine. The remains were dominated by the dwarf birch (*Betula nana*), willow (*Salix reticulata*) and dryad (*Dryas octopetala*); according to Szafer (1946), this composition suggests a severe arctic climate in the discovery area during the lifetime of the Starun rhinoceros.

After V.N. Sukachev this trend of investigations was led by Professor B.A. Tikhomirov. He took part in digging out a mammoth, found in the Shrenk River Basin, northwestern Taimyr (Fig. 1) and in studying present flora and vegetation in the area of the find (Tikhomirov, 1950a). Conditions of burial and the geology of the area were studied by Popov (1950, 1957). Palynological analysis of the deposits which both encompassed the mammoth skeleton and built up terraces in the vicinity of the burial site was carried out by Zaklinskaya (1954). Fossil mosses were examined by Zenkova (1950), Savich-Lyubitskaya and Abramova (1954).

Sporo-pollen analysis of plant remains in the food of the Berezovka mammoth was made as late as the late 1950s (Tikhomirov and Kupriyanova, 1954; Kupriyanova, 1957). Analyses showed that the pollen and spores were excellently preserved in the food remains of the animal even after almost 60 years in storage, hence the method showed promise for this kind of research. This was supported by spore and pollen analysis of food remains, preserved between the teeth and in the oral cavity of the woolly rhinoceros, discovered on the Kholbui River, Yakutia as early as 1877; analysis was performed by Metelseva (Garrut et al., 1970). In spite of the fact that the head of the animal had been kept at the museum for more than 90 years, plant remains, as well as pollen and spores, stuck between the teeth, though rare, were relatively well preserved.

Sporo-pollen analysis of a 2 m thick strata enclosing a mammoth skeleton at the village of Chekurovka in the lower part of the Lena River showed not only living conditions of the mammoth but also the character of later variations of vegetation and, hence, of environmental conditions in the area (Korzhev and Fedorova, 1962).

Seventy years passed after the find of the Berezovka mammoth and in 1968 a fossil horse was discovered on Balkhan Creek, a right tributary of the Elga River, which is tributary to the Indigirka River (Fig. 1). By efforts of B.S. Rusanov, P.A. Lazarev and O.V. Egorov, the find was dug out and delivered to Yakutsk and then to the Leningrad Zoological Institute. Plant remains from different sections of its gastrointestinal tract were analysed at the Komarov Botanic Institute.

The next summer (1969) an expedition was organized to the area of find. The expedition enlisted N.V. Lovelius (chief, dendrochronologist), Zh.M. Belorusova (geomorphologist), M.V. Sokolova (florist), V.P. Bibikov (geobotanist), V.I. Kozhanchikov (carpologist), and V.N. Adamenko (climatologist); they studied environmental conditions of the area where the fossil horse was found, and its burial conditions. Furthermore, the scientists collected herbarium, seeds, pollen, samples of wood and sediments from strata enclosing the horse and thus obtained material, required to identify plant remains in the food of the horse and to reconstruct paleogeographic conditions of its lifetime.

Even results of preliminary examinations (Tikhomirov and Kultina, 1973) confirmed the fact that pollen and spores practically do not decompose in the gastrointestinal tract (Tikhomirov and Kupriyanova, 1954; Kupriyanova, 1957) reflecting the composition of plants, both directly eaten by the animal and those growing in the vicinity of its habitat.

The year 1971 witnessed two extremely valuable finds: a complete mammoth skeleton with a well preserved frozen monolith of internal organs (Fig. 1) in the Shandrin River of the lower Indigirka River and a relatively well preserved corpse of a fossil bison female (Fig. 1) in the middle part of Indigirka River, Mylakhchin District. B.S. Rusanov, A.P. Lazarev and their mates dug up both skeletons from permafrost and delivered them to the freezing pit of the Institute of Permafrost at Yakutsk. In January, 1973, the two unique finds were transported to Novosibirsk and examined by a group of scientists, including paleontologists, anatomists, microbiologists, parasitologists, botanists, and geologists. Two scientists from the V.L. Komarov Botanic Institute of the USSR Academy of Sciences - N.V. Lovelius and the author - participated in the study. A program of botanical studies, developed by Professor B.A. Tikkomirov and the author, was submitted for consideration of the Commission. According to the program, the following studies should be fulfilled: 1) analysis of macroremains of vegetative parts of plants, preserved in the gastrointestinal tract of the mammoth (stems, leaves roots, rootstocks and the like); 2) carpological analysis; 3) palynological analysis of different parts of the intestinal tract; 4) radiocarbon analysis of the food remains.

The program envisaged the organization of complex expeditions to the area of the two finds where geologists and botanists planned to perform joint investigations. For some reasons, the expedition has never taken place. However, in 1974 a group of scientists of the Botanic Institute undertook investigations in the area of burial of the Shandrin mammoth. Examinations of the content of the gastrointestinal tract of the mammoth, based on studies and collections, made during the expedition, allowed reconstruction of floristic composition and the character of vegetation for the mammoth's lifetime (Solonevich et al., 1977; Gorlova, 1982); furthermore, the investigations served as the basis for the reconstruction of main climatic characteristics, first performed for the area (see Chapter 4.3).

Two more fossil animals were found in 1977. In June a wholly preserved corpse of the baby mammoth (Fig. 1) was the first discovered in the midstream of Kirgilyakh Creek, on the upper part of the Kolyma River; it has become universally known under the name of Dima. This unique find was conveyed to Magadan. The Presidium of the USSR Academy of Sciences at once established a commission under the chairmanship of Academician N.A. Shilo with the aim of examination of the mammoth. The members of the Commission were representatives of a number of academic institutes, namely, the Institute of Microbiology; the Institute of Evolutionary Morphology and Ecology of Animals; and the biology chair of the Moscow University. In accordance with the instructions of the Presidium of the USSR Academy of Sciences, the author was the representative of the Botanic Institute in the Commission. When the Commission completed the work, the group of botanists, led by the author, moved from the area of Berezovka mammoth burial to the Kirgilyakh Creek area where the botanists studied present flora and vegetation, collected herbaria, took surficial samples for palynological analysis in key types of vegetation cover.

Baby mammoth Dima outshone another very important paleontological find of 1977—remains of some parts of the corpse of the mammoth of an early type, as defined by L.I. Alekseeva (oral communication); it was found in the Bolshaya Lesnaya (Large Forested) Rassokha River Basin, a right tributary of the Novaya River, in southeastern Taimyr (72°35′ N) (Fig. 1). No gastrointestinal tract of the mammoth has been preserved, but joint investigations, undertaken in the area, including detailed palynological study of corpse-enclosing deposits, permitted reconstruction of both living conditions of the mammoth and major stages of evolution of habitats in this part of Taimyr for the last 53,000 years (see Chapter 6.2).

And at last, in early September, 1979, it became known that a mammoth corpse was found in the middle part of the Yuribei River on Gydansky Peninsula (Fig. 1). A group of specialists was gathered at once with the objective of digging it out and conducting an examination of it. The group included a geologist, paleontologists, an anatomist-morphologist, a palynologist-florist, a geocryologist, and other specialists, as well as a "Trud"s correspondent. As early as September 18 the group of specialists of the USSR Academy of Sciences arrived to the area of the find and started working. In the process of excavation they dug up much of the skeleton, four feet containing soft tissues, separate more or less well preserved parts of soft tissues, wool and part of the internal organs. A gastrointestinal tract, tightly filled up with remains of plants, which had been eaten shortly before the death, was complete and well preserved. At the site of excavation, the scientists took samples from all parts of the gastrointestinal tract for all kinds of botanical analysis and for radiocarbon dating.

From the brief review above, it can be seen that starting from the early finds of fossil animals in this country, a strong emphasis was placed not only on examination of food remains they contained, but also on study of enclosing deposits. Since the late 1960s (i.e. the time of fossil horse discovery) multidisciplinary researchers of academic and non-academic institutions have participated in joint investigations; this has allowed a number of problems of both purely academic and applied value to be tackled successfully.

Results of studies of all the finds noted above were published in Russian in different years and in various publications but, primarily, in such small numbers of copies, that they quickly became rarities. The author took upon herself the difficult task of summarizing and critically commenting on the earlier publications.

Chapter 2

Material and Methods

2.1. Material

This monograph is based on the study of:

I. Food remains of the following herbivorous fossil animals which perished in the late Pleistocene and Holocene in various areas of north Siberia: 1) Selerikhan horse, upper Indigirka River; 2) Shandrin mammoth, lower Indigirka River; 3) Mylakhchin bison, middle Indigirka River; 4) Kirgilyakh (Magadan) mammoth, upper Kolyma River; 5) Yuribei mammoth, middle Yuribei River (Fig. 1).

II. Samples of deposits in which some of the above animals were buried; samples were taken from: (i) all the three terraces at the burial site of the Vereshchagin mammoth remains on the Bolshaya Lesnaya Rassokha River, southeastern Taimyr; (ii) terrace II above the floodplain of the Berelekh River at the site of the Berelekh "cemetery" of fossil animals in the lower Indigirka River; (iii) Pleistocene and Holocene deposits in the middle Novaya River Basin, southeastern Taimyr, and other sites. In all, the author palynologically examined 180 samples, taken directly from: deposits in which the animals were buried; strippings near the repository, and natural exposures, including peat bogs.

III. Surficial samples, 30 in number, taken in key types of vegetational cover in the areas where the fossils were found.

2.2. Methods

Pollen and spores, extracted from food remains of the fossil animals, were examined by the method of paleopalynoflora analysis and the method known as sporo-pollen analysis, combined with anatomomorphological and carpological analyses; in addition, they were dated by the radiocarbon method. Pollen and spores, extracted from deposits, were studied by the method of sporopollen analysis in conjunction with the geologo-geomorphological method and radiocarbon analysis. Surficial samples, taken at key sites of present vegetational cover, were examined with the aid of sporopollen analysis and geobotanical methods.

Food remains of all the above mentioned fossil animals were examined by the complex technique which required a special approach: for anatomomorphological, carpological and palynological analyses, samples were simultaneously taken from the same sections of gastrointestinal tracts; this allowed comparison of data, obtained by different methods, and evaluation of the potentialities of each. Such an approach showed that percentages of major plant groups - herbs, shrubs, and low shrubs, established, for example, in the stomach of the Selerikhan fossil horse from macroremains and pollen, are similar and, hence, they may be considered as reflecting true percentages of the above plant groups in animal feeding. On the other hand, the palynological method provided a more comprehensive picture of taxonomic plant composition (Tikhomirov and Kupriyanova, 1954; Kupriyanova, 1957; Tikhomirov and Kultina, 1973; Ukraintseva (Kultina), 1977; and others), due to the fact that pollen and spores appear to be practically unaffected by gastric juice, while tissues of some plants, especially those of meadow forb representatives, are almost completely digested which makes their identification practically impossible. In the best case, only the family or subfamily could be determined. The carpological method is very efficient in identification of, for example, sedges (Egorova, 1977), whereas their pollen are hardly ascribed a generic identification.

Differential study of the contents of different sections of gastrointestinal tracts contributes towards the understanding, for instance, by palynological spectra, of the character of the plant communities which served as pastures for animals or were resorted to by them shortly before their death.

When preparing for palynological analysis of food remains of the fossil animals, samples of deposits and surficial samples, a separation methods, proposed by Grichuk (1940) was used. The above laboratory treatment provided precipitates containing pollen and spores which were subjected to acetolysis (Erdtman, 1960).

2.2.1. Paleopalynoflora method

The problem to be tackled - was as follows: to determine, as completely as possible, the composition of plants which grew in the vicinity of the fossil animal ranges with the aid of pollen and spores, separated from their food remains.

The studies performed showed that the task could be successfully fulfilled with the help of paleopalynoflora method only. To this end, a collection of permanent preparations was gathered for each fossil. On the basis of organic precipitates, containing pollen and spores of plants eaten by the animals, the study of the collection provided the most complete understanding of the composition of plants which once grew in the areas of finds. Experience accumulated in the above studies showed that the number of permanent preparations required for the study, is determined by wealth of local paleoflora composition: the richer paleoflora, the greater number of preparations should be made for a more detailed insight into its composition. Thus, a collection containing 150 preparations was made for paleoflora richest in taxonomic respect which was synchronous to the Selerikhan horse. A collection of 40 preparations was gathered in the result of study of pollen and spores, separated from food remains of the Shandrin mammoth. However, the examination of the first 20 preparations showed that pollen and spores of new taxa ceased to be found. But, taking into account the fact that in working with species determination it is not expedient to note more than 2 to 3 spores or pollen grains in one preparation, the number of pre-parations, required for the study, was doubled. Collections of preparations on all the discoveries are stored at the Palynological Laboratory of the V.L. Komarov

Botanic Institute, USSR Academy of Sciences, Leningrad.

The examination of the fossil forms, their identification and description were performed on MBI-3 and MBI-15 microscopes with the use of a MC-51 comparison microscope. Pollen and spores were photographed on Bush microscope by the author with the participation of B.T. Shapkov, and on a MBI-15 microscope by the author. Measurements were made with the aid of a OMB-15 screw eyepiece micrometer. Pollen and spores, recognized in food remains of the fossil animals, were described and documented with microphotos (Ukraintseva, 1977, 1982a)

As a rule, descriptions were made on the basis of study of no less than 10 forms of each species. In the cases when pollen and spores of some plants were infrequent, the number of samples on which the species was determined and described, was indicated in the description (Ukraintseva (Kultina), 1977.

At the final stage, a list of all plant taxa in the rank of species, genus, and family which were established in studying; the collection of permanent preparations, was compiled. The list gives notions of: (i) the taxonomic composition of plants which grew in the animal's lifetime in both the area of its death and in adjacent areas; (ii) the character of paleobiotopes which served as pastures for the animal; (iii) the season when it died. Such a list may be regarded as local paleoflora whose time of existence is determined by radiocarbon analysis of food remains and skeletal bones of the animal.

The studies showed that in the case of such finds the composition of local paleofloras can be revealed more or less completely; this depends, as shown below, on a number of factors of both objective and subjective character (see Chapter 4). Nevertheless, paleopalynoflora analysis only, but not sporo-pollen analysis, can reveal a detailed composition of local paleofloras; but the sporo-pollen analysis has some advantages over the former. That is why it is expedient to use sporo-pollen analysis at the following stage, since its data are vital for solutions of some concrete problems, in particular, to reveal zonal patterns of the character of vegetation in the areas where the animals died; to determine percentages of certain groups of plants in their nourishment, etc.

2.2.2. Method of sporo-pollen analysis

This method is universally adopted and widely used. Interpretation of data, obtained with the method, is known to be based on the correspondence principle between the composition of vegetation and produced sporo-pollen spectra, collected from marine, continental, lacustrine, palustrine and other deposits.

Food remains of herbivorous fossil animals were found also to be good collectors of pollen and spores of both plants which served as forage for the animals and other plants which grew ln the vicinity of their ranges. Pollen and spores are practically not digested in intestinal tracts of the animals under the action of gastric juice and enzymes, and therefore both the intestinal contents of the animals (Tikhomirov and Kupriyanova, 1954; Tikhomirov and Kultina, 1973; Ukraintseva (Kultina), 1977; Ukraintseva et al., 1978; and others) and their excrements (Spaulding and Martin, 1979; Hansen, 1980; Thompson et al., 1980; Agenbroad et al., 1984; Davis et al., 1984) can be used for sporo-pollen analysis.

Palynologists reconstruct vegetation of the past and analyze flora on the basis of revealing and comparing pollen spectra. So far the term "sporo-pollen spectrum" has been differently interpreted in literature (Sladkov, 1967). The author adheres to the most precise definition of the notion, given by Grichuk and Zaklinskaya (1948). In his opinion,sporo-pollen spectrum is a whole collection of pollen and spores, both from the present soil surface and found in fossil state; the collection is expressed as percentages of the components. Spectra of the contents of gastrointestinal tracts of herbivorous animals were formed primarily by pollen and spores of the plants, eaten by the animal, and by pollen and spores, accumulated during flowering on other plants and in water basins (streams, lakes, rivers). In such spectra, a "factor of redeposition" is, in fact, ruled out, and redeposited forms, if any, are easily discernible. Adventitious (foreign) pollen and spores are also readily recognized because of their restricted part in a pollen spectrum.

In studying food remains of the fossil animals by sporo-pollen analysis, it was necessary to understand: (i) whether the composition of sporo-pollen spectra of gastrointestinal content adequately reflects the role and percentage of certain plant groups in forage of the animals and, if adequately, (ii) the extent of adequacy; and (iii) zonal patterns of paleovegetation of their ranges. These features must have been controlled by: 1) percentages of main fodder plant groups, determined from their macroremains in samples under study; 2) composition of sporo-pollen spectra of animal-embedding deposits or sediments, synchronous to its existence; 3) spectrum composition of surficial samples, taken at key sites of main types of present vegetation of the areas where the fossil animals or their remains were found; 4) composition of spectra of food remains of present herbivorous animals.

In this connection, the author adopted a single technique of estimation of the composition of sporo-pollen spectra of: (i) samples taken from different sections of gastrointestinal tracts of fossil animals; (ii) samples of deposits in which the animals were embedded; and (iii) surficial samples. The composition of each spectrum, obtained in study of all the above-mentioned sample types, was divided into four groups, corresponding to four groups of life forms of plants: 1) pollen of trees; 2) pollen of shrubs and low shrubs; 3) pollen of herbs and dwarf shrubs; 4) spores of sporophytes. In addition, pollen of three more subgroups, such as grasses, sedges, and forbs were recognized in the group of grass pollen. The technique adopted by the author was reflected in construction of histograms and pollen diagrams. This approach allowed establishment of a common criterion which can be reliably used for both appraisal of participation of these or those plant groups in nutrition of the animals and reconstruction of zonal patterns of paleovegetation. This criterion is a common spectrum composition, calculated from four, but not three component groups, as adopted in most papers on sporo-pollen analysis. Such an approach was found to lead to a more reliable recognition of both zonal and local components of spectra; their percentages gave a more valid notion of zonal patterns of paleovegetation and of the role and percentage of certain groups – shrubs, low shrubs, herbs, including grasses, sedges, herbs, in forage of the animals.

It should be noted that the author first carried out palynological study of Holocene deposits of Taimyr (Kultina et al., 1974) and Yakutia (Belorusova et al., 1977) by using this approach; this allowed tundra and forest spectra to be reliably differentiated.

2.2.3. Method of paleoclimatic reconstructions

Climate influences the life of herbivorous animals both, directly (air temperature, humidity, etc.) and indirectly, determining the distribution of plants, including forage crops, the qualitative composition of forage, its seasonal and weather dynamics. Knowledge of the character of climate can give an objective notion of living conditions of fossil animals, and their ecology. Therefore climatic reconstruction is one of the main tasks in studying habitats.

Plants once growing in the ranges of particular fossil animals can serve as reliable indicators of paleoclimate and its variations. Hence, their composition is one of foundations for paleoclimatic reconstructions. As early as 1914, V.N. Sukachev endeavored to determine the character of paleoclimate of the Berezovka River Basin, synchronous to the life of Berezovka mammoth, judging from plants, ascertained in its food remains, and plants, found in deposits of the steep bank of the Berezovka River near the site of its death. Trunks of the larch (*Larix* sp.), woody birch (*Betula* sp. ex sect *Betula*), and alder (*Alnus* sp.), found under the mammoth carcass and in the deposits of the steep bank, led V.N. Sukachev to the conclusion that the climate of the area during the lifetime of the mammoth was not more severe than that of the present. The above-mentioned species grow now in the Berezovka River Basin and in other areas of the Kolyma River. Was the climate warmer than that of today? V.N. Sukachev could not answer the question because of inadequate data.

A more complete composition of plants, revealed by palynological analysis of food remains of the Berezovka mammoth led L.A. Kupriyanova (1957) to the conclusion that in the past the climate of the Berezovka River Basin was warmer than that of the present, as some plants, such as the pine (*Pinus sibirica*), pea-shrub (*Caragana jubata*) and others, occurring then in the Berezovka River Basin, now grow much farther south and southwest of the area.

Paleofloristic data have long been used in successful reconstruction of quantitative characteristics of some elements of paleoclimates. These data usually permit reconstruction of quantitative characteristics of the following elements of paleoclimate: mean temperature of July, the hottest month; mean temperature of January, the coldest month; mean annual temperature; mean annual precipitation; the duration of frost-free period (Iversen, cited from Grichuk, 1969; Szafer, 1946, 1954; Zagwijn, 1963; Baryshnikov and Malaeva, 1967; Klimanov, 1976, 1982; Klimanov and NikiIorova, 1982; Byrashnikova et al., 1982; and others).

The author used zonal vegetation types, as indicators of paleoclimates of the areas where the animals dwelled, reconstructed with the aid of sporo-pollen analysis in the process of examination of their food remains and the deposits where the animals were buried. To estimate the character of both present and past climates, the author used the following elements: 1) mean month and mean annual air temperatures; 2) sum of air temperatures above 0°C; 3) sum of air temperatures below 0°C; 4) mean annual precipitation; and 5) precipitation for a period of positive temperatures.

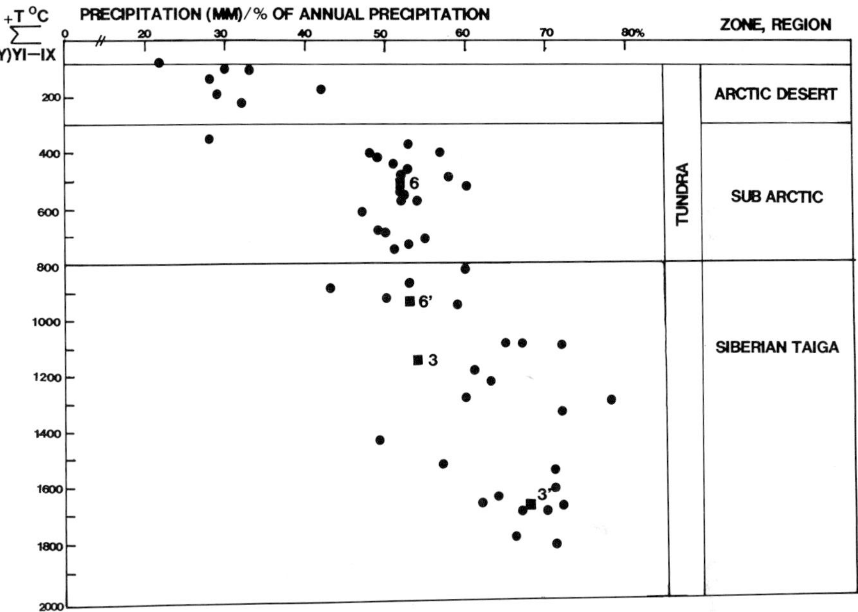

Figure 3. Correlation between positive temperature sums and precipitation during the growth and vegetation zoning in Siberia.

Sums of air temperatures above 0°C and precipitation for the period of air temperatures above 0°C, expressed as a percentage of total annual precipitation, were used to reveal conditions of growth of zonal vegetation types. Climatologists and biologists have widely used the method of sums of air temperatures since Réaumur times in both original and variously modified forms (Budyko, 1967; Young, 1971; Romanova, 1978; Tuhkanen, 1980). The method also gained recognition in agriculture and forestry (Selyaninov, 1961; Davitaya, 1964); besides, it is used in phytogeography (Tuhkanen, 1980). Nevertheless, the method has not found proper application in reconstructions of paleoclimates of the past yet. But a distinct correlation, observed between heat (sum of air temperatures above 0°C) and moisture provision (precipitation for the period of positive air temperatures) and zonat vegetation types suggests that these climate characteristics show promise for estimation of past paleoclimates on paleo-phytogeographic gound, since with this approach "the possibilities of skewed estimation of the climate will become minimal" (Borisov, 1975, p. 291).

S.S. Savina and N.A. Khotinsky (1982) were the first in Russia who employed the method of sums of air temperatures for the purposes of reconstruction of Holocene paleoclimates; they used sums of air temperatures above 5°C and 10°C, mean air temperatures of January and July, and vaporability values (Eo) to estimate heat conditions of plant growth.

Paleoclimatic reconstructions, carried out by the author, showed that the application of data on air temperatures of the hottest and coldest months of a year and data on sums of air temperatures above 5°C and 10°C for estimation of conditions of plant growth in the high latitude areas did not give a complete insight into the character of provision of vegetation with warmth in these latitudes. This is suggested by analysis of quantitative characteristics of ambient air temperature from data of weather stations, situated in tundra and forest (northern type) zones. According to multi-year observations of 10 stations, the mean air temperature of July (August) did not exceed 5°C; the number of stations made up 18% of 54 stations whose observational data were used; 16 stations (30%) fixed the air temperature in July (August) not higher than 10°C. Hence, data on 26 stations (48%) of all the stations) would not have been used if traditional sums of air temperatures above 5°C and 10°C had been used for estimation of heat provision of the areas.

First of all, quantitative characteristics of the above-mentioned elements of present climate were obtained for each area of paleoclimatic reconstruction; the characteristics were calculated by interpolation of mean multi-year data of no less than 3-6 stations which were the nearest to a particular area under study (Reference book of the climate of the USSR, 1966, 1968, 1970, 1972, 1973). Sums of air temperatures were obtained by multiplying mean month temperatures, calculated by interpolation, by the number of days in appropriate month (Table 1). Characteristics obtained allowed construction of climagraphs (Fig. 4). Such climagraphs appear to be graphic examples of the annual course of heat and moisture provision of zonal vegetation types.

Further, usual mapping of the ranges of dominants and subdominants, recognized within reconstructed paleovegetation of a certain "time section" of a particular area under study permitted us to reveal an area of their common growth at present, which was regarded as an "area-analogue". Quantitative characteristics of the above-mentioned elements of climate were obtained for a recognized "area-analogue" also by interpolation of mean multi-year data of some weather stations, situated within the area; the characteristics were regarded as "reconstructed elements of paleoclima" (REP) for the area in question; they described its heat- and moisture provision during a time span in the past which is reconstructed. Differences, revealed in quantitative characteristics of the above-mentioned elements of present climate and paleoclimate (Table 2), allowed us to determine the character and directivity of climate variations in a particular area under study, as compared to the present. Climagraphs which can be considered as paleoclimagraphs were also constructed from the reconstructed elements of paleoclimate.

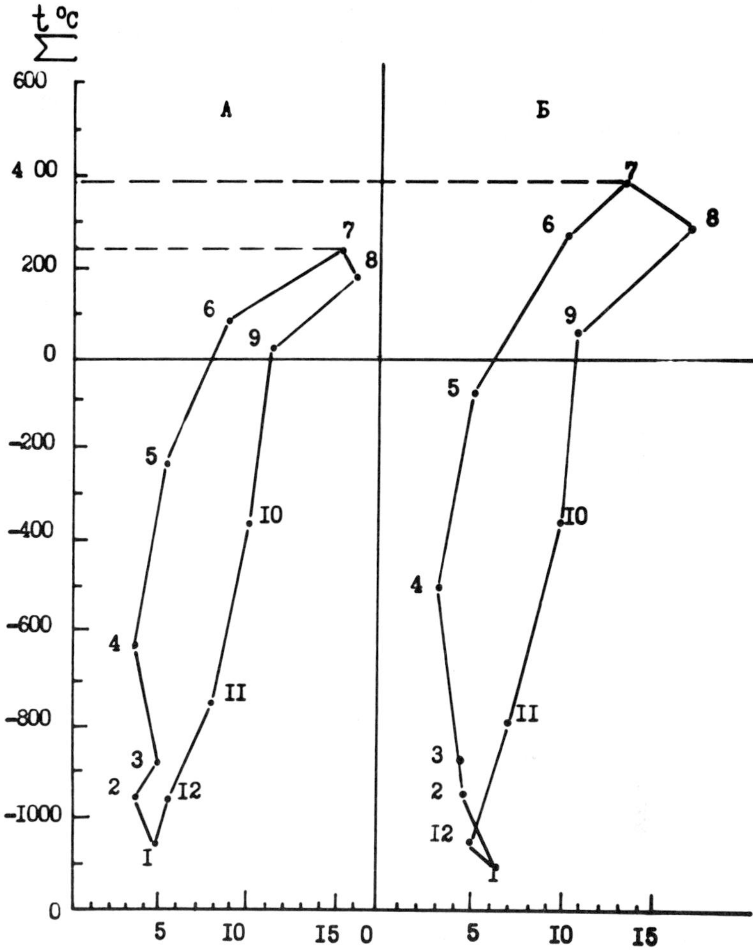

Figure 4. A) Climagraph characterizing the annual temperature and moisture conditions in the middle Shandrin River Basin. Constructed from multi-year data of three meteorological stations, nearest to the area where the mammoth was found; B) Paleoclimagraph characterizing climatic conditions of the area 40, 350±880 yr B.P.; 1-12 months.

Thus, fulfilled paleoclimatic reconstructions suggest that in high latitudes and, particularly in mountainous areas, it is advisable to use, for estimation of the character of paleoclimate, in addition to mean month characteristics of air temperature and precipitation, also sums of temperatures above 0°C and precipitation for a period of positive temperatures, i.e. the characteristics which determine the distribution of zonal vegetation types.

Data, obtained for each area under study, give a notion of the character, general directiveness and range of climate variations in some area of western Siberia, Taimyr, Yakutia and the North East of the USSR during the last 53,000 years.

Table 1

Annual course of temperatures and precipitation in the mid-Shandrin River Basin
lower Indigirka River[1]

	Months												Year	
	1	2	3	4	5	6	7	8	9	10	11	12		
	Temperatures													
Xt	-34.4 -34.4	-28.8	-21.0	-7.8		2.9	7.8	5.5	0.4	-11.7	-24.9	-31.2	-15. 0	513.0
Σ	-1066.0	-963. 0	-894. 0	-630.0	-243. 0	87.0	242.0	172.0	12.0	-363. 0	-747. 0	-967. 0		
	Precipitation, mm/% of sum of annual precipitation													
X	11	7	10	7	11	17	28	30	21	19	14	11	186	96.0
%	5.0	4.0	5.0	4.0	5.0	9.0	15.0	17.0	11.0	10.0	8.0	5.0		52.0

1) Calculated by interpolation of mean multi-year data of three weather stations, the nearest to the site where the mammoth was found.

Table 2

Values of heat- and moisture provision of the lower Indigirka River Basin
in the present and in the past

Area	Temperature, °C			t°C months (V)VI-IX	Sum of annual precipitation mm	Precipitation for months (V)VI-IX: mm/% of sum of annual precipitation
	July	January	mean annual			
	Instrumental data					
Mid-Shandrin River, lower Indigirka River	8.0	-31.0	-15.0	513	185	96.0 52.0
	Reconstructed for 40,350±880 yr B.P.					
	12.0	-34.0	-13.0	936	277	146.0 53.0
	Difference					
	4.0	3.0	2.0	423	92	50.0 54.0

Calculated by interpolation of data of three weather stations, the nearest to the site where the mammoth was found.

Figure 5. Mammoths (*Mammuthus primigenius* Blumenbach). Drawing by K.K. Flerov.

Chapter 3

"Mammoth" Faunal Complex: Brief Ecologo-Geographic Characteristics

Mammoth, bison, and Chersky horse are representatives of "glacial", as defined by V.I. Gromov, or "mammoth" fauna, which has been recently described in detail (Kuzmina, 1977; Vereshchagin and Baryshnikov, 1980; Vereshchagin, 1979). According to their data, main components of this fauna in the Arctic zone of Eurasia were: mammoth, Chersky horse, woolly rhinoceros, Pleistocene Arctic hare, glacial gopher, Kamchatka marmot, narrow-skulled vole, collared lemming, reindeer, primeval bison, Baikalian yak, musk ox, bighorn sheep, northern saiga, wolf, Arctic fox, polar bear, glutton (wolverine), cave lion, and cave bear. Remains of some of the above animals are more or less common in late Pleistocene deposits of northern Eurasia and in those of high and middle latitudes of North America. However, bones, teeth, skulls, tusks, more or less complete skeletons and even frozen carcasses of the woolly mammoth *Mammuthus primigenius* are the most common finds, and that is why the faunal complex of this period was named the "mammoth" complex.

The above-enumerated animals are dwellers of open landscapes. In the course of evolution they developed morphofunctional adaptation to living under cold and dry climatic conditions. For instance, the woolly mammoth had thick hair-covering consisting of thin down hairs, 15-16 cm long, and guard hairs, up to 1 m long; long intermingled hairs hung down from the lower part of chest and from the belly in the form of skirt (Fig. 5). Its legs, but for feet, trunk, ears and tail were completely covered with hairs. The mammoth's ears were 10-12 times smaller than those of the African elephant and 5-6 times smaller than ears of the Indian elephant (Vereghchagin and Baryshnikov, 1980). Similar to feet of its predecessor, *Mammuthus trogontherii* and all mastodons, the feet of the mammoth were aserial and therefore adapted to walk on snow and unsteady soil as a wide middle carpal bone of a foot rested not only upon the capitale bone, but also upon two adjacent lower bones; this aided in an even distribution of the great animal's weight upon the feet thus making them more steady and firm (Garutt, 1951).

The mammoth's heart was found to be reminiscent of that of modern elephants (Shilo et al., 1983); nevertheless, the mammoth' circulatory system shows the tendency toward doubling venous vessels serving as a deponent of blood which is typical of archaic or extremely adapted and not mobile animals (Ivanova, 1981).

Most of the above-listed animals were herbivorous mammals. The mammoths were real grassophages. They has special organs for grasping (Flerov, 1931) and grinding rigid grasses (Gromov, 1948). Flerov (1931) showed that the "bilobate" structure of the mammoth's proboscis end was related to the means of eating: it consisted of a very branched lower lobe and an upper digital process and, hence, represented a suitable organ to pick grass. According to Gromov (1950), the evolution of the masticatory apparatus from the primitive *Archidiskodon (Elephas) meridionalis* to the mammoth led to: an increase in the height of the crown and in the number of tooth plates; the appearance of folded enamel; and a decrease in its thickness. After Gromov (1950), these changes in the masticatory apparatus were caused by a change in the composition of food from more succulent and soft to coarser, which may have been associated with the transition of the mammoth's predecessors from forest to open areas, dominated by rigid herbs.

Naturally, a notion of reproduction of these extinct animals can be provided by indirect data only. Vereshchagin and Baryshnikov (1980) came to the conclusion that the net reproduction of the mammoth was probably lower as compared to the present elephants; this is suggested, in their viewpoint, by an age composition of the Berelekh mammoth paleopopulation in the lower Indigirka River. Young mammoths (under 10 years old) and foetal individuals respectively accounted for 30% and 2% of remains of 140 individuals of the Berelekh cemetery.

The musk ox is a contemporary of the mammoth; it has survived to the present in the north of Canada, in Greenland and on Spitsbergen, i.e. in the area where climatic and landscape conditions are related to the presence or proximity of glaciers. The musk ox is of square build; its neck, legs, ears, tail are very short;

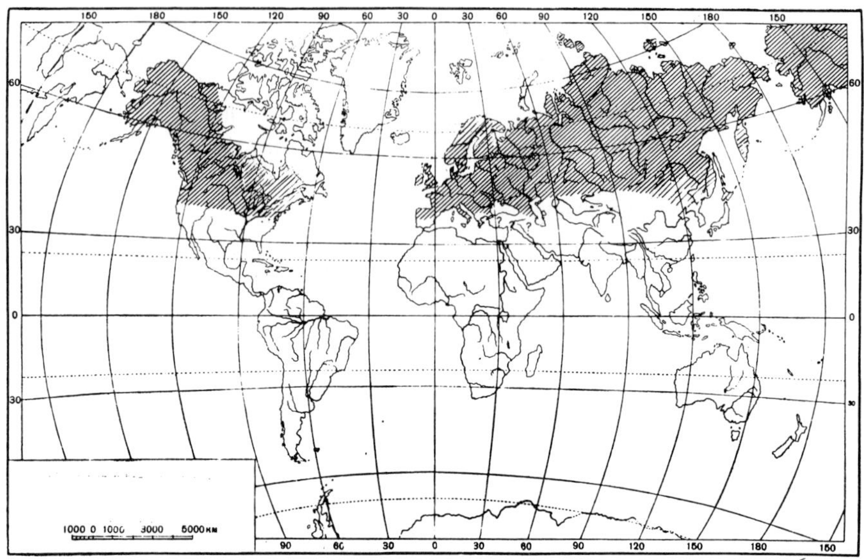

Figure 6. Range of mammoth (*Mammuthus primigenius* Blumenbach) in the Pleistocene. Modified from Flerov et al., 1955. After Agenbroad, 1984.

small lateral hoofs or feet are arranged high above the ground; wool is thick, guard hairs hang down almost to the ground; they are 60-90 cm long on the chest and sides. Fur hair, 10-15 cm long, accounts for 60% of its wool. Blood circulation is reduced in the animal's legs and their supercooling does not harm it. The animal is indifferent to bitter and long frosts but it suffers from wet warm winters, accompanied by deep snow and thaws with subsequent glazed frost. The musk ox has probably derived from northeastern Siberia. A lineage of its ancestral forms is known from there (Sher, 1971). The oldest remains of the musk ox are known from the Mindel gravel at Sussenborn, Germany (Soergel cited from Vereshchagin and Baryshnikov, 1985). In the middle and late Pleistocene, the range of musk ox swept almost the whole of Europe and northern Asia (Fig. 6). Musk oxen lived at Chelyuskin Cape as late as 3000 years B.P. Earlier dated finds have not yet been reported for Taimyr.

The emplacement of cold tolerant species of fauna and flora was related to cold spells in the late Pliocene (Lazukov, 1973) and the development of thick ice sheets in the Pleistocene (Gerasimov, 1961; Markov et al., 1961; Markov et al., 1968; Velichko, 1973).

Paleozoological evidence suggests that the faunal complexes of the Eopleistocene and early Pleistocene of Europe contained no marked admixtures of cold tolerant elements. A wider distribution of cold tolerant "mammoth" fauna (after V.I. Gromov) in both Europe and Siberia occurred during the Riss (Dnieper) glaciation (Velichko, 1973). According to Ivanova (1965), *Mammuthus trogontherii* was completely supplanted by the whooly mammoth during the vast glaciation. Kahlke (1976) believes that the mammoth appeared in Europe no

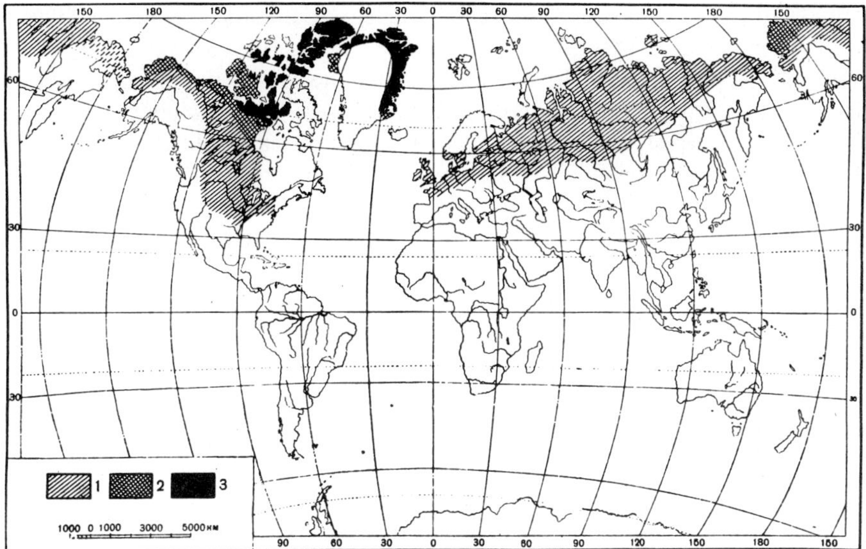

Figure 7. Range of musk ox (*Ovibos moschatus* Limm) in the Northern Hemisphere. After Flerov et al., 1955. 1 Pleistocene, 2 19th century, 3 at present.

earlier than the middle Pleistocene, however, it was not a leading form among the elephants in the faunal complex.

The acme and wide distribution of "mammoth" fauna throughout extra-tropical Eurasia took place during the second half of Pleistocene or the glacial period proper. After Kahlke (1976), in late Pleistocene time corresponding to the last glaciation maximum, the situation drastically changed in the Far East, as well as in the west. In this area, the population of paleoloxodons were replaced by those of *Mammuthus primigenius*. At the time the mammoth had the widest distribution in eastern Asia and settled not only over the Bering Land Bridge, but also over the Tatar Land Bridge (Kahlke, 1976).

According to the present concepts, the formation of the core of "mammoth" fauna occurred in the northeast of the Asian part in the early Pleistocene (Sher, 1970, 1971; Vereshchagin, 1971). According to Sher, "it is in the north of Eurasia, particularly on the Primorsky lowlands, that the core of the future periglacial upper Palaeolithic fauna was probably emplaced in the lower Pleistocene, since sufficiently severe climatic conditions, similar to those of the present hypo-Arctic, had already existed there by that time" (Sher, 1970, p. 523).

The middle Pleistocene witnessed the acme of early "mammoth" fauna (Vereshchagin, 1979; Vereshchagin and Baryshnikov, 1980). At the time, the mammoth (its earlier form), large horse, ass, *Elasmotherium sibiricum*, gigantic camel (*Camelus knoblochi*), gigantic deer (*Megaloceras giganteus*) and red deer (*Cervus elaphus*), musk ox, and long-horned bison settled over the plains of Siberia and eastern Europe. For the first time the moose (*Alces*) and reindeer became widely settled and became rather abundant; the saiga dwelled on drainage divides. In the late Pleistocene, the late "mammoth" fauna distributed in various areas of Eurasia; the remains of representatives of this fauna occur at present from England in the west to Alaska in the east; they are known as the "upper Palaeolithic" or "Khvalynskoe" theriocomplex. Woolly rhinoceros, three or four species of horse, Pleistocene ass, as well as Asiatic wild ass in southern Siberia and in the Transbaikalian area, were indicator species of the fauna. Red deer (maral, Manchurian deer, gigantic deer) widely settled in river valley in mid-latitudes. Reindeer, saiga, and musk ox had extensive ranges. Mammoths, musk oxen, reindeer and other representatives of the "mammoth" complex inhabited the Arctic with the mammoth and musk oxen dwelling in the circumpolar areas (Fig. 7). The southern limit of woolly mammoth distribution in Europe and North America reached 39°40°N (Flerov et al., 1955; Agenbroad, 1982). In the Territory of China, the mammoth dwelled slightly to the east, reaching 35°N (Kahlke, 1976; Liu Tung-sheng and Li Xing-guo, 1984). The most northerly finds of mammoths are known from Severnaya Zemlya (Makeev et al., 1979).

The "mammoth" fauna reached the acme and wide distribution in the Pleistocene, and at the end of the period it underwent substantial changes. The mammoth, woolly rhino, Chesky horse, primeval bison, cave bear and some other animals died out. According to radiocarbon dating, the last representatives of the genus *Mammuthus* perished in both Siberia and America (Agenbroad, 1984) about 10,000 yr B.P. The genuine bison, saiga, yak, and musk ox sharply reduced the ranges (Fig. 6). In Siberia, the musk ox substantially outlived the mammoth: as late as 2920 ± 50 yr B.P. (GIN-2945, L. Sulerzhitsky, oral communication) the musk oxen inhabited Cape Chelyskin. On Spitsbergen, they survived to our days. In the early 1970s a very small population of the animals lived only there. By the December of 1982, a single cow remained on Svalbard Island (E. Alendal, written communication).

Factors such as "the sum total of which overcomes such giants as the elephant and rhino that were able to develop warm clothing in the course of moving northward, and such patient animals that were able to content themselves with frugal food and stand considerable cold as the horse" (Pavlova, 1924, p. 39) remained unclear for a long time. Nevertheless, the scientists have long ago come to the conclusion that the causes of extinction of certain groups of organisms, both plants and animals, cannot be understood with no insight into their living conditions, their ecology. Vegetation is one of the main factors characterizing living conditions of herbivorous animals. Plant remains, once eaten by animals and preserved in more or less digested state in their gastrointestinal tracts, are the source of extremely valuable paleobioecological information (Ukraintseva, 1979, 1981b, 1985).

Chapter 4 deals with the results of complex study of food remains of some representatives of the "mammoth" faunal complex and enclosing deposits which contributed greatly towards the understanding of living conditions of both concrete individuals and the "mammoth" fauna as a whole.

Chapter 4

Living Conditions of Representatives of the "Mammoth" Faunal Complex in Siberia (as derived from Paleobotanic Data)

Long before the Berezovka mammoth with plant remains preserved in its gastrointestinal tract was found, A.F. Middendorf wrote: "What complete idea would we obtain of living conditions of the mammoths and reasons for their extinction if the stomach contents of the best preserved carcasses were to be subjected to microscope analysis" (Middendorf, 1861, p. 855).

Up to now nine fossil animals—representatives of the "mammoth" faunal complex—have been found in various areas of the north of Siberia; the content of the gastrointestinal tracts had been preserved in six of them, namely, the horse, bison, four mammoths; five of the fossils were found in the Kolyma and Indigirka Basins (Fig. 1). In addition, an entire "cemetery" of mammoths and some of their associates was dug out on the Berelekh River, the lower Indigirka River.

Each new find has contributed to our knowledge of living conditions of the animals of the "mammoth" faunal complex such as: the composition of plants growing in the vicinity of the sites where the animals dwelled; the character of vegetation dominating this or that area; the composition of plants eaten by the animals and percentages of different plant groups in their feed; habitats which served as pastures for the animals; the time, season and circumstances of their death; taphonomic and geological conditions of burial of the animals; and, lastly, the character of paleoclimates peculiar to each of the areas where the animals were found.

Results of the study of each find were widely published. Here the author summarizes firm data on all the discoveries, including her own research obtained during the investigations. Owing to efforts of many investigators, living conditions of all representatives of the "mammoth" faunal complex, found during the past 80 years,[1] have been studied in sufficient detail. New data allow a new approach to be applied to the problems of the causes of extinction of some groups of animals in the Pleistocene. A separate chapter is devoted to the problem. It is not merely a tribute to "the problem old as the hills" (Vereshchagin, 1979, p. 168), but the result of multi-year investigations and earnest interest to this complex and acute problem.

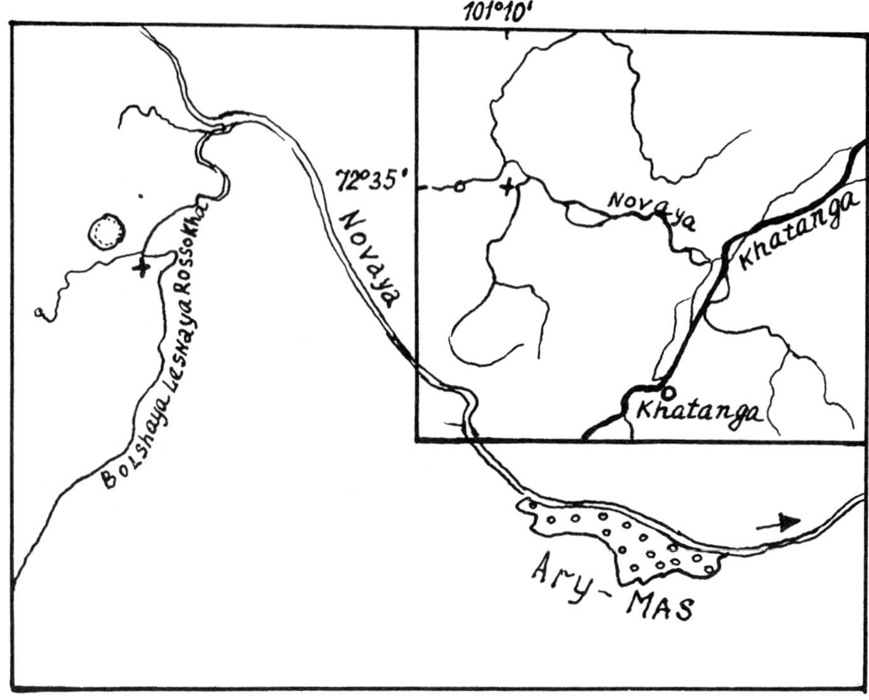

Figure 8a. Site where the Vereshchagin mammoth was found (marked "+").

4.1. Vereshchagin (Khatanga) Mammoth

Mammoth remains, including the head with trunk and tusks, a right femur void of muscles and ligaments, crus and foot coupled in hide with ligaments, and two ribs (Vereshchagin and Nikolaev, 1982), were found in the summer of 1977 on the left steep bank of Bolshaya Lesnaya (Large Forested) Rassokha River, the right tributary of the Novaya River, in the lower Khatanga River Basin, southeastern Taimyr, 15 km from the river mouth (Fig. 8a). The remains of this animal occurred in permafrost sand, 5 m from the water edge (in the low-water period) and 105 m above the key peat bed (Fig. 8b). In late July-August, 1978, the peat retained a peculiar smell of the gastrointestinal content which was not that of putrefaction. This suggests that the rest of the body, including entrails, had been recently gone with water when the terrace had been broken (Ukraintseva et al., 1981).

In July-August, 1978, an integrated study of the area was undertaken. The party of the Polar Expedition, V.L. Komarov Botanical Institute, led by the author, studied recent flora, and other vegetation in the vicinity of the site and enclosing deposits. The site where the remains were found is situated 20-25 km north-west of Ary-Masurochishche — the most northerly forest tract in the world (Norin, 1978a). The area under study lies in the tundra zone. The vegetational cover is represented by: 1) shrub tundras; 2) forb - undershrub tundras; 3) "spot-

Figure 8b. Left bank of the Bolshaya Lesnaya Rassokha River, mammoth's burial site. Photo by
V.V. Ukraintseva.

Figure 9. *Larix gmelinii*, the larch. At the edge of alder forest, right bank of the Bolshaya Lesnaya
Rassokha River. Photo by V.V. Ukraintseva.

ty" undershrub tundras; 4) polygonal bogs; 5) floodplain shrubs (Ukraintseva
and Kozhevnikov, 1981). The most abundant are the shrub tundras, associated
with terraces II and III above the flood-plain and around lakes. Forb — undershrub

and "spotty" undershrub tundras are fragmentary. The right, low elevation river bank is the north-westernmost position of the range of larch in the area. Some specimens of the larch (*Larix gmelinii*), reaching 1.5-1.6 m in height and 10-12 cm in diameter, were encountered there on the margins of an alder scrub forest 1 km in length and 0.5 km in width; these young trees had already born fruits (Fig. 9). On the left, high elevation bank, the larch occurred as beds of elfin woods only.

Outcrop No. 1, whose deposits embedded the mammoth's remains, exposes the following beds[2] (in ascending order):

Bed	**Thickness, m**
1. Sand light- and dark-grey, fine-grained, horizontally bedded	0.50
2. Clay dark-grey with seams of organic matter	0.10
3. Sand yellowish – light grey, quartz, medium-grained, well washed, rare gravels and pebbles	1.95
4. Peat unripe with thin (up to 1 mm) bands of sand and silty particles	0.45
5. Sand yellowish – light grey, inequigranular, well washed, with bands of dark-green sand and inclusions of coal crumbles and organic material (small twigs, bark, leaves)	2.55
6. Sand yellowish – light grey, medium-grained, in places fallow (ferruginate), with inclusions of poorly rounded pebbles and thin (1 mm) bands of vegetable debris	1.20
7. Sand dark-grey, medium-grained, horizontally bedded, in places showing evidence of strong ferrugination	1.65
8. Sand yellowish-grey, quartz, medium-grained, similar to bed 3	0.80

Net apparent thickness 9.65 m.

Forty seven specimens were taken for palynological analysis from the above-enumerated deposits upon their thorough cleaning up; specimens 6 and 10 were taken from the peat bed for botanical and radiocarbon analysis, respectively. Surficial samples were collected for palynological analysis at key sites of the main vegetation types. All the materials served as the basis for paleogeographical reconstructions of the area under study.

According to radiocarbon analysis, the accumulation of the key peat bed (Fig. 10, Bed 4) was completed over 34, 730 yr B.P. Botanical analysis, carried out by M.S. Boch, showed that peat of this bed is chaff, unripe (it contains up to 5 % humus); it is almost completely composed of sedge fiber (*Carex aquatilis* ssp. *stans*) (Table 3), which is typical of present polygonal bogs in Taimyr and in the tundra zone as a whole.

Many nuts of *Carex* sp., rare seeds of *Juncus* sp., *Andromeda polypholia*, and sclerotia of fungi were washed and identified by M.G. Kipiani in the

Table 3

Results of Botanical analysis of peat from outcrop No. 1

| | SPECIMEN NO. | | | | | | | | | |
COMPOSITION OF PLANTS, %	1	2	3	4	5	6	7	8	9	10
Carex aquatilis ssp. stans	95	90	100	80	70	100	90	100	90	100
Eriophorum sp.		5		5	10					
Drepanocladus vernicosus				15	20		10			
Drepanocladus sp.	5								10	
Unidentified			5							

specimen taken by N.K. Vereshchagin from this peat bed (Vereshchagin and Nikolaev, 1982, p. 7).

An extremely poor species composition of macroremains suggests that it reflects the composition of vegetation taking part in the structure of pools in the polygonal bogs which had already started developing, at that time in the area under study.

At present, the polygonal bogs are second to shrub tundra in distribution in the study area. They are associated with the floodplain; terrace I above the floodplain; and the former bottoms of lakes of terraces II and III. More or less wet polygons may be as large as 10 X 10 m, 10 X 12 m, and, rarely, 10 X 20 m in area. *Carex aquatilis* ssp. *stans* is the dominant species of the pools, whereas *Carex rariflora, C. chordorhiza, Eriophorum polystachion,* and *Juncus arcticus* are subordinate. *Comarum palustre, Saxifraga hirculus, Pedicularis albolabiata, Betula nana* ssp. *exilis, Salix reptans* are rare. *Betula nana* ssp. *exilis, Cassiope tetragona, Pyrola* sp., *Calamagrostis holmii, Arctagrostis latifolia* and some other specimens of forb are associated with the drier mounds. The lichen-moss layer is characterized by a high degree of coverage (about 100%) in the depressions and on the mounds.

Palynological analysis of specimens No.s 10 through 13 from the peat bed (Fig. 10) allowed us to reveal both a more complete composition of flora of the time of formation of the bed and to trace changes in the character of vegetation of this period in the lower part of Bolshaya Lesnaya Rassokha River.

Sporo-pollen spectra of the four above-mentioned specimens are dominated by Quaternary pollen and spores (91.0-82.0%). Redeposited pollen and spores of Mesozoic-Neogene age (hereafter marked as Mz-N for brevity) range within 9.0-18.0%.

Quaternary pollen and spores are dominated by pollen of herbaceous plants [NAP(non arboreal pollen)], accounting for 49.7-77.3%. The herb group

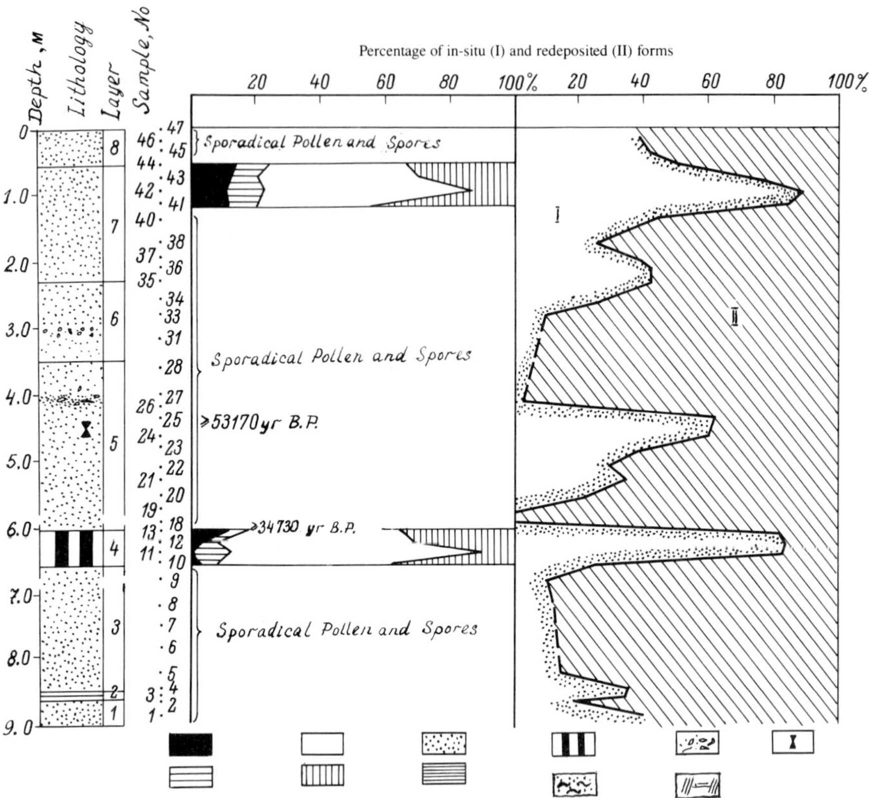

Figure 10.   Spore-pollen diagram of outcrop No. 1, exposing Upper Pleistocene deposits on the left bank of Bolshaya Lesnaya Rassokha River, at mammoth's burial site. The position of mammoth's remains in the outcrop is marked "X".

is dominated by pollen of sedges (*Cyperaceae*) (75.0-79.0%); pollen of grasses (*Poaceae*) and the pink family (*Caryophyllaceae*) constitutes (1.5) 9.6-17.0%, and 1.0-6.2%, respectively. Sporadic grains represent pollen of the following plants: *Juncus* sp., *Polygonum viviparum, P. bistorta, Rumex arcticus, Stellaria* sp., *Papaver* sp., *Tofieldia* sp., *Saxifraga nivalis, Saxifraga* sp., *Valeriana capitata, Artemisia* sp.1, sp.2, Nardosmia sp., *Ericaceae, Armeria arctica, Dryas punctata.* Spores account for 10.5-38.0% with spores of true amounting to 95.0-100.0%, but they are represented by no more than three species. There were noted sporadicspores of *Sphagnum* sp., *Huperzia (Lycopodium)* cf. *selago.*

Pollen of shrubs makes up 5.0-11.0%; in fact, it is represented by rare grains of the birch (*Betula exilis*) and alder (*Alnus fruticosa*). Pollen of the willow (*Salix* sp.1, sp. 2) is sporadic.

Pollen of trees in specimens Nos. 10 through 12, is represented by occasional grains of *Larix* sp., *Picea obovata, Pinus sibirica, Betula* sp., total 0.6 to 4.3%.

It should be noted that spectra of specimens Nos. 10, 11, and 12, taken from the peat bed (Fig. 10),are similar to those of the present polygonal bogs of the area, but differ in composition from spectra of present shrub tundra. This suggests that shrub tundra, at present the most widespread type of vegetation in the area, may have been fragmentarily distributed only in the vegetational cover of the lower Bolshaya Lesnaya Rassokha River during the formation of the lower part of peat bed. At that time the vegetational cover of the area was dominated by polygonal bogs. However, the composition of component species of both flowering plants and mosses was poorer, as compared to that of the present. Even mosses which made up the moss layer were composed of no more than three species, as evidenced by the composition of both macroremains and spores.

In spectrum of specimen No. 13 (depth 6.0-6.05 m), arboreal pollen reaches its relative maximum, 11.5%, although it is also represented by rare pollen grains of *Larix* sp., Siberian spruce (*Picea obovata*), cedar pine (*Pinus sibirica*), Scotch pine (*Pinus sylvestris*), mountain pine (*Pinus pumila*), large woody birch (*Betula* sp.), speckled alder (*Alnus incana*). In spite of the fact that in specimen No. 13 pollen of trees is obviously dominated by long distance-transported pollen of spruce, pine, woody birch, the amount of pollen of larch has markedly increased in it. In addition, the spectrum of this specimen contains pollen of *Menyanthes trifoliata*, spores of *Lycopodium annotinum* and *Equisetum* cf. *boreale*.

Taking into account a generally poorer preservation of fossil larch pollen and its lower pollen productivity as compared to the above-mentioned trees, it may be inferred that sedge larch forests and open woodlands were widely developed in the period of formation of the upper part of peat bed in the lower Bolshaya Lesnaya Rassokha River. The northern limits of spruce and cedar pine may have been close to the present ones. This indicates a climate like that of the present or somewhat warmer since forests, although light ones, reached a treeless area (at present) and probably extended farther north.

The pollen spectra of the specimens, taken from beds 1 through 3 which underlie the peat bed (Fig.10), are dominated by redeposited pollen and spores accounting for 59.0-92.0%; this points to an important part played by the processes of denudation in the period of the formation of the beds in the area.

Pollen and spores of Quaternary age constitute 8.0-41.0%; they are represented by occasional grains of the following species of trees (*Abies sibirica*, *Pinus sibirica*, *Betula* sp.); shrubs and low shrubs (*Alnus fruticosa*, *Betula exilis*, *Salix* sp.); grasses (*Carex* sp., *Poaceae*, *Artemisia* sp., *Taraxacum* sp. and others). They also contain scarce spores of the mosses (*Bryales* spp.) and sporadic spores of *Sphagnum* sp. and *Huperzia selago*. The composition of spectra of these specimens indicates that the time of formation of the peat bed was preceded by severe climatic conditions when no stable vegetational cover existed in the area and this caused strong denudation of previously deposited sediments.

Twenty six specimens taken from sand which overlies the peat bed (beds 5 through 8, Fig. 10) were also studied. Twenty two specimens (Nos. 18 through 40, 45 through 47) are characterized by: (i) a poor saturation with Quaternary pollen and spores; (ii) the predominance of redeposited Mz-N forms (Fig. 10) in spectra; (iii) the presence of long-distance transport pollen of the following trees: primarily, the Siberian pine (*Pinus sibirica*), Siberian spruce (*Picea obovata*), birch (*Betula* sp.). Percentages of Quaternary pollen and spores and redeposited forms (Fig. 10) in the composition of spectra of the sequence suggest unstable conditions of sedimentation at the time of its formation; processes of denudation dominated at some time (depth interval 5.75-6.0 m) or became stable at another (4.25-4.75 m) or again denudation dominated (1.30-4.0 m).

Pollen of grasses and spores of mosses (*Bryales* sp.) are few in the composition of palynological spectra of specimens, collected from depth interval 3.0-3.9 m. Pollen grains of the birch (*Betula exilis*) and alder (*Alnus fruticosa*) are sporadic whereas long-distance transported pollen of the Siberian Pine (*Pinus sibirica*) is the most abundant.

In depth interval 4.0-4.75 m where the mammoth's remains were buried, spectra of two specimens (Nos. 24, 25) are dominated by Quaternary pollen and spores accounting for 62.0-65.0% of the sum total of all the forms calculated. Mz-N redeposited forms amount to 35.0-38.0% (Table 4). Spores total 42.4-54.0% in the composition of Quaternary spectra with those of *Huperzia selago* prevailing; rare spores of the true mosses (*Bryales* sp.), bog mosses (*Sphagnum* sp.), club-mosses (*Lycopodium* sp.) were noted. Pollen of grasses make up 27.0-28.0%, but it is represented in fact by occasional grains of the grasses (*Poaceae*), sedges (*Cyperaceae, Ericaceae*), pink (*Caryophyllaceae*), serpent grass (*Polygonum bistorta*), sorrel (*Rumex arcticus*), buttercup (*Ranunculus* sp.), French honeysuckle (*Hedygarum hedysaroides*), dandelion (*Taraxacum* sp.) and others. Pollen grains of *Betula exilis* and *Alnus frutocosa* are rare. Each preparation contains six to sixteen grains of long-distance transported pollen of Siberian cedar.

Specimens, taken from depth intervals 5.75-1.5 m (beds 5 through 7) and 0.5-0.0 m, contain few pollen and spores, represented, if present, by rare pollen grains of grasses, scarce spores of mosses, and scanty grains of long-distance-transported pollen of Siberian cedar.

Four specimens only (No.s 44 through 41) from sand (Bed 7, upper part) are sufficiently impregnated with pollen and spores. The composition of their pollen spectra is dominated by pollen and spores of Quaternary age (51.0-89.0%). Redeposited Mz-N pollen and spores account for 11.0-49.0% (Fig. 10).

Pollen of grasses is most abundant (33.8-64.1%) in the composition of Quaternary pollen and spores with pollen of the sedges (*Cyperaceae*) and heather family (*Ericaceae*, cf. *Cassiope tetragona*) accounting for 55.4-78.2% and 15.0-28.0%, respectively. Pollen of grasses (*Poaceae*) constitutes 1.5-4.0% only. Separate pollen grains of *Eriophorun* sp., *Juncus* sp., *Sparganium* sp., *Polygonum bistorta, Rumex arcticus, Saxifraga hirculus, Saxifraga* sp., *Stellaria*

Table 4
**Results of palynological analysis of specimens from outcrop No. 1,
bed 5, which the mammoth's remains were buried**

Plants	Specimen Nos., depths in meters		
	sp. 23, 4.15-4.20	sp. 24, 4.35-4.40	sp. 25, 4.55-4.60
1	2	3	4
Pinus siberica	10	16	6
P. pumila	-	-	1
Betula sp. (ex sect. Fruticosae)	-	-	1
Betula exilis	3	7	-
Betula sp. ex sect. Nanae	-	-	2
Alnus fruticosa	-	8	2
Poaceae	-	4	3
Cyperaceae	2	2	4
Polygonum bistorta	2	4	2
Rumex arcticus	-	1	-
Caryophyllaceae	-	1	-
Ranunculus sp.	-	1	-
Papavaer sp.	-	1	-
Hedysarum hedysaroides	-	1	-
Ericaceae	5	7	4
Valeriana capitata	1	-	-
Artemisia sp.	-	1	-
Asteraceae	-	2	-
Hepaticae	1	-	1
Sphagnum sp.	-	-	1
Bryales sp.	2	11	2
Botrychium sp.	1	-	-
Polypodiaceae	2	-	1
Hyperzia selago	21	22	18
Lycopodium alpinum	2	4	1
Lycopodium sp.	-	1	1
Total amount of pollen and spores	50	100	50
of trees	10 20.0%	16 16.0%	6 12.0%
shrub and low shrub	3 6.0%	15 15.0%	6 12.0%
grasses	10 20.0%	27 27.0%	14 28.0%
spores	27 54.0%	42 42.0%	24 48.0%
Quaternary age	50 34.0%	100 66.0%	50 62.0%
Polygonum bistorta	85 65.0%	52 34.0%	31 38.0%

sp., *Caryophyllaceae, Dryas punctata, Polemonium* sp., *Ranunculus* sp., *Gentiana* sp., and *Artemisia* sp. were noted. Spores (*Bryophyta* and *Pteridophyta*) make up 29.6-46.1%. Spores of no more than three species of true mosses are predominant (65.3-74.6%) Spores of fir club-moss account for 2.0-18.2% reaching a maximum of 52.3% at a depth of 7.80-7.85 m. Spores of bog moss (*Sphagnum* sp.) and common club-moss (*Lycopodium clavatum*) are rare (they account for 8.8% at a depth of 8.0-8.25 m). Two specimens contain spores of *Selaginella rupestris.* Pollen shrubs of [*Betula exilis, Betula* sp. (sect. *Nanae*), *Alnus fruticosa, Salix* sp.] accounts for 9.0-12.0%; pollen of these plants generally shows evidence of underdevelopment. Pollen of trees [*Larix* sp. *Picea obovata, Pinus sibirica, Betula* sp. (sect. *Betula*)] totals 11.0-13.0%.

Spectra of the four specimens described above are very similar in composition to that of specimen No. 13, taken from the upper Part of peat bed, but differ from it in a greater amount of heather pollen. Percentages of main components, trends of curves of the total content of pollen and spores, the composition of palynoflores of the specimens, taken from depth interval 1.25-0.5 m, indicate that during the period of their formation, cassiope-sedge and sedge-cassiope larch forests were distributed in the area under study.

The composition of palynological spectra of specimens Nos. 1 through 23 (Bed 5, Fig. 10) suggests that very severe climatic conditions preceded the time of the mammoth's death. At that time the area under study was either void of vegetational cover, or it was so sparse that an extremely small amount of pollen and spores produced by occasional plants, got in the rocks on which they grew. Ground which had poor turf cover was readily subjected to denudation in summer when snow and frozen grounds actively thawed. This is clearly evidenced by a high percentage of redeposited forms in spectra. At that time the area was probably dominated by plant aggregations of the polar desert type which are presently distributed only on Taimyr in the Cape Chelyuskin area. Vegetational cover there ranges from 0 to 10-15%. Flowering plants are rare. No more than 10 species of flowering plants take part in the structure of the vegetational cover. The number of moss species in plant communities ranges between 5 and 12 with one species usually dominating (Matveeva, 1979). Only July and August have positive temperatures (1.5° and 0.8°C) but even in these months temperature can drop below 0°.

Percentages of Quaternary and redeposited pollen and spores (Fig. 10) suggest that in the period of mammoth's death the processes of denudation ceased; this was associated, among other causes, with an increase in density of vegetational cover. At that time meadowlike grass-forb aggregations developed with the participation of *Poa* sp., *Poaceae, Cyperaceae, Polygonum bistorta, Rumex arcticus, Caryophyllaceae, Ranunculus* sp., *Ranunculaceae, Papaver* sp., *Hedysarumhedysaroides, Valeriana capitata, Artemisia* sp. *Asteraceae, Huperzia selago* and others in grass stands; they occupied steep slopes of southern, southwestern and southeastern aspects; sedge-grass communities were spread on river and lake banks. Shrubs of *Betula exilis* and *Alnus fruticosa*, as well as mosses were subordinate in the vegetational cover, as suggested by a significant amount of pollen and spores of these plants in the pollen spectra characterizing the time of formation of mammoth-enclosing beds.

Outcrop No. 1 which exposes peat bed 4 below the mammoth (Bed 5) is represented by erosion scarps of terraces II and III above the floodplain; the terraces are very close to each other because of undermining and destruction of terrace II. By the present, terrace II (Karginsky) had been practically destroyed and the Zyrianka (=early Wisconsin) mammoth-enclosing beds were exposed. At the burial site, the Karginsky terrace has been preserved as a narrow step (1.5-2.0 m wide) hardly recognized in exposure, but distinctly marked by a ledge peat (Ukraintseva et al., 1981, fig. 4, line A-A1). There is also the possibility of a small sand sequence in depth interval 7.75-8.50 m (Bed 7, upper part), as suggested by the character of palynological spectra. Deposits underlying the peat bed were destroyed by water-jet pumping when digging and cleaning up the outcrop to take samples for palynological analysis.

Thus, integrated investigations showed that only a narrow near-bank fragment of terrace II of Karginsky (middle Wisconsin) age insets the older (Zyrianka), terrace III scarp at the burial site. Naturally, part of the section which overlies the peat bed and exposed deposits of terrace III yielded a higher radiocarbon value, as compared to peat. The accumulation of peat (over 34,730 yr B.P.) coincided in time with the Malaya Kheta phase (42,000-35,000 yr B.P.) of Karginsky warming-up, according to the scheme, proposed by N.V. Kind (1974), i.e. it took place at the close of second warm interval within the Karginsky Interglacial.

Karginsky sediments accumulated at different absolute heights, but the most favorable conditions for accumulation was under continental environments and further erosion took place in river valleys of the central North Siberian lowlands which suffered no major continental glaciation in Sartan time (Main Wisconsin). As mentioned above, in Sartan time the emplacement of the Novaya River Valley was associated with a drastic decrease in base level of erosion and with undercutting and destruction of the Karginsky terrace which exposed deep horizons of previously accumulated sediments (Belorusova and Ukraintseva, 1980). Erosion which continued into the Holocene was so vigorous and deep that it practically destroyed the Karginsky terrace at the site of the mammoth's burial.

The lithology of Bed 5 (Fig. 10) where the mammoth was found indicates it was accumulated under conditions of intensified erosion. Coal crumbles, fragments of organic matter in Bed 5, and a low impregnation of the specimens with Quaternary pollen and spores, and, at the same time, a high percentage of redeposited Mz-N forms point to intensive erosional processes, sparse vegetational cover and cooling (Ukraintseva et al., 1981).

The time of life and death of the mammoth is dated by the Zyrianka Ice Age (=early Wisconsin) when glaciers (primarily mountain and valley ones) descended from the Byrranga and the Putorana Mountains, but did not close in the center of the North Siberian lowlands. At that time, watershed plains were open areas with sparse vegetational cover, and the periglacial vegetational complex of the river valley contained meadowlike grass-forb aggregations and sedge-grass communities on banks of rivers and lakes.

Conclusions

1. The mammoth perished in Zyrianka time as suggested by both radiocarbon dating and geological-geomorphological analysis of the area in proximity of the mammoth's burial site.

2. Outcrop No. 1 exposing the peat bed above which the mammoth was buried is represented by scarps of terraces II and III remnants above the floodplain; the terraces are very close to each other because of the undermining and destruction of terrace II.

3. Palynological characteristics of the Zyrinka complex of mammoth-enclosing deposits indicate that the time of animal's death was preceded by severe climate, responsible for poor flora of flowering plants and mosses and for sparce plant cover. At the time vegetational aggregations of the polar-desert type became spread in the area under consideration.

4. At the time of mammoth's death, about 53,000 yr B.P., meadowlike grass-forb communities were common in the Bolshaya Lesnaya Rassokha River Basin, whereas sedge-grass communities occurred on the river and lake banks which served as pastures for animals of the "mammoth" faunal complex.

5. The Karginsky terrace has been practically destroyed there. It remains as a very narrow ledge (1.5-2.0 m wide), distinctly marked by the peat bed, and as a small sand bed (Bed 7, upper part, 1.25-0.5 m), as indicated by the composition of palynological spectra.

6. The formation of the marker peat bed was completed about 34,470 yr B.P., i.e. at the last stages of second Karginsky warm interval. During the formation of the lower part of peat bed, polygonal bogs may have been widely distributed in the area under study. The formation of the upper part of peat bed took place under more favorable climatic conditions. At this stage, sedge-larch forests, similar to the present larch forests of the Ary-Mas massif, (the most northerly forest "island" in the world on the Novaya River) had already occupied the lower Bolshaya Lesnaya Rassokha River Basin.

4.2 Kirgilyakh mammoth (DIMA)

On July 23, 1977, a team of gold diggers discovered a corpse of the Kirgilyakh (Magadan) baby mammoth at Kirgilyakh Creek, a left tributary of the Berelekh River (Fig. 11), in the upper course of Kolyma River (Shilo et al., 1983). The corpse proved to be one of the best preserved specimens of fossil mammals recently found in Siberia; this allowed: (i) study of the morphology of these extinct animals and the anatomy of internal organs; (ii) biochemical and cytologic examinations of muscles and brain; (iii) study of blood vessels and blood cells; (iv) attempts at obtaining cell culture and chromosome preparations

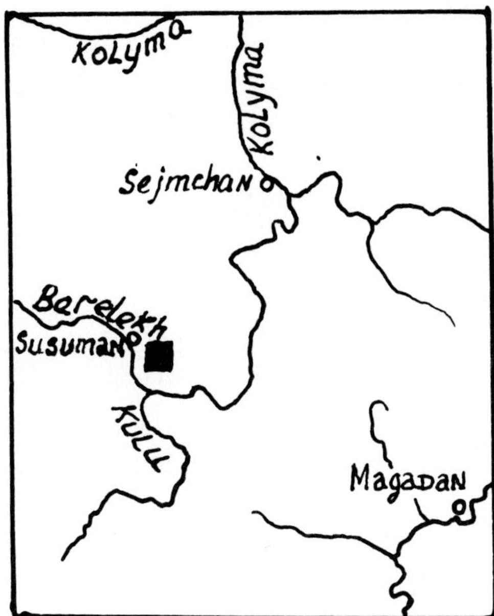

Figure 11.   Location of the Kirgilyakh mammoth (marked black square).

(Magadan baby-mammoth, 1981; Shilo et al., 1983). But what was environment in the area in the period of mammoth's death? Due to efforts of many scientists, geologists and geomorphologists, paleobotanists and cryologists, paleogeographers and even mineralogists, this vital question was also answered.

In addition to the corpse of baby mammoth, separate bone remains of other typical representatives of late mammoth fauna, namely, horse, woolly rhinoceros, reindeer, and primeval bison were also found in the Kirgilyakh Creek Valley (Vereshchagin and Lazarev, 1977).

The Kirgilyakh Creek Basin is situated on the Upper Kolyma highland. In geomorphological respect, it is a typical fragment of middle- and low-mountain relief (Fig. 12). Absolute heights of the interfluves can be as high as 1180-1200 m, whereas relative heights do not exceed 200-400 m. Neither glacial landforms nor drift have been observed near the burial site and in the vicinity of the discovery (Titov, 1982; Shilo et al., 1983).

In botanical-geographical relationship, this area is quite typical of the inland part of the southern Magadan region (Ukraintseva and Kozhevnikov, 1979). The most extensive areas there are occupied by different types of more or less thinned-out larch forests with an understory in which birch, alder and mountain pine are common (Fig. 13). They cover gentle mountain slopes up to the crests, progressively giving way to mountain pine which forms the upper limit of the forest. The southern and southeastern slopes are covered by cowberry-larch forests containing *Betula platyphylla* which does not grow in the valleys. *Betula exilis, B. midendorffii,* and *Salix xerophila* are common in understory

Figure 12. Berelekh River Valley where Kirgilyakh Creek falls into it (on the left) and general view of middle-height relief. Photo by V.V. Ukraintseva.

Figure 13. Larch forest in the vicinity of Kirgilyakh mammoth's burial site. Mountain pine, dwarf birch and alder in the undergrowth. Photo by V.V. Ukraintseva.

of the larch forest. *Dasiphora friticosa* is rare. Besides low shrubs, *Calamagrostis lapponica, Stellaria ciliatocephala, Polygonum tripterocarpum, Pedicularis labradorica, Taraxacum ceratophorum* and other plants dwell in the

Figure 14. Poplar-chozenia-larch forest, dry flood plain of the Berelekh River. Photo by V.V. Ukraintseva.

ground cover. At some localities of northwestern slopes, *Alnus fruticosa* and *A. kamtschatica*, growing up to 3-4 m high, as well as large and high shrubs of *Pinus pumila* are abundant in the undergrowth (Fig. 14). Larch forests also occur on valley bottoms, in particular, on low terraces above the floodplains, which usually flood. On terraces above the floodplain, the larch forests are generally conterminal to moist willow-chozenia forests with an undergrowth, composed of *Rosa acicularis, Ribes triste* and *R. dikuscna* in which *Lactuca sibirica* can be encountered. As swampiness increases, the willow-chozenia small forests give way to willow-dwarf shrub formations, made up of *Betula exilis, Salix myrtilloides,* and *S. fuscescens,* and *S. pseudo-pentandra* growing up to 3 m high. Dry sites of valley floor above the floodplain, flooded for brief time spans, are under poplar-chozenia-larch forests with an undergrowth composed of *Salix xerophila, S. schwerinii, Ribes triste, R. dikuscha, Pinus pumila, Juniperus sibirica* (rare), *Rosa acicularis* and others (Fig. 14). In the forests there are moist and dry localities which show xero- and mesomorphic habits of ground cover respectively. The floristic composition of the forests is relatively rich. Either grass cover, (made up of *Poaurssulensis*), or forbs (including *Rosa acicularis, Urtica angustifolia, Aconitum delphinifolium,* and *Astragalus frigidus* are formed in places under the canopy of the forests. As one moves farther away from stream channels, the forests pass into brushwood (willow stands, dwarf shrubs) or hillock bogs, dominated by *Eriophorum vaginatum* or *Carex lugens*. *Rubus chamaemorus, Calamagrostis lapponica, Stellaria palustris* and other plants grow in the hillock bogs. In proximity of these bogs there are sphagnum

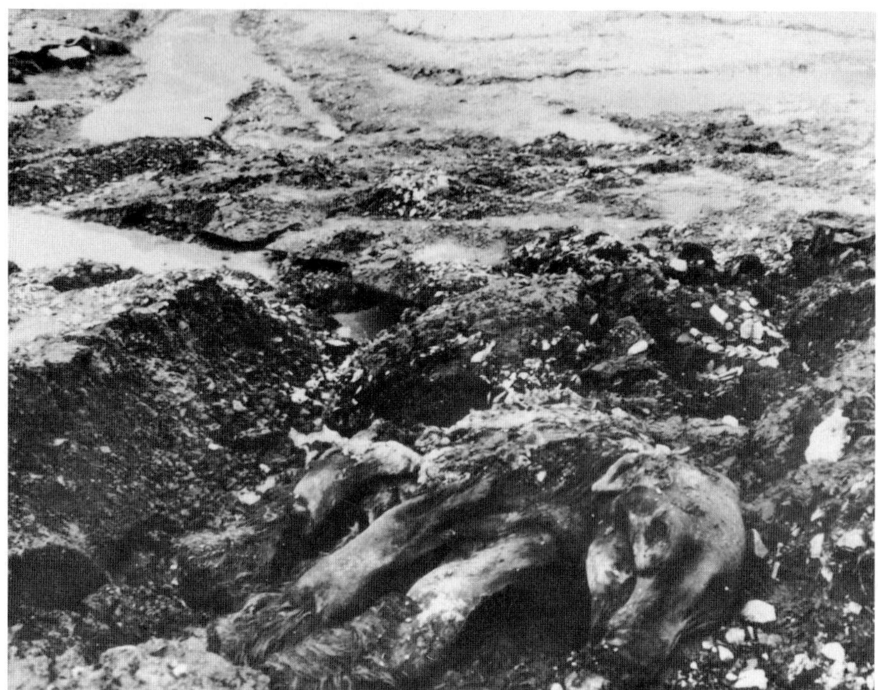

Figure 15. Position of the Kirgilyakh mammoth *in situ* when exposed by excavation. Photo by S. Stelmashenko.

bogs containing *Oxycoccus microcarpa*, *Smilacina trifolia*, and *Rubus chamaemorus*. Some localities, conterminal to the bogs may be classified as wet meadows. *Luzula multiflora, Erigeron elongatus, Tanacetum boreale, Ptarmica alpina* and others usually occur there. "Steppoids" were found on the southern and southeastern slopes over the right bank of Berelekh River. These habitats are represented by *Arenaria tschuktschorum, Chamaerhodos erecta, Draba cineres, Potentilla viscosa, Allium strictum, Atragenea chotensis, Poa attenuata* ssp. *bothyoides, Aquilegia parviflora, Dracocephalum palmatum, Saxifraga multiflora, Calamagrostis purpurascens, Campanula rotundifolia* ssp. *langedorddiene, Silene repens,* and *Viccia macrantha* sp. "Steppoids" occupy the most thoroughly warmed localities as is also evident from contiguous vegetation: small insular groves of birch (*Betula platyphylla*), which are evident in other areas, are usually conterminal with the localities occupied by "steppoids"; *Salix xerophila* is more common.

Three river terraces are recognized in the Kirgilyakh Creek Valley: terrace I is 2 m high, its ^{14}C age ranges from 6,500 to 1,170 yr B.P. Terrace II is 7 m high, alluvium is 5-7 m thick; the age of alluvial soil is 29,000-30,000 yr B.P. Terrace III is 11 m high; the thickness of alluvium is 3.7 m; age is 42,500 yr B.P. (Titov, 1982; Shilo *et al.*, 1983). The mammoth was found at the level of terrace III in loose sediments of terrace-swale facies, 1.8-2.0 m below the sur-

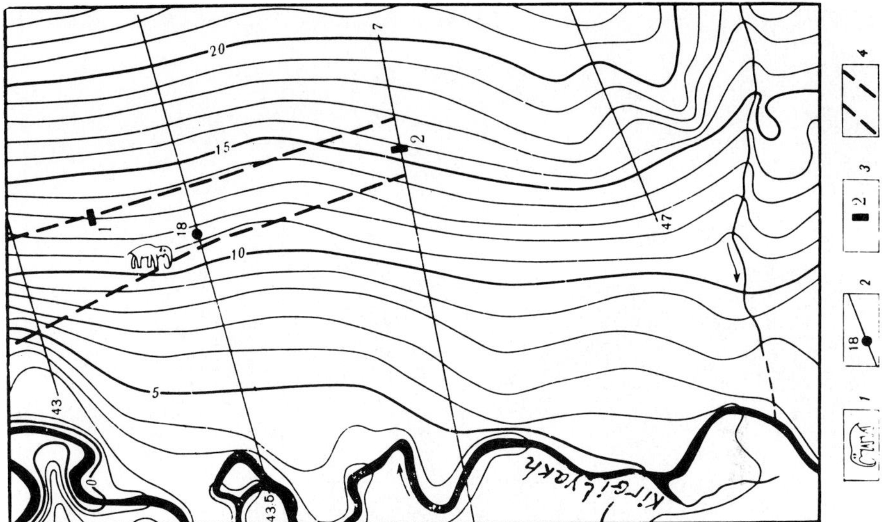

Figure 16. Position of the mammoth. After Shilo et al., 1983: 1) place of discovery with actual orientation of the animal when found; 2) lines along which the corpse position was allocated; 3) studied sections 1 and 2.

face. The animal was buried in a "young" heterogeneous sequence; the ^{14}C age of the sequence falls in the range of 30 to 3,000 yr B.P. (Shilo *et al.*, 1983).

When it was found, the mammoth lay on its left side with its head upstream and legs to the stream channel (Fig. 15). The rear part of the body lay somewhat higher than the fore-part and head. A protruding part of the right side was partially cut off by the bulldozer blade as this part was overlain not by ice, but by loady rock. The body was practically hairless. Some hair was preserved on the feet, trunk and ears (Fig.s 16, 17). Plenty of wool was frozen into ice below the mammoth; the body lay almost at the intersection of two ice veins.

A series of radiocarbon dates, obtained on skin fragments, slices of intestine and muscle tissues, showed that the baby mammoth died about 40,000 yr B.P. (Arslanov *et al.*, 1981; Shilo and Lozhkin, 1981; Shilo *et al.*, 1983). The animal was no older than 6-12 months as suggested by the following facts: (i) premolar teeth (pd2 and pd3) had been slightly wornout; (ii) milk tusks had just started cutting; (iii) permanent tusks had been formed as caps only and sat behind the roots of milk tusks (Vereshchagin, 1981). The animal was 104 cm in height; the body length was 74 cm in an oblique line; the height of chest above ground was 52 cm. Its live weight must have been about 100-115 kg, and the corpse weighed 61 kg.

Its slightly worn premolar teeth suggest the baby mammoth was fed on not only its mother's milk, but on vegetation also. However, dissection revealed no plant remains in its gastrointestinal tract. Its simple stomach about 1 liter in capacity contained 50 grams of dark-brown earthy mass. The small intestine about 315 cm in length was quite empty. The large intestine about 132 cm in

Figure 17. Position of the mammoth after partial ice thawing out from under it. Photo by A.V. Lozhkin.

length was filled with earthy mass similar in consistency and color to that found in stomach (Vereshchagin, 1981). The substance which filled the intestine was composed of the following major components: (i) mineral; (ii) soluble organic matter, and (iii) insoluble organic matter. The biogenic group which accounts-for 0.1 of net solid matter (Shilo, *et al.*, 1983) was represented by fine roots and fragments of sedge epidermis, with bits of willow epidermis. Bark fragments of willow (*Salix* spp.) and birch (*Betula* sp.), scraps of moss stems were also encountered (Ukraintseva, 1981a). A sample, taken from the large intestine, contained no less than three species of sedges (*Carex* sp.), as determined from nut fragments; two nuts of *Rumex acetosella,* one fruit of *Ranunculus* aff. *flamule* showing good preservation; fruit fragments of *Ranunculus* sp.; one fruit of *Potentilla* sp., as well as remains of true mosses, namely, some offsets of *Pleurozium* sp., two leaves of *Mnium affine* Bland. *emend* Toumik, some offsets of *Calliergon* sp., one sprout of *Helodium* sp., some offshoots of *Bryales* indet.; megaspores of *Salaginella* aff. *rupestris* (*sibirica*)(Nikitin, 1981).

All the samples, taken from the gastrointestinal tract, as well as samples taken by A.V. Lozhkin from deposits beneath, and above the mammoth, appear to be abundant in pollen and spores. Belaya and Kisterova (1978) who examined the samples came to the conclusion that sporo-pollen spectra of all the samples

Table 5

Results of palynological analysis of the samples studied.

Plants	Contents of the						Underlying bed		Overlying bed	
	stomach		rectum		large intestine					
	sample Nos									
	204, 1-5		206, 1-5		201, 4-7		210, 1-5		205, 1-5	
1	**2**		**3**		**4**		**5**		**6**	
	abs.	%	abs.	%	abs.	%	abs.	%	abs.	%
Larix sp.	1	0.7	-		1	1.0	-		-	
Picea sp. x)	4	2.8	1	2.0	-		2	1.7	-	
Pinus sibirica x)	2	1.4	1	2.0	-		1	0.8	1	
P. sylvestris x)	-		-		2	2.0	-		-	
P. pumila	13	8.9	2	4.0	15	15.0	24	20.7	3	
Pinus sp.	15	10.2	2	4.0	21	21.0	17	14.6	-	
B. platyphylla	7	4,	1	2.0	-		2	1.7	1	
Betula sp. (sect. *Betula*)	-		-		3	3.0	-		-	
Betula sp. (sect. *Fruticosae*)	9	6.1	1	2.0	3	3.0	-		2	

To be continued

Table 5 (Cont)

1	2		3		4		5		6	
	abs.	%	abs.	%	abs.	%	abs.	%	abs.	%
B. *exilis*	26	17.7	11	20.5	5	5.0	12	10.3	8	
Betula sp.	57	38.5	23	49.0	40	40.0	50	43.4	10	
Alnus fruticosa	6	4.1	1	2.0	6	6.0	4	3.4	11	
Salix spp.	6	4.1	6	11.5	4	4.0	4	3.4	11	
Populus sp.	1	0.7	-		-		-		-	
Poaceae (Gramineae)	10	3.5	12	5.6	9	4.0	10	7.5	13	3.8
Cyperaceae	99	34.2	91	40.1	63	27.7	52	36.2	177	52.0
Allium schoenoprasum	2	0.7	1	0.5	1	0.4	2	1.4	-	
Polygonum viviparum	8	2.8	11	5.1	10	4.3	2	1.4	7	2.1
P. bistorta	-		-		1	0.4`	-		-	
Rumex sp.	-		-		1	0.4	2	1.4	-	
Chenopodiaceae	2	0.7	-		-		2	1.4	1	0.3
Minuartia sp.	-		-		-		1	0.7	-	
Stellaria sp.	-		-		6	2.6	-		1	0.3
Caryophyllaceae	34	11.8	17	8.0	14	6.1	10	7.5	47	13.8

To be continued

48

Table 5 (Cont)

1	2		3		4		5		6	
	abs.	%	abs.	%	abs.	%	abs.	%	abs.	%
Thalictrum foetidum	3	1.0	-		2	0.8	1	0.7	4	1.2
Ranunculus sp.	22	7.6	19	8.9	22	9.5	14	9.8	18	5.3
Papaver sp.	1	0.8	-		-		-		-	
Brasicaceae (Cruciferae)	7	2.5	2	0.9	5	2.2	-		5	1.5
Saxifraga spp.	3	1.0	2	0.9	3	1.3	-		2	0.6
Potentilla sp.	-		-		-		-		1	0.3
Sanguisorba officinalis	3	1.0	-		-		1	0.7	-	
Fabaceae (Leguminosae)	-		-		-		1	0.7	-	
Geraniaceae	-		-		-		-		1	0.3
Chamaenerion angustifolium	1	1.0	2	0.9	-		1	0.7	3	0.9
Apiaceae (Umbelliferae)	-		-		1	0.4	-		-	
Ericaceae	4	1.4	4	1.9	1	0.4	2	1.4	-	
Gentiana tenella	1	0.3	-		-		-		3	0.9
Polemonium acutiflorum	1	0.3	-		1	0.4	-		1	0.3
Valeriana capitata	12	4.2	-		8	3.5	4	2.8	7	2.1
Artemisia sp.	14	4.9	17	8.0	29	12.6	-		13	3.8

To be continued

Table 5 (Cont)

1	2		3		4		5		6	
	abs.	%	abs.	%	abs.	%	abs.	%	abs.	%
Aster sp.	-		-		1	0.4	-		-	
Asteraceae (Compositae)	6	2.0	3	1.4	6	2.6	9	7.0	11	3.2
Dicotyledoneae indet.	54	18.8	32	15.0	`46	20.0	28	19.7	25	7.3
Sphagnum spp.1 & 2	23	14.3	2	2.8	36	21.6	19	27.0	2	1.6
Bryales spp. 1-3	49	30.2	23	32.4	42	25.0	15	21.4	25	20.2
Botrychium lunaria	-		1	1.4	1	0.6	-		-	
Dryopteris fragrans	-		-		-		1	1.4	1	0.8
Polypodiaceae	-		-		12	7.2	-		-	
Equisetum sp.	2	1.2	-		2	1.2	2	2.8	-	
Lycopodium cf. annotinum	-		-		2	1.2	-		-	
Huperzia selago ssp. *arctica*	-		-		1	0.6	-		-	
Selaginella sibirica	66	41.3	33	49.4	44	26.0	30	43.2	74	60.3
Hepaticae	20	12.4	11	14.0	26	15.6	-		20	16.2

To be continued

Table 5 (Cont)

1	2		3		4		5		6	
	abs.	%	abs.	%	abs.	%	abs.	%	abs.	%
Sporites indeter.	1	0.6	-		-		3	4.2	1	0.8
Total amount of pollen and spores	595		333		496		328		505	
trees	13	2.2	3	0.9	6	1.2	3	1.3	2	0.5
shrubs and undershrubs	134	22.6	46	13.8	94	18.8	113	34.4	40	7.9
herbs\	287	49.2	213	64.0	230	46.0	142	42.9	340	67.2
sporophytes (Bryophta, Pteridophyta)	161	27.0	71	21.3	166	34.0	70	21.4	123	24.4

x) Long-distance
transported pollen

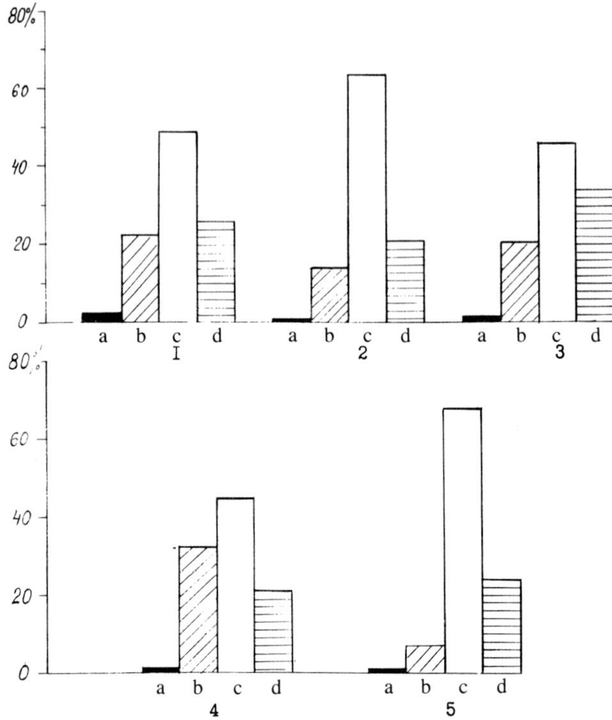

Figure 18. Palynological spectra (general composition) of samples of the gastrointestinal tract of Kirgilyakh mammoth and enclosing deposits: 1) stomach, 2) rectum, 3) large intestine, 4) underlying bed, 5) overlying bed; pollen of: (a) tree species, (b) shrubs and under-shrubs, (c) grasses; (d) spores of mosses and ferns.

belong to the same type in both the presence of sporophytes, treeshrub and undershrub-grass components and in floristic composition. The author (Ukraintseva, 1981a) showed that spectra of the samples, taken from the intestinal tract and deposits beneath the mammoth, are practically identical in the percentage of pollen of trees, shrubs and undershrubs, grasses, spores (Bryophyta + Pteridophyta), and in taxonomy. The spectrum of the sample from the deposits overlying the mammoth differs somewhat from those of the samples, taken from the intestinal tract and the beds beneath the mammoth[3] (Table 5; Fig. 18). Facts mentioned above led the author to the conclusion that shortly before its death the baby mammoth had drunk water from a pool, saturated with particles of the surficial soil layer, probably mixed up a little. Almost equal proportions of "underdeveloped" pollen in the sample from the intestinal tract (17.9% on average) and in the sample from the underlying bed (19.7%) is another evidence for the fact that soil had gotten into the mammoth's intestinal tract when it had been alive. This is supported by mineralogical analysis of the soild residue of the intestinal contents and the deposits enclosing the mammoth; analysis showed that substance of the intestinal tract contained the most complete set of minerals typical of the enclosing deposits (Table 6).

Table 6

Mineral composition of solid residue of mammoth intestinal contents
(samples 1-d, 2, 3-d) and enclosing deposits (sample 4)

(from Shilo et al., 1983)

| | Sample | | | |
Minerals	1-dX	2	3-d	4
Amphiboles	x	x	x	x
Andalusite	x	-	-	-
Apatite	x	-	-	x
Barite	x	-	-	x
Biotite	x	x	-	-
Cassiterite				x
Chalcopyrite	-	-	-	x
Chlorite	x	x	x	x
Epidote	x	x	x	x
Garnet	x	x	-	-
Hematite	-	-	x	-
Ilmenite	x	x	x	x
Magnetite	x	x	-	-
Pyrite	x	x	x	x
Pyroxene	-	-	x	-
Quartz	x	x	x	x
Rutile	x	x	x	x
Sphene	x	-	x	-
Topaz	-	-	x	-
Tourmaline	x	-	x	-
White mice	x	x	x	-
Wolframite	x	-	-	-
Zircon	x	x	x	x

Note: x - the presence of a mineral in identifiable amount.

Thus, results of palynological and mineralogical analyses permit us to consider substance (soil) which filled the mammoth's intestinal tract as a peculiar "averaged surface sample", "taken" by the baby mammoth with the aid of its trunk from the nearest earth surface. Thorough analysis of a set of geological-geomorphological, palynological and radiocarbon data suggests that the "sample was taken" from Kirgilyakh terrace III with an age of 42,500 yr B.P. (Titov, 1982). It is precisely this terrace where the baby mammoth perished and where it was buried 40,000-41,000 yr B.P. (Shilo et al., 1983). Hence, plant remains (macroremains, pollen, spores), found just below the mammoth in the underlying deposits and in the intestinal tract, reliably and more or less completely show the composition of plants growing at that time in the area and give a notion of the character of flora and vegetation of the whole terrain.

Let us start from the fact that the rear, slightly raised part of the baby mammoth's body, rested on a tussock of the sedge (*Carex* sp.). The workers saw it when they lifted the corpse—besides, there were some leaves of willow and a small sod of mosses, composed of *Sphagnum angustifolium, S. russowii* and *Calliergon stramineum.*[4] The plant remains (mammoth's contemporaries) were well preserved. The leaves were pale greenish in the center of tussock and dried up and died out around the periphery. Stems and cauline leaves became colorless, but their structure was well preserved.[5] Soil between the roots and lenticules of clay partings which surround the sedge tussock yielded separate sprigs and leaves of the following moss species: *Pogonatum urnigerum, Polytrichum alpinum, P. alpestre (P. structum), Ditrichum flexicaule, Ceratodon purpureus, Vistichium capillaceum, Dicranella* sp., *Dicranum congestum, Encalypta,* sp., *Pottia* sp., *Tortula macronifolia, T. ruralis, Trichostomum* sp., *Tortella fragilis, Tortella* sp., and *Rhacomitrium canescens, Pohlia cruda, Pohlia* sp., *Ivlnium spinosum, M. ambiguum, M. rugicum, Aulacomnium turgidum, A. palustre, Philonotis* sp., *Pseudoleskeella tectorum, Thuideum abietinum, Th. philibertii, Campylium protensum, C. chrysophyllum, Drepanocladus fluitans, D. aduncus, Calliergon giganteum, Cirriphyllum cirrosum, Entodon concinnus, Hypnum cupressiforme, Rhytidium rugosum* (Abramov and Abramova, 1981). Some of the above-mentioned mosses were also found in samples, taken by I.A. Dubrovo in a quarry close to the mammoth's burial site. A sample, taken at a distance of 2-3 m from the burial site at a depth of 2 m, contained *Polytrichumalpestra, Bryum* sp., *Myurella tenerima, Rhytidium rugosum* and others, whereas a sample, taken at 1.2 m yielded *Polytrichumalpestre, Mniumrugicum, Calliergon stramineum, Drepanocladus fluitans* and others; in addition, sprigs of *Sphagnum teres,* possibly admixed with scanty *S. russowii,* were noted in abundance. A sample, collected at the level of mammoth burial, but at a distance of 20 m from the burial site, yielded nine moss species with four of them being plants of dry and stony habitats; they are *Encalypta* sp., *Thuideum abietinum, Entodon concinnus, Rhytidium rugosum.*

Palynological analysis of the samples, taken from the intestinal tract and the deposits lying directly under the mammoth, extended the list of plants considerably, which are mammoth contemporaries. Pollen and spores of over 50 taxa of plants which grew at that time in the burial site area and its vicinity and long-distance transported pollen (*Picea* sp., *Pinus sibirica, P. sylvestris, Betula* sp. ex sect. *Betula*) had gotten into the mammoth's intestine together with water and soil. Palynological spectra of samples, taken from the intestine and the deposits underlying the mammoth proved to be dominated by pollen of herbs (Table 5, Fig. 18), ranging from 49% to 64%. Spores of mosses and ferns make up 21.3-34.0%; pollen of shrubs and undershrubs constitutes 13.8-22.6%; pollen of tree species is represented by rare grains of the larch (*Larix* sp.), spruce (*Picea* sp.), large woody birch [*Betula platyphylla, Betula* sp. (sect. *Betula*)], and pine (*Pinus sylvestris*) totalling to only 0-9-2.2% of the sum total of pollen calculated. Although pollen of herbs is represented by a rather enlarged list (Table 5), however, the group is dominated by pollen of sedges which accounts for 27.7-40.1%; pollen of grasses ranges from 3.5 to 7.5%; forbs are represented by pollen of the pinks (*Caryophyllaceae*), *ranunculi* (*Ranunculaceae*), jointweed (*Polygonaceae*) mustards (*Brassicaceae*), *Polemoniaceae* (*Polemonium acutiflorum, Polemonium* sp.)] and compositae Asteraceae; pollen of the Ericaceae is represented by sporadic grains and totals only 0.4-1.9%. 15% to 20% of "underdeveloped" and, hence, unidentifiable pollen was noted as pollen of herbs.

The spectrum of the sample, taken directly from under the mammoth is actually identical to those of the samples of the intestinal tract in percentages of pollen of trees, shrubs, low shrubs, herbs, in percentages of spores of sporophytes and in taxonomy (cf. data presented in Table 5 and Fig. 18). The similarity is emphasized by about equal proportions of "underdeveloped" pollen in the spectrum of the sample taken from the underlying bed (19.7%) and in the samples of the gastrointestinal tract (average 17.9%).

Ice fragments, from cut off ice vents under the mammoth[6], contained a low count of pollen and spores of the following plants: sedges (*Carex* sp.), grasses (*Poaceae*), two species of Greek valerin (*Polemonium*) *acutiflorum, Polemonium* sp.), selaginella (*Selaginella rupestris*), pinks (*Caryophyllaceae*), compositae (*Asteraceae*), and rare spores of true mosses. The spectrum of the sample, taken from the overlying beds (Table 5, sample 205, 1-5), is also dominated by pollen of herbs, accounting for 67.2% of all pollen, but pollen of shrubs and under-shrubs, namely, that of mountain pine (*Pinus pumila*), birch (*Betula exilis)* and other low shrubby birches [*Betula* spp. (sect. *Nanae*)], alder (*Alnus fruticosa*), willow (*Salix* sp.) drops to 7.3%. Arboreal pollen is represented by occasional grains of the pine (*Pinus sibirica*) and woody birch (*Betula* sp.). The propor-tion of spores is about equal to that in spectra of the intestinal tract and the underlying bed; the fact that its qualitative composition varies is of importance. The quantity of spores of the bog mosses (*Sphagnum* spp.) drastically drops to 1.6% of the spectrum, whereas the amount of spores of *Selaginella rupestris* increases to 60.3%. It is important that the spectra of this sample contains no spores of such plants as the stiff club-moss (*Lycopodium annotinum*), fir club-moss (*Hyperzia selago*), as well as spores of ferns of the family Polypodiaceae, recorded in the spectrum of the sample, collected from the mammoth's large intestine. The amount of "underdeveloped" pollen sharply drops in the spectra and comes to only 7.3%.

The sample, taken from the bed overlying the mammoth, yielded 16 species of freshwater diatoms, dominated by *Pinnularia lata* (Breb.) W. Sm. and *P. Borealis* Ehr.; *Pinnularia alpina* W. Sm., *Eunotia praerupta* (Ehr.) W. Sm. et var. *bidens* Grun., *Navicula mutica* Kutz., *Cymbelle heteuropleure* var. *minor* Cl. and *Hantzsehia amphioxys* (Ehr.) Grun. are minor; *Ceratoneis arcus* (Ehr.) Kütz., *Tabelaria fenestrata* (Lyngb.) Kütz., *Eunotia exigua* var. *compacta* Hust., *Stauroneis javanica* Grun., *Navicula mutica* var. *vetricosa* (Kütz.) Cl., *Pinnularia microstrauron* var, *brebissonii* (Kütz.) Hust., *Neidium bisuleatum* (Lagerst.) Cl., *Cymbella perpusilla* A.Cl. are rare. The above-mentioned diatoms dwell on moistened soil, in pools and on moss cushions in the northern alpine areas.[7]

Spectra of the gastrointestinal contents and the bed underlying the mammoth turned out to be similar to those of the lower sequence of recent Kirgilyakh terrace III (bed of lower buried soil in depth interval 5.0-5.7 m, Fig. 19) for which a marked maximum of herb pollen and undershrubs was ascertained (Lozhkin, 1983, cited from Shilo et al., 1983). It should be noted that the maximum of pollen of herbs is very characteristic of spectra of the intestinal tract and the bed underlying the mammoth (Ukraintseva, 1981a).

The spectra described above of the intestinal contents, the bed underlying the mammoth and the synchronous spectra of the bed of lower buried soil in terrace III of Kirgilyakh Creek were found to differ basically from the subrecent spectra of key types of persent vegetation (Ukraintseva, 1981b) (Table 7; Fig. 20). The latter were dominated by pollen of shrubs and undershrubs [*Pinus pumila, Alnus fruticosa, Salix* spp., *Betula exilis, Betula* sp. (sect. *Nanae*)], accounting for 67.0-74.0% of the total composition of spectrum. Pollen of tree species constituted (3.8) 7.8-14.3%; it is mainly composed of pollen of the larch (*Larix cajandri*) (4.3) 6.8-16.2%; pollen of the woody birch is common to only (0.5) 1.1-2.9%. Pollen of herbs ranges from (8.4) 17.7% to 24.4%. Spores of mosses account for only 2.3-4.9%. It should be noted that mosses play an essential role in the ground layers of larch forests. A minor role of their spectra of surface samples can be explained by suppression of the process of sporogensis and the predomination of vegetative reproduction.

A low percentage of larch pollen in the composition of the above-characterized spectra also confirms its poor preservation in fossil state (Vaskovsky, 1957; Piavchenko, 1968, 1982; Karaevskaya, 1972, and others). As a result, the role of shrubs and undershrub pollen in the spectra is overestimated and hence the role of know forest types of spectra is skewed. The above characterized "skewed" spectra reflect the composition of monodominant forests, composed of larch, or larch forests with minor woody birch. They contain pollen of the following main groups of plants (in descending order): shrubs + undershrubs + herbs

Figure 19. Spore-pollen diagram of deposits of Kirgilyakh Creek terrace III above the flood plain (a section 5 km from the creek mouth). After Shilo et al., 1983: 1) soil, 2) loam with detritus, 3) sand with detritus, 4) deposits containing detritus and pebbles, 5) sandy loam containing pebbles and detritus, 6) coarse gravel containing sand and plant remains, 7) loam containing detritus and pebbles, 8) coarse gravel, 9) ice; pollen of: 10) trees and bushes, 11) grasses and forbs; 12) spores, 13) the number of micro-remains.

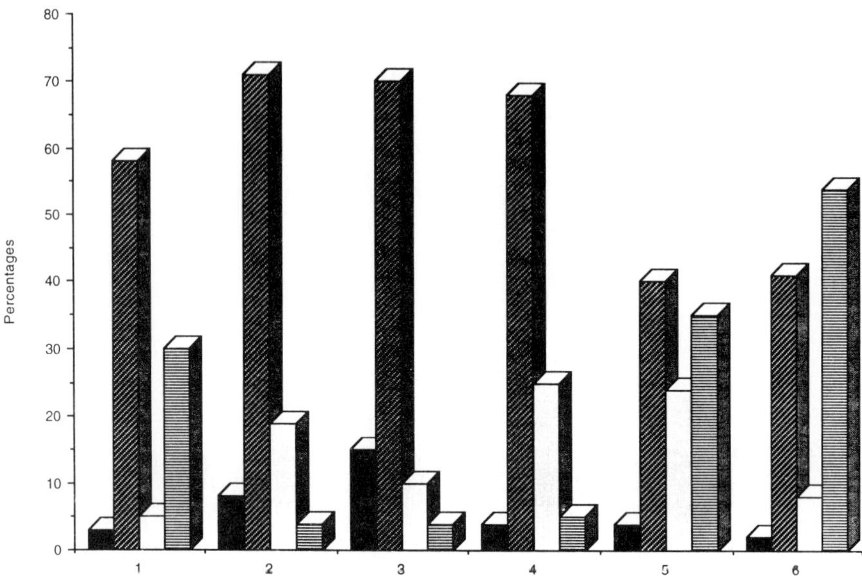

Figure 20. Palynological spectra (general composition) of surficial samples taken at key sites of main types of vegetation in the Kirgilyakh Creek Basin. After Ukraintseva, 1982: 1) moss-hillock dwarf shrub formation on the sides of right bank of Kirgilyakh Creek (vicinity of mammoth's burial site), 2) cup moss shrub larch forest, the same place, 3) larch forest with abundance of cowberry on the sides of left bank of Kirgilyakh Creek, 4) wild rosemary-dwarf shrub larch forest, the same place, 5) sites of the "steppoid" type on the southeastern sides of Berelekh River Valley, 6) mixed poplar-chozenia-larch forest, Berelekh River flood plain.

+ trees (larch ± birch ± poplar) + spores of mosses.

It is evident from comparisons performed (Table 5; Fig. 18 and Table 7; Fig. 20) that spectra of the gastrointestinal tract; the bed underlying the mammoth; and synchronous spectra of the bed of lower buried soil in the section of Kirgilyakh terrace III (Fig. 19) reflect a treeless type of vegetation which dominated in the area about 41,000-40,000 yr B.P. This is supported by species composition of mosses, found at the mammoth's burial site (Abramov and Abramova, 1981), and by the composition of fruits and seeds (Nikitin, 1981).

According to Belaya and Kisterova (1978), hypnum moss and grass-hypnum moss bogs, dwarf willow-sedge-hypnum moss and grass-forb aggregations predominated in the vegetational cover in the Kirgilyakh Creek area during the mammoth's lifetime; aggregations of debris slopes were subordinate; only individual slope localities and the valley were occupied by larch forests. Nikitin (1981) suggests the presence of paludal tundra there. Abramov and Abrarova (1981) arrived at the conclusion that vegetation at that period resembled drier mountain tundra in character.

Analysis of all the paleobotanical data, obtained by the above authors' independent of one another, reveals that all available data show a rather good cor-

Table 7

Results of palynological analysis of surface samples of key sites in main types of vegetation in the Kirgilyakh Creek area (Upper Kolyma River).

Sample Number	1		2		3		4		5		6	
	Moss hillock dwarf shrub		Scrub and moss larch forest with *P. pumila*		Cowberry larch forest with *Pinus pumila* and *Alnus fruiticosa*		Wild rosemary and dwarf shrub larch forest with *P. pumila* and *A. fruticosa*		Sites of "steppoid" type (margin of larch forest)		Poplarchozenia larch forest with *P. pumila*	
	abs.	%	abs.	%	abs.	%	abs.	%	abs.	%	abs.	%
1	2		3		4		5		6		7	
Larix cajanderi	3	2.5	12	6.8	34	16.2	8	4.3	4	5.7	3	2.1
Pinus pumila	21	17.2	100	57.2	82	38.9	31	16.8	57	82.7	109	76.2
Pinus sp.	10	8.2	3	1.7	-		-		-		-	
Betula platyphylla	2	1.6	5	2.9	1	0.5	2	1.1	-		-	
Betula sp. (ex sect.) *fruticosae*	-		7	4.0	1	0.5	5	2.7	2	2.3	1	0.7
Betula exilis	8	6.6	3	1.7	11	5.2	29	15.4	2	2.3	5	3.5
B. middendorfii	-		-		1	0.5	-		-		-	
Betula sp. (ex sect. *Nanae*)	22	18.0	16	9.1	5	2.5	12	6.4	1	1.4	6	4.2

To be continued

Table 7 (cont.)

1	abs.	%	abs.	%	abs.	%	abs.	%	abs.	%	abs.	%
	2		3		4		5		6		7	
Alnus fruticosa	56	45.9	27	15.5	66	31.4	90	48.0	3	4.3	11	7.7
Alnus camtschatica	-		-		-		2	1.1	-		-	
Chozenia arbutifolia	-		-		-		-		-		4	2.8
Salix sp.	-		2	2.1	8	3.8	9	4.8	8	1.4	4	2.8
Salix sp.	-		-		1	0.5	-		-		-	
Poaceae	-		-		3	-	22	33.8	4	-	-	
Cyperaceae	2	-	3	-	4	-	3	4.6	6	-	2	-
Allium sp.	-		-		-		-		2	-	-	
Urtica sp.	-		2	-	-		-		-		-	1
Chenopodiaceae	-		1	-	-		-		-		-	
Caryophyllaceae	-		-		-		3	4.6	2	-	-	
Pulsatilla sp.	1	-	-		-		-		-		-	
Thalictrum foetidum	-		-		-		1	-	-		-	
Ranunculaceae	-		-		-		-		3	-	1	-
Hedysarum sp.	1	-	-		-		-		-		-	
Chamaenerion sp.	2	-	-		-		-		-		-	1
Galium verum	-		-		-		-		1	-	-	

To be continued

Ericaceae	10	-	29	-	11	0	34	52.4	1	-	18	-
Labiatae	-		-		-		-		11	-	-	
Valeriana capitata	-		-		-		1	1.5	-		-	
Artemisia sp.	2	-	1	-	2	-	2	3.1	1	-	1	-
Aster sibiricus	-		-		-		-		-		2	-
Dicotyledoneae indeter.	1	-	3	-	-		-		14	-	-	
Hepaticae	-		-		14	-	-		-		-	
Sphagnum sp.	32	53.4	-		2	-	1	-	3	5.4	8	5.5
Bryales sp.	18	30.0	2	-	5	-	12	-	50	89.2	10	6.9
Bryales sp.	5	3.4	1	-	-		-		-		-	
Dicranum sp.	-		1	-	-		-		-		-	
Botrychium lunaria	-		-		-		-		1	1.8	113	85.5
Lycopodium sp.	1	1.6	-		-		-		-		1	0.7
Selaginella siberica	2	2.3		-	-		-		2	3.6	2	1.4
Polypodiaceae	2	3.3	1	-	-		-		-		-	

To be continued

Total amount of pollen and spores of:	201		219		238		266		171		313	
trees	5	2.5	17	7.8	34	14.3	10	3.8	4	2.3	4	2.2
shrubs and undershrubs	117	58.0	158	72.2	176	73.9	178	66.9	66	39.6	136	39.6
herbs	19	9.5	39	17.8	20	8.4	65	24.4	45	26.3	26	8.3
sporophytes (Bryophyta, Pterydophyta)	60		5		8		13		56		144	49.9

relation and point to (i) the tundra and/or forest-tundra type of vegetation and (ii) its heterogeneous and mosaic character of vegetational cover.

"Mosaic" vegetation of the Kirgilyakh Creek Basin was composed of communities of moist (paludal), mesophytic and dry mountain tundras, petrophytic communities of poorly soil-covered and debris slopes. Palynological data suggest that woody plants were primarily represented by larch which tended to grow in the river valley and flood plain and occurred as isolated trees, groups of trees or small forests. Large bushes of *Alnus fruticosa* and *Pinus pumila*, which form more or less thick undergrowth of recent forests and separate zones at the upper forest boundary (Ukraintseva and Kozhevnikov, 1979), were also poorly represented; this confirms a very limited distribution of woody plants at that time.

Sedge marshes and sedge meadows appear to have been the major type of vegetation, combined with dwarf shrub formations and willow stands. Lowshrub-moss meadows and sedge-grass-forb communities were developed at better drained localities. A large volume of sedge pollen points to rich sedge vegetation.

Along with sedges (*Carex* spp.) and cotton-grass (*Eriophorum* spp.) various riverside and aquatic plants (*Potamogeton, Alisma, Myriophyllum*), creeping willows (*Salix* sp.), buttercups (*Ranunculus* sp.), cinquefoil (*Potentilla* sp.), starwort (*Stellaria* sp.), valerian (*Valeriana capitata*), and other flowering plants, as well as mosses (*Aulacomnium palustre, A. turgidum*) and other types grew in overmoistened swamp tundras on lower terraces.

Higher river terraces and southern hillside aspects were dominated by dry tundras. Sedges and grasses were also essential elements of this type of tundra; forbs were composed of a small set of taxa (*Minuartia* sp. *Draba* sp., sp., *Papaver* sp., *Cruciferae* and others), whereas a set of xerophilous mosses was sufficiently broad (*Encalipta* sp., *Pottia* sp., *Tortula mucronifolia, Rhacomitrium canescens, Mnium spinosum, Hypnum cupressiforme* and others). Bryoflora contained a group of mosses of more southerly association which partially grew on stony soils; the author believes that the group, on the one hand, reflects local conditions near the burial site and, on the other hand, suggests that petrophytic communities were rather widely distributed under suitable conditions at the time of mammoth's death, i.e. 41,000-40,000 yr B.P. A high frequency of occurrence of *Selaginella rupestris* in spectra of the mammoth's gastrointestinal tract and the enclosing rocks (Table 5) is a forcible argument for such a conclusion.

It cannot be ruled out that short-grass meadows and, possibly, sites of steppe-like character, containing *Thalictrum foetidum, Aster* sp., *Polemonium* sp., *Selaginella rupestris* and mosses (Rhitidium rugosum, *Thuidium abietinum, Tortula ruralis*) and others were developed on southern hillside exposures.

The distribution of tree species and their associates, in particular, bushes (*Pinus pumila, Alnus fruticosa, Juniperus sibirica*), and some thermophilic herb species is bounded, as suggested by reconstructions of main climatic characteristics, by lower warmth and moisture provision (Table 8 Fig. 21). At that time, mean July temperatures probably did not rise higher than 9-10°C, and

Table 8

Values of heat- and moisture-provision of the middle Berelekh River Basin
(upper Kolyma River at present and in the past)

Locality, area	Temperatures, C			Precipitation		
	July	January	mean annual	t months (V)VI-IX	of annual precipitation, mm	Precipitation for months (V)VI-IX: mm/% of of annual precipitation
1. Susuman	13.5	-39.8	-13.6	1090	238	155/65%
2. Mid-Berelekh River calculated for the present	13.4	-39.5	-13.5	1096	266	186/69%
reconstructed for 41,000-40,000 yr B.P.	9.0	-29.0	-10.0	680	280	132/47%
Deviation	4.4	10.5	3.5	416	14	54/20%

1) Calculated by interpolation of data of four weather stations, namely, Susuman, Berelekh (Kolyma), AGRES, Frolych, the nearest to the site where the mammoth was found (Reference book of the climate of the USSR, 1966, issue 33, part 2).

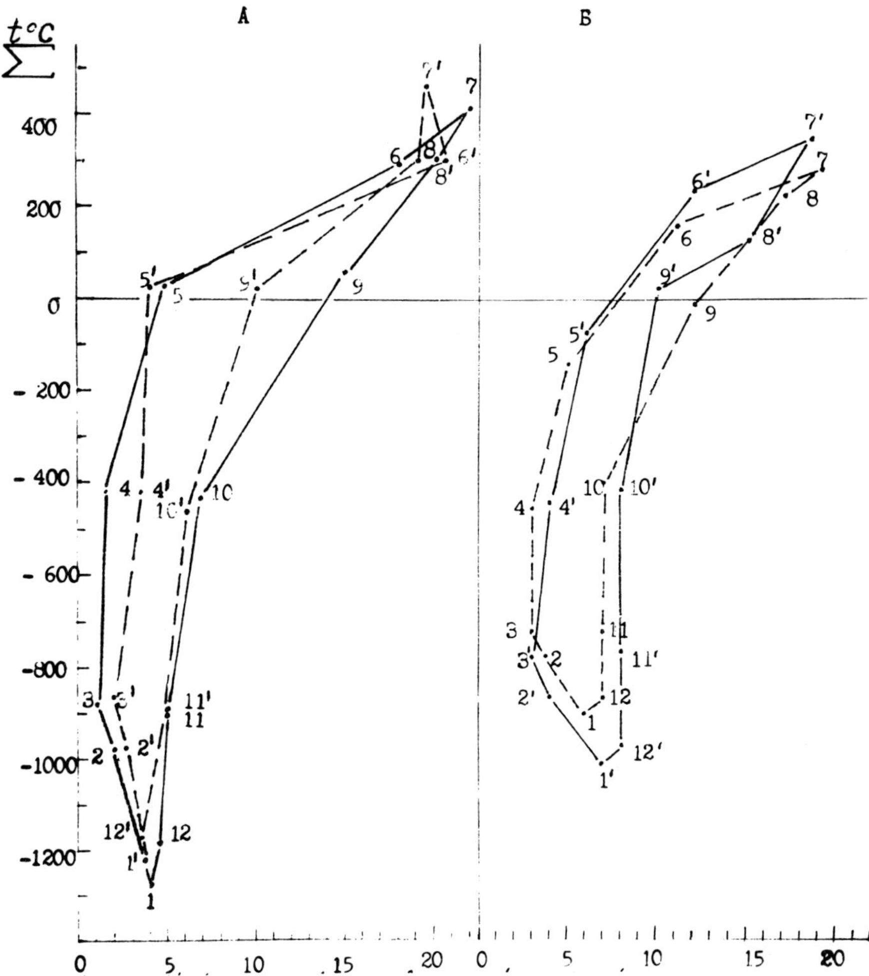

Figure 21. A) climagraph characterizing temperature and moisture conditions in the vicinity of the settlement of Berelekh (solid line) and in the area of the mammoth's burial (dashed line). Constructed from data of the meteorological stations: Berelekh, AGRES, Susuman, Florych; B) paleoclimagraphs characterizing probable temperature and moisture conditions in the vicinity of Kirgilyakh Creek (dashed line) and Berelekh River Basin (solid line) 41,000-42,000 yr B.P.

the sum of temperatures above 0⁰C did not exceed 80⁰, i.e. the threshold of temperature which, when obtained, leads to a wide distribution of tree species, especially larch.

The period when the area under study was dominated by open treeless landscapes was found to last for about 6,000 years, from 45,000 to 39,000 yr B.P. It was named the Kirgilyakh cooling (Shilo *et al.*, 1983). This cold interval proved to correspond to the first cold interval within the Karginsky Interglacial in Siberia.

A drastic change in the vegetational cover of the area, caused by a new warm wave about 38,000 yr B.P., is fixed by palynological spectra of deposits of Kirgilyakh terrace III in depth interval 3.8-1.0 m (Shilo et al., 1983). At this stage, diverse tundras gave way to forest formations, made up of larch with minor woody birch, when present. In the forest undergrowth, the role of mountain pine and shrub alder sharply increased and that of peat mosses decreased. This was associated with greater warmth, which at that time may have been close or analogous to the current warm temperatures in the area.

Conclusions

1. The baby mammoth, discovered on Kirgilyakh creek, died at the age of 6-12 months in time interval 41,000-40,000 yr B.P.

2. Its premolar teeth slightly worn out suggest that the mammoth was fed not only on mother's milk, but also on vegetation, however, no remains of vegetative food were found in dissection of its gastrointestinal tract.

3. Soil which filled the mammoth's gastrointestinal tract was composed of deposits of Kirgilyakh terrace III; the deposits had been formed in depth interval 5.0-6.0 m as suggested by an obvious analogy in palynological spectra of samples, taken from the gastrointestinal tract, the bed underlying the mammoth and the bed in lower sequence of deposits of Kirgilyakh terrace III, as well as by analogy in mineral composition of the samples, taken from the intestinal tract and the deposits enclosing the mammoth.

4. Taxonomic composition of palynological spectra of the samples, taken from the intestinal tract, the bed underlying the mammoth, the bed of lower strata of deposits of Kirgilyakh terrace III, as well as the composition of seeds, and moss remains under the burial more or less completely show the composition of local paleoflora in the vicinity of mammoth's burial site.

5. Many of plants (both flowering and mosses), typical of the recent-flora of the area, were represented there as early as 41,000-40,000 yr B.P.; they include larch, mountain pine, alder, most species of herbs and mosses, but the phytocenological role of some of them, and, especially, larch and large bushes, differed from their present role.

6. In time interval 45,000-39,000 yr B.P., the landscape of the area where the mammoth was found and adjacent territories was dominated by various tundras, such as moist (paludal), mesophytic and dry mountain tundras; petrophytic communities of poorly soil covered and debris habitats were rather common. Sedge marshes and sedge meadows appear to have been the major type of flood-plain vegetation; they were accompanied by shrub willows, dwarf shrub formations, as well as sedge-grass-forb communities at better drained sites.

7. However, arboreal plants did not disappear completely at that time. They were represented by separate trees, small groups of widely spaced trees, and probably, light forests; they grew in river valley and flood plains; large bushes (*Alnus fruticosa* and *Pinus pumila*) were also scanty in the vegetational cover.

Figure 22. Shandrin River terrace II above the flood plain near the mammoth's burial site. Photo by V.V. Ukraintseva.

8. At that time a wider distribution of forest formations was restricted by lower temperature and moisture.

9. Palynological data suggest that the basic change in vegetational cover, caused by warming, took place about 38,000 yr B.P. At this new stage, different tundras gave way to forest formations, made up of the larch (*Larix gmelinii*) with minor woody birch (*Betula* sp.), if present. In the undergrowth of the forests, the role of mountain pine and shrub alder sharply increased. The role of herb communities was reduced.

10. Thus, integrated studies, carried out in the region in connection with the baby mammoth's find, enabled us to follow the dynamics of character of its environment during the past 45,000 yr; this is of prime theoretical and applied importance.

11. This find has not only great paleobiogeographical, but common biological significance.

4.3. Shandrin mammoth

In the summer of 1972 D.D. Kuzmin and A.M. Struchkov, residents of the settlement of Chokurdakh, found a mammoth's skull and tusks on the right bank of the mid-Shandrin River (right tributary of the lower Indigirka River). News of the find reached B.S. Xusanov and P.A. Lazarev, scientists of the Institute of Geology, Yakutsk Branch of the Siberian Departments, USSR Academy of Sciences. In August, 1972, they arrived at the site and made excavation which

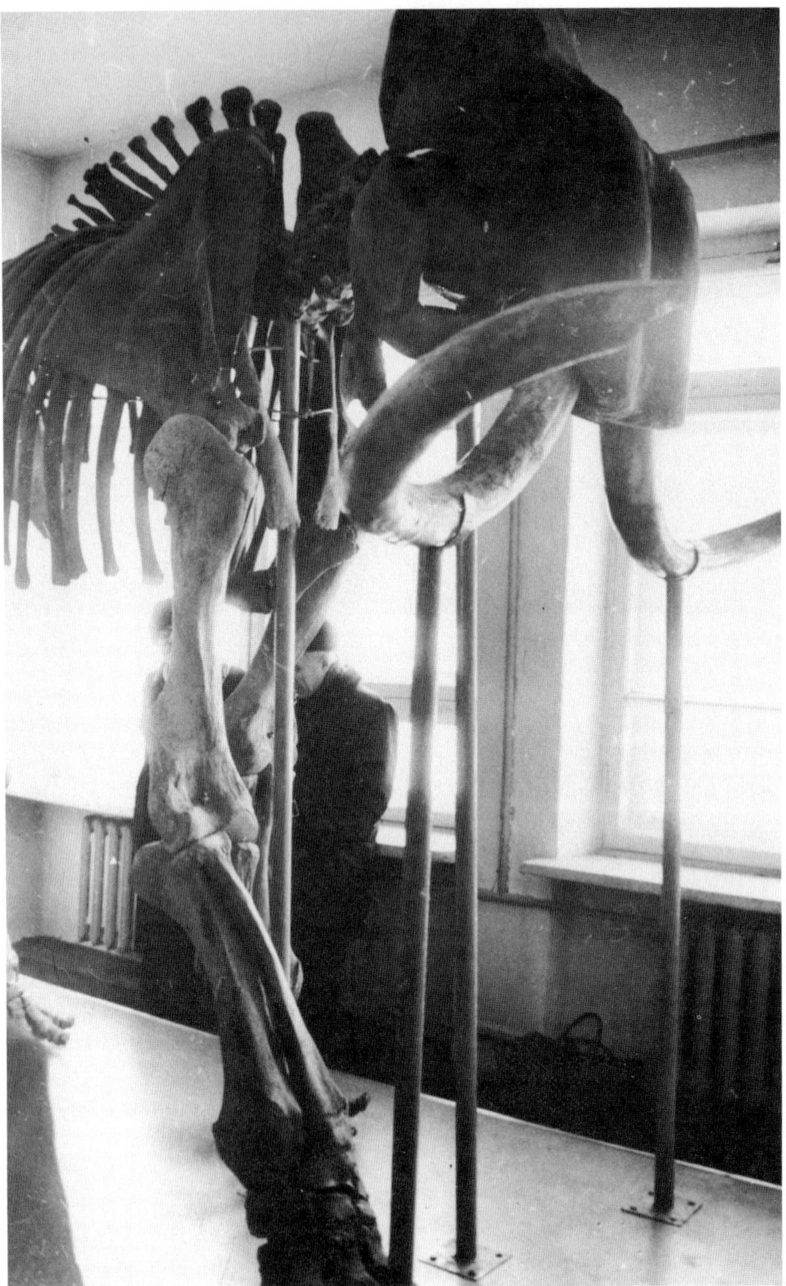

Figure 23. Skeleton of Shandrin mammoth exhibited at the Institute of Biology, Novosibirsk, 1973. Photo by N.V. Lovelius.

revealed a mammoth skeleton, buried at the base of terrace II above the flood plain (Fig. 22), in anatomical order. The skeleton proved to be that of an old, though not particularly big, male mammoth, 60-70 years old (Vereshchagin, 1975). It was embedded in laminated river loam interbedded with coarse gravel. Under a cover of ribs and broad pelvis bones, the internal organs could be seen as a well preserved frozen monolith. The skeleton bones and monolith were recovered from permafrost and transported first to Yakutsk where they were placed a permafrost pit. Then the skeleton was brought to Novosibirsk where P.A. Lazarev mounted it (Fig. 23) at the Novosibirsk Biological Institute, Siberian Department of the USSR Academy of Sciences. The monolith of gastrointestinal tract was also brought to Novosibirsk in January, 1974, and examined by a team of specialists, including paleontologists, anatomists, microbiologists, parasitologists and geologists. N.V. Lovelius and the author participated in the work as representatives of the Komarov Botanical Institute of the USSR Academy of Sciences.

A program for botanical examination, elaborated by Professor B.A. Tikhomirov and the author, was presented to the commission. This program planned to carry out: 1) examination of macroremains of plant vegetative parts preserved in the mammoth's gastrointestinal tract (stems, leaves, roots, bark, rhizomes, etc.); 2) carpological analysis; 3) palynological analysis of the gastrointestinal tract contents. The program also proposed a special expedition to the site of the mammoth's discovery to investigate the mammoth-enclosing sediments and the recent flora and vegetation there.

After thorough examination of the monolith of frozen gastrointestinal tract (Fig. 24) it was sawed up into eight segments (Fig. 25); this allowed reconstruction of the structure of abdominal organs of the extinct animal for the first time in history (Yudichev and Averikhin, 1982) and estimation of the mass of intestinal tract contents by weight; this totalled 291 kg plus 25 kg of decomposed intestinal parts. One of the eight segments of monolith (Fig. 25) clearly displays loops of the small and large intestines. Examination of the monolith its subsequent investigation by Yudichev and Averikhin (1982) revealed the following portions of internal organs: 1) parts of the abdominal wall; 2) remains of the diaphragm which preserved its thin fibrous structure; 3) the pancreas preserved in the anterior lower monolith segment (Fig. 25); 4) fragments of the spleen and kidneys; 5) the stomach with walls being very thin, but solid, was preserved in fragments only; 6) the small intestine; 7) the large intestine fully preserved with a mummified wall 2 mm thick. The muscle tissue and submucous layer displayed extensive hemorrhages. Biochemical examination of the large intestine revealed a considerable amount of protein.

About 15 kg of food mass from different sections of the gastrointestinal tract was given over to the Komarov Botanical Institute in Leningrad.

The site of mammoth's discovery is in the Yana-Indigirka subprovince, East Siberian Province of subarctic tundras (Aleksandrova, 1977). Information on the vegetation and flora of the lower Indigirka River can be found in works

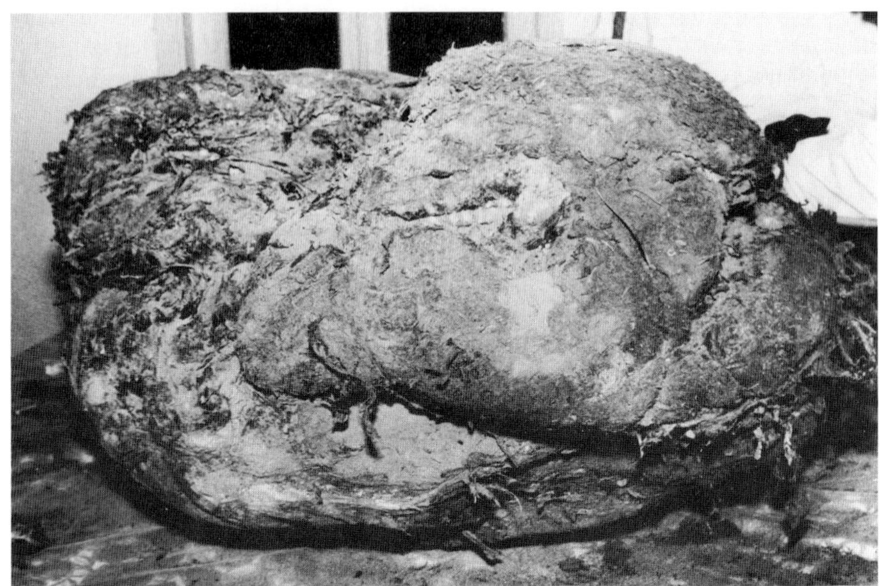

Figure 24. Frozen gastrointestinal tract of Shandrin mammoth. Photo by N.V. Lovelius.

by Aleksandrova (1963), Tyrtikov (1958), Shchelkunova (1968), Andreev (1974), Boch and Tsareva (1974), among others.

In 1974 a team of workers of the Komarov Botanical Institute undertook a study of the flora and vegetation in the area of mammoth discovery (A.A. Korobkov florist-geobotanist; V.V. Ukraintseva palynologist-florist; O.M. Afonina bryologist-florist). The work done by the team showed that the vegetational cover of the area is dominated by various types of low shrub hummock tundras (Fig. 26), including forb-sedge-bush and cotton-grass-bush tundras with abundant *Ledum decumbens; Betula exilis* and *Vaccinium idaea* are second in abundance with hillocks being made up of *Eriophorum vaginatum*. Cotton-grass-birch-herb tundras with *Eriophorum vaginatum* occupy large areas of terrace II above the flood plain. Occurring between hummocks are *Betula exilis, Salix reptans, Ledum decumbens, Vaccinium vitis-idaea, Pyrola grandiflora, Cassiope tetragona*. Grasses are represented by *Arctagrostis arundinaceae, Festuca brachyphylla, Poa arctica,* etc.; sedges are made up of *Carex lugens, Luzula confusa*. Herbs include *Dryas punctata, Stellaria* sp., *Saxifraga hieracifolia, Parrya nudicaulus, Valeriana capitata, Nardosmia frigida, Senecio* sp. and other plants. Lichens in the soil layer are primarily represented by *Peltigera aphtosa* and *Cetraria* sp.; true mosses are composed of *Aulacomnium turgidum, Polytrichum strictum, Dicranum* sp., polygonal and sedge-hypnum bogs, as well as moss-herb hummock tundras are common on terrace I and in depressions of terrace II and III above the flood plain.

Current climatic conditions of the mid-Shandrin drainage basin can be

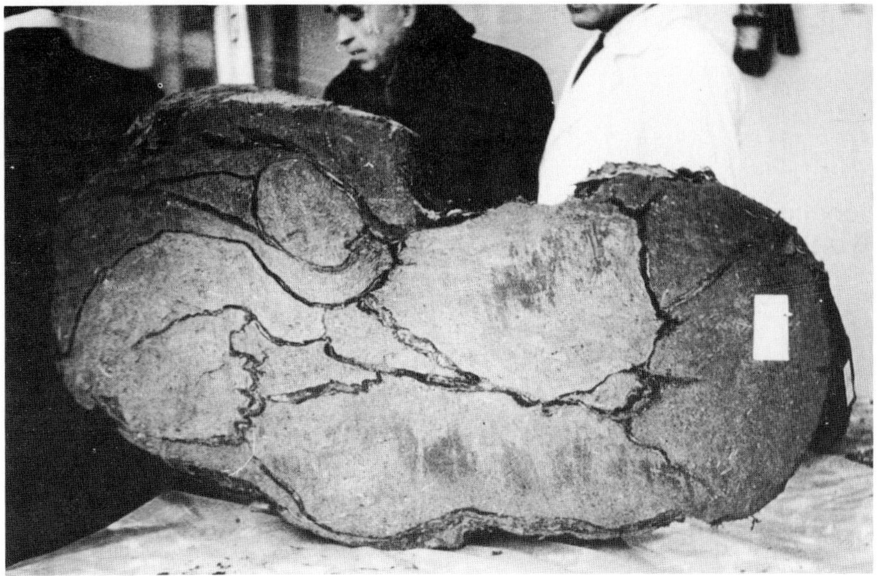

Figure 25. Gastrointestinal tract of Shandrin mammoth (in section). Photo by N.V. Lovelius.

Figure 26. Cotton grass-bushy forb tundras in the vicinity of mammoth's burial site. Photo by V.V. Ukraintseva.

inferred from the data on annual temperature and precipitation distribution, obtained by interpolating actual observations of three meteorological stations, nearest to the site of mammoth find (Table 1). A climatograph based on the data

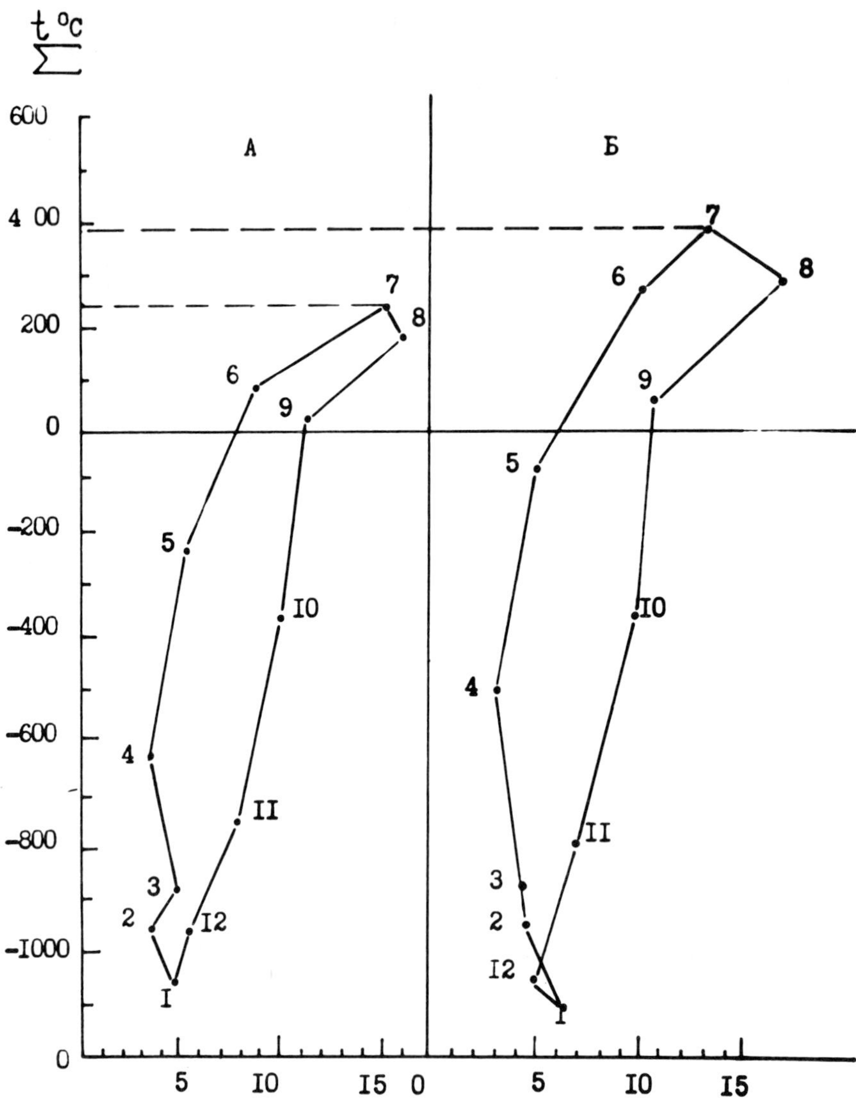

Figure 27. A) Climagraph characterizing present climatic condition of the middle Shandrin River Basin. Constructed from average multi-year data of three nearest meteorological stations; B) paleoclimagraph characterizing climatic conditions of the Shandrin River Basin 40, 350±880 yr B.P.

shows annual warmth and moisture proportions in the area (Fig. 27). Analysis of firm data and those calculated for the area (Table 1) shows low warmth supply and abundant precipitation in the vegetation period in the lower Indigirka and the Alazei Basins. Mean temperature of July, the warmest month in the area, does not exceed 6.6°-9.7°; the sum of plus temperatures varies within 42.3°-46.3°

(in the Arctic tundra area) 51.4⁰ (in the subarctic tundra area) reaching 70.9⁰ near the settlement of Chokurdakh, located the tundra and forest tundra boundary. Precipitation within the vegetation period accounts for 49-55% of the total annual precipitation, a factor contributing to the extensive formation of various types of marshy tundras and swamps in three areas of Primorsky lowland (Sheludyakova, 1938; Boch and Tsareva (1974), marshy tundras and swamps occupy as much as 70% of the area near the settlement of Polarny.

What were the vegetation, flora and climate patterns of the area at the mammoth's lifetime, i.e. $40,350 \pm 880$ yr B.P., as determined by radiocarbon dating of plant remains extracted from the mammoth stomach (Arslanov et al., 1980)? To provide an answer to the question, let us present the results of the investigation of its gastrointestinal contents. It must be noted that it was a firmly compressed mass of plant fragments varying in length (up to 8-10 cm) and diameter (up to 0.5 cm). Analysis of portion of the mass by N.G. Solonevich showed that it was composed of specimens of different plant groups, basically, herbs, namely, their epidermal tissues and vascular-fibrous clusters. The epidermal structure indicates that they were mainly low shrubs of the family Ericaceae as suggested by the ramification pattern. There occurred intact or semi-destroyed leaves of *Vaccinium vitis-idaea* and *Salix* sp., as well as small amounts of needles, bark, remains of seed scales of *Larix gmelinii* (Rupr.) Rupr. (= *Larix dahurica* Turcz.). The bulk of needle mass was partially destroyed, lacking either the base or the cap with the maximal needle length not exceeding 107 cm. Abundant moss remains occurred as short sprigs or stem apexes as well as single intact or semi-destroyed leaves and leafless stem fragments. Basically, the species represented were closely related in ecology. Remains of *Polytrichum commumis* Hedw., P. *strictum* Sm. and *Aulacomnium turgidum* (Wehlenb.) Schwaegr., as well as of bog mosses were more common. The latter group yielded the following species: *Sphagnum angustifolium* C. Jens., sp. *girgensohnii* Russ., *Sphagnum* sp. of the section *Subsecmda*, and *Sphagnum* sp. of the section *Palustria*. Remains of true mosses such as *Tomenthypnum nitens* (Hedv.) Loeske and five-six other species were less common. In addition, some semi-destroyed leaves of liverwarts were also found.

Sporo-pollen analysis of forage mass from different parts of the tract allowed identification of the plants on which the animal grazed shortly before its death (Solonevich et al., 1977).

An averaged palynological spectrum is composed predominantly of spores belonging to some species of true and bog mosses (77.0%). Herb and low shrub pollen accounts for 19.4% the group being dominated by grass and sedge pollen. Small quantities of pollen of *Dryas punctata* s.1., *Ledum* sp., *Saxifraga* sp., *Chamerion dahuricum, Valeriana capitata, Artemisia vulgaris, Artemisia* sp. and others were also detected; these plants currently grow in the area. Pollen of tree species, shrubs and low shrubs accounts for only 3.6%; this includes pollen of larch (dominant), alder, willow, birch. The overall list of plants, identifies from plant macroremains, pollen arid spores, amounts to 50 taxa of flowering plants and sporophytes, the species, genus and family of which were ascribed. Boreal and hypoarctic plants are predominant (Table 9). The composition of plants and frequency of their occurrence in forage mass showed that shortly before its death the mammoth fed on sedges, cotton-grasses and grasses and some herbs whose epidermal tissues are readily digestible and therefore they are rare to absent in forage mass (*Pedicularis* sp., *Saxifraga* sp., *Valeriana capitata* and others), whereas their pollen occur in low amounts. Low shrubs (*Dryas punctata, Vaccinium vitis-idaea, Cassiope tetragona, Ericaceae*), branches of the birch (*Betula exilis*) and willow (*Salix* sp.,), thin sprigs of the larch (*Larix gmelinii*) were also eaten. Besides, the mammoth fed on true and bog mosses, presumably abundant in the soil cover. Remains of some of these plants are common in the intestinal contents (Solonevich et al., 1977). As established by Gorlova (1982) from a small sample of gastric contents, provided by the author, moss remains account for only 1.0%, which seems quite probable. In the author's opinion, the above sample reflects local composition of plants on which the animal grazed shortly before it died. Presumably, the animal was in the recumbent position and unable to rise on its feet, therefore it ate plants, mainly herbs that grew around it. An unique photo, taken by Sukachev (1914), displays a similar situation: herbal remains remained unswallowed with clearly seen toothprints of the Berezovka mammoth (Sukachev,1914, Table 1). A poor set of plants that served as food for the Shandrin mammoth is noteworthy. This might be due to a scarcity of species composition of the communities, frequented by the animal shortly before its death or the fact that it perished either in early spring when only few flowering plants had been in blossom (they may have been snow-covered) or in late autumn when most flowering plants had shed their blossoms, or they may have been snow-covered again. Dissection showed that "the mammoth had died in early spring of asphyxia owing to acute meteorism of the stomach and intestines caused by consumption of enormous quantities of poorly digestible fodder such as old dry grass, turf, bush branches" (Yudichev and Averikhin, 1982, p. 37).

Abundance of macroremains, pollen and spores of plants of moist and swampy habitats such as sedges, cotton grasses, grasses, for example, *Arctophila fulva*, true and bog mosses in the forage mass of the mammoth's gastrointestinal tract indicates that shortly before the death the mammoth had lived on moss-herb communities. These including shrubs and low shrubs in open woodland or in forest-tundra, composed of the larch (*Larix gmelinii*), bark, needles, small cones, seed scales, pollen of which was detected in forage mass (Solonevich et al., 1977; Gorlova, 1982). The present-day equivalent of the past paleocommunities is probably the forest-tundra communities in the Yercha River Basin, their north boundary lying 200-250 km south of the mammoth's burial site. Exploration of the basins of Yercha's left tributaries, namely, the Bolshoi Tabagychan and the Verkhnii Tugychak, in August, 1974, revealed the following types of larch forests and open woodlands in the area: 1) dwarf birch and willow; 2) dwarf birch and alder; 3) dwarf alder-willow-birch; 4) cotton grass-sedge; 5) cotton grass-

Table 9
List of plant identified in the gastrointestinal contents of Shandrin mammoth
(after Solonevich et al., 1977)

Ord. No.	Plant	Frequency of pollen, spores	occurence of macro- remains
1	2	3	4
1.	*Larix gmelinii (=L. daharica)*	+	+
2.	*Picea obovata*	(x)	-
3.	*Pinus pumila*	(x)	+
4.	*Betula exilis*	+	-
5.	Betula sp. ex sect. Nanae	+	+
6.	*Alnus fruticosa*	+	-
7.	*Salix* spp.	+	+
8.	*Poa arctica*	+	-
10.	*Arctophila fulva*	+	-
11.	*Festuca sp.*	+	+
12.	Poaceae	+	+
13.	*Eriophorum sp.*	+	+
14.	*Carex spp.*	+	+
15.	Cyperaceae	+	+
16.	*Minuaritia sp.*	+	-
17.	Caryophyllaceae	+	-
18.	*Ranunculus spp.*	+	-
19.	*Saxifraga spp.*	+	-
20.	*Parnasia sp.*	+	-
21.	*Dryas punctata s.l.*	+	-
22.	Rosaceae	+	-
23.	*Ledum sp.*	+	-

To be continued

Table 9
(continued)

1	2	3	4
24.	*Cassiope tetragona*	+	-
25.	*Vaccinium bitis-idaea*	+	+
26.	Ericaceae	+	+
27.	Labiatae	+	-
28.	*Pedicularis sp.*	+	-
29.	*Valeriana capitata*	+	-
30.	*Valeriana sp.*	+	-
31.	*Companula sp.*	+	-
32.	*Artemisia vulgaris*	+	-
33.	*Artemisia spp.*	+	-
34.	*Crepis sp.*	+	-
35.	Asteraceae	+	-
36.	Dicotyledoneae indeter.	+	+
37.	*Sphagnum angustifolium*	-	+
38.	*S. girgensonii*	-	+
39.	*Sphagnum* sp. (ex sect. Subsecunda)	-	+
40.	*Sphagnum sp.*	+	+
41.	*Sphagnum sp.*	+	+
42.	*Sphagnum spp.*	+	+
43.	*Tomenthypnum nitens*	-	+
44.	*Dicranum sp.*	+	-
45.	*Bryales spp.*	+	+
46.	Polypodiaceae	+	-
47.	*Equisetum spp.*	+	+

x) Long distance tetrasported pollen grains

cassiope; 6) lichen-sedge-cotton grass. The open woodlands are characterized by many swamps' a relatively poor and uniform composition of the herb-
undershrub layer, and abundant true and bog mosses in the ground cover. The
paleobotanical data (Solonevich et al., 1977; Gorlova, 1982) above suggest that
in the mammoth's lifetime, i.e. 40,350$\pm$880 yr B.P., the area was dominated
by paludal open woodlands, composed of larch, and probably forests, made up
of *Larix gmelinii* or *L. dahurica*. The light larch forests alternated with shrub
and low shrub breaks, while the river flood plain was basically occupied by low
shrub-moss meadows with herb-grass communities and lowland cotton grass-
sedge marshes presumably occupying large areas there. It should be noted that
Tomskaya (1981, p.159-162) interprets the spectra of two samples, taken from the
Shandrin mammoth's intestine, as tundra-steppe ones; this is quite inconsistent
with conclusions made by Solonevich and coworkers (1977) and Gorlova (1982).

The mammoth's death falls on the Karginsky Interglacial, namely, its optimum (Kind, 1974) . The existence of forests in optimal phases of the Karginsky Interglacial in the present treeless areas of North-east Siberia is suggested
by data presented by (i) Gitterman (1972) and N.O. Rybakova (Kaplina et al .,
1980) on the Bolshoi and Maly Anyui, lower Kolyma River basin; (ii) Lozhkin
(1975) in the Indigirka and Alazei interfluve; and (iii) Kind (1974) on the adjacent areas. Recent larch formations (their north version) of the Indigirka and
Kolyma Basins can be treated as equivalents of the above monodominant forest
paleoformations that existed in different terrains of the Primorsky lowlands during the Karginsky Interglacial. A notion of environments necessary and sufficient for their growth, related to warmth and moisture proportions in the vegetation period, can be provided by film data of meteorological observations in the
areas, for instance, at the settlements of Vorontsovo and Druzhina on the Indigirka River and Chersky and Kolymskaya on the Kolyma River (Table 10).
Information obtained by interpolating the observational data of the four
meteorological Stations (Table 10) rather reliably reflect environments necessary
and sufficient for growth of the monodominant larch paleoformations, developed
in northern parts of the Primorsky lowlands during the Karginsky Interglacial.
This was associated with a higher temperature in the area at that time. Calculations performed showed that about 40,000 yr B.P., the solar input in the mid-
Shandrin River Basin was twice that of today (Table 11); like today, over half
annual precipitation (53 %) fell the period when temperatures were above 0°C
and were almost doubled recent precipitation. The summer was warmer, the
deviation from current mean July temperatures measured 4°C. It was precisely this proportion of warmth and moisture that promoted growth of forests in
these currently treeless areas. A climagraph, based on data calculated using
paleophytogeographic evidence, graphically displays warmth-moisture proportion in the pre-Shandrin River Basin in optimal phages of the Karginsky Interglacial (Fig. 27), suggesting more favorable conditions of the area in the past.
Nevertheless, permafrost soils existed, although the level of their thawing was
no doubt lower than that of today. This is evident from a large "cold content"

Table 10

Annual course of the temperatures and precipitation in the Indigirka and Kolyma River Basins.   Calculated by interpolation of data of four meteorological stations, namely, Vorontsovo, Druzina, Kolumskaja, Cherskiy.

Meteostation locality	Temperature, C			Σ t°C for months (V)VI-IX	Σ of annual precipitation	Σ Precipitation for months (V)VI-IX (in %% from months I-XII)
	July	January	mean annual			
Kolumskaja	11.7	-34.8	-13.4	870.7	317	53.0
Vorontsovo	11.7	-37.8	-14.1	953.4	232	59.0
Cherskiy	12.0	-33.0	-11.6	986.4	323	43.0
Druzina	13.8	-39.0	-13.4	1191.6	263	61.0
Indigirka-Kolly River Basin	12.0	-34.0	-13.0	936.0	277	54.0

Table 11

Annual course of the temperatures and precipitation in the Indigirka River Basin at present and in the past.

Region	Temperature, C			$\Sigma\,t°C$ for months (V)VI-IX	Annual precipitation	Precipitation within months (V)VI-IX (MM/% of the annual sum total)
	July	January	mean annual			
The Mid-Shandrin River. At present.	8.0	-31.0	-15.0	513	185	
40,350±880 years ago	12.0	-34.0	-13.0	936	277	
Difference	+4.0	+3.0	+2.0	+423	+92	

1) By interpolating data of three meteostations, nearest to the mammoth find site
2) By interpolating data of meteostations Vorontzovo, Kolymskaya, Chersky.

(sum of minus temperatures) and its seasonal trend (Fig. 27). This is in agreement with taphonomic observations (Vereshchagin, 1975). According to Vereshchagin (1975), a good preservation of the gastrointestinal tract was ensured by the ribs, and a complete decomposition of the wool cover, skin and muscles was due to the fact that the dead animal carcass underwent repeated thawing and freezing in the summer and winter seasons, respectively, but the internal organs, particularly, the stomach and intestines, filled with raw forage mass, never thawed, a factor preventing their decomposition.

Conclusions

1. A strict anatomical order of the mammoth's skeleton in the enclosing sediments rules out its displacement, therefore the discovery can be used to substantiate the age of the sediments.

2. The sediments of Shandrin terrace II above the flood plain which enclosed the mammoth's skeleton, were formed about 40,000 yr B.P., i.e. during the Karginsky Interglacial, as established by ^{14}C dating of forage mass from the intestines and muscle tissues of the mammoth. This is also supported by geologo-geomorphological data, obtained in the area of mammoth's discovery.

3. During the mammoth's lifetime, i.e. $40,350 \pm 880$ yr B.P. this currently treeless area (province of subarctic tundras) was dominated by larch forests and open woodlands alternating with shrub and low shrub breaks; the Shandrin River Valley was presumably occupied by vast areas of cotton grass-sedge, herb-grass, and, predominantly, grass communities, herb-moss bogs, as well as bottom and undershrub-moss meadows.

4. Shortly before its death the mammoth grazed on highly moistened or heavily paludal sites in open larch woodlands. Its fodder was mainly composed of sedges, cotton grasses, grasses, low shrubs (dryad, red bilberries, cassiope and other plants), sprigs and leaves of larger shrubs and undershrubs, including alder, willow, low birch, as well as young shoots of larch. The mixture also included mosses abundant in the ground cover.

5. The present paludal forests and open woodlands whose northern boundary lies in the Yercha River Basin, 200-250 km south of the mammoth's burial site, appear to be equivalents of vegetation growing in the pra-Shandrin River Basin at the time of mammoth's death.

6. The larch forests growing during the Karginsky Interglacial in the currently treeless terrains were due to higher radiation influx in the regions which was almost twice that of today at these latitudes, as suggested by the sum of mean monthly air temperatures above 0^{0}C.

4.4. Selerikhan fossil horse

The corpse of a fossil horse was discovered at the Selerikhan gold mine, on the upper Indigirka River, when driving a horizontal drift. The animal rested at a depth of 8-9 m from the surface in a bed of loam with inclusions of scattered pebbles, unrounded material, 2-3 cm in size, and vegetational remains

(Belorusova, 1977; Lazarev, 1982).

Morphological study of fragments of the corpse and postcranial skeleton points to a separate species of the horse of high arctic latitudes (Vereshchagin and Lazarev, 1977). The animal is named the Chersky horse in Russian, and *Equus lenensis* Russanov, 1968, in Latin. The Chersky horse was a bay (coffee-colored) stallion with coal-black mane and tail. In growth and exterior, it was short with a height of about 134-136 cm at the withers. As suggested by some characters in the structure of skull and skeleton, the Chersky horse resembled the present Yakut horse. The geographical distribution of this species in the late Pleistocene was confined to the northeastern regions of Siberia, Beringia and, probably, Alaska (Vereshchagin, 1977).

Radiocarbon analysis of plant remains from the horse's stomach showed that it died $38,590\pm1,120$ yr B.P. (Arslanov and Chernov, 1977). The date of death as indicated by analysis of muscle tissue is 35,000 yr B.P. Another specimen of muscle tissue, given by N.K. Vereshchagin to Professor I. Harrington (National Museum, Ottawa) yielded a date of $33,800\pm2,100$ yr B.P. According to Arslanov and Chernov (1977), the date of $38,590\pm1,120$ yr B.P. is the most reliable value as obtained with a relatively small statistical error.

Detailed geological-geomorphological study of the area of the find, performed by Belorusova (1977), showed that the horse lay on the bottom of an ancient buried valley of Balkhan Creek at an elevation of 710 m. The elevation of the burial site above present water level in the Bolshoi Selerikhan River is about 110-120 m and corresponds to an erosional terrace 110-120 m high, well defined on the Bolshoi Selerikhan riverside. The river basin is situated on the central Yana-Oimyakon Plateau (Fig. 28) and is associated with a planation surface ranging from 800-900 to 1000 m which is, at the same time, a 200 m terrace of Bolshoi Selerikhan Creek.

This geomorphological level is of great interest as its surface is downcut by present and ancient valleys of Balkhan and Badran creeks whose alluvial deposits enclosed the horse.

Evidence of glacial activity has not been found, either in relief or in the lithological section of Quaternary deposits of the Bolshoi Selerikhan Basin. This led Belorusova (1977) to the conclusion that the area under study had not been covered by ice of either mountain or continental glaciation. In the upper Indigirka River Basin, only mountainous massifs of the Chersky Highland, Suntar-Khayata Range, and their immediate mountain flange may have undergone glaciation (Vaskovsky, 1959; 1967; Glushkova, 1982).

The burial site lies in the zone of mountain larch forests (Sheludyakova, 1938). The Selerikhcqn Creek Valley (Fig. 29) and surrounding hill slopes are well forested. According to Sokolova (1977), the vegetational cover is represented by:(1) flood plain forests, composed of *Populus suaveolens, Chozenia arbutifolia* and subordinate *Larix gmellnii;* the forests have a thick shrub layer, made up of *Salix schwerinii, S. pseudopentandra, Alnus fruticosa* and a herb layer (*Hedysarum obscurum, Lathyrus pilosus, Galium verum, Aster sibirica* and

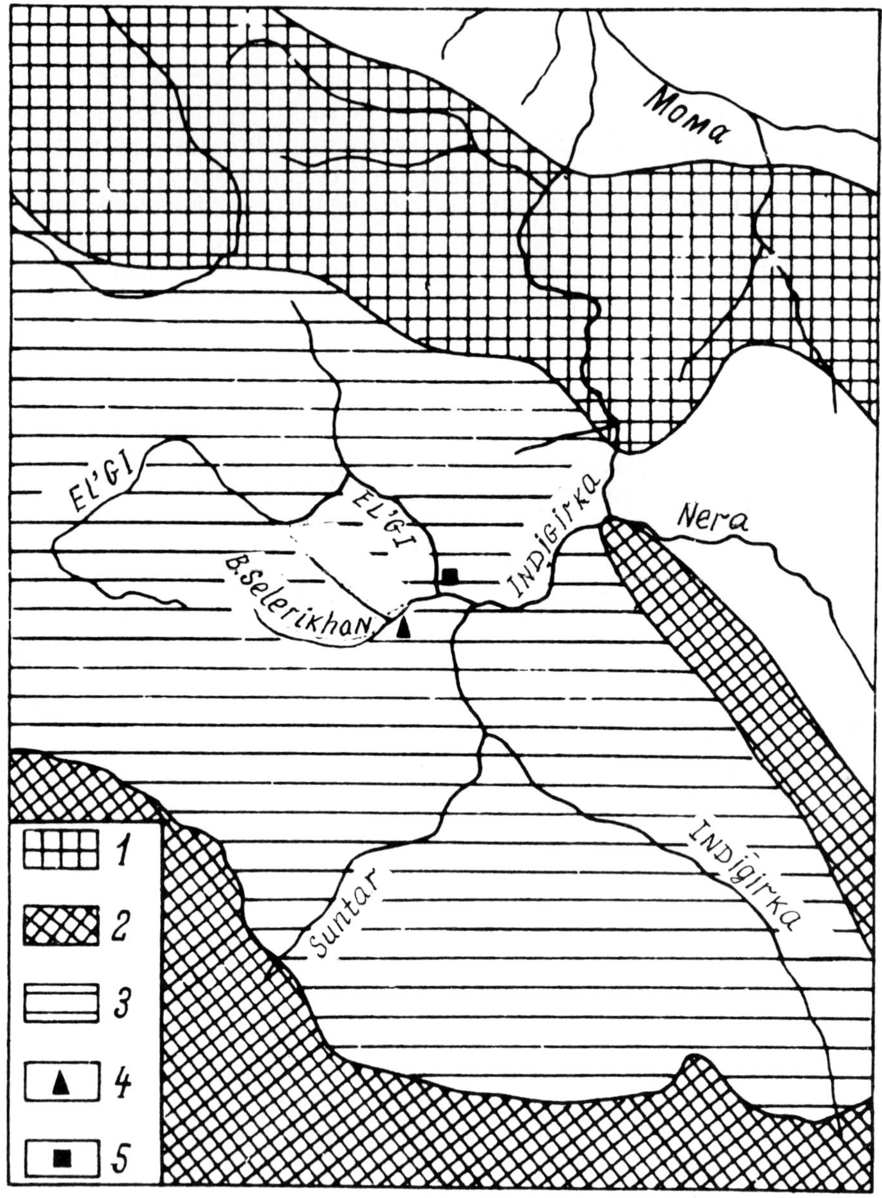

Figure 28. Morphostructural scheme of the area under study: 1) Chersky Upland, 2) Suntar-Khayata Range, 3) Oimyakon Highland, 4) fossil horse's burial site, 5) rhinoceros' burial site. After Belorusova, 1977.

others); (2) river terrace arid hillside forests, composed of *Larix gmelinii* - larch forests with a cowberry-moss layer; moist larch forests with a lichen layer; dry larch forests with a herb layer; (3) sites covered by steppe vegetation on steep,

Figure 29.   Larch taiga in the Bolshoi Selerikhan River Basin. Vicinity of the horse's burial site. Photo by N.V. Lovelius.

hill slopes of southern aspect, dominated by *Helictotrichon krylovii;* (4) bogs with the most extensive sedge hillock marshes and subordinate sedge-cotton grass swamps. Small areas are under forb-grass meadows, situated at slightly raised sites of the river terrace. The list of flora includes 264 species of plants, assigned to 139 genera and 156 families (Sokolova, 1977). The flora is dominated by boreal (115 species) and hypoarctic (108) species, accounting for 43.7 % and 41.3 %, respectively. The number of arctic and arcto-alpine species is as low as 40 only (15 %).

Plant remains which filled the gastrointestinal tract of the horse were poorly masticated and digested. This suggests that the horse died suddenly before it had time to digest the plants eaten before (Tikhomirova and Kultina (Ukraintseva), 1973). According to Vereshchagin (1977), a sudden death of the animal is also inferred from the attitude of the corpse: it lay in the deposits with its back up, neck raised, forehead slightly thrown up, the far foreleg raised, with the near foreleg stretched and tilted, and hind legs half-lowered. The animal must have tried its best to rear up and get out of a trap. Neither thoracic nor abdominal cavities were cut in the course of burial as evidences by the absence of mineral

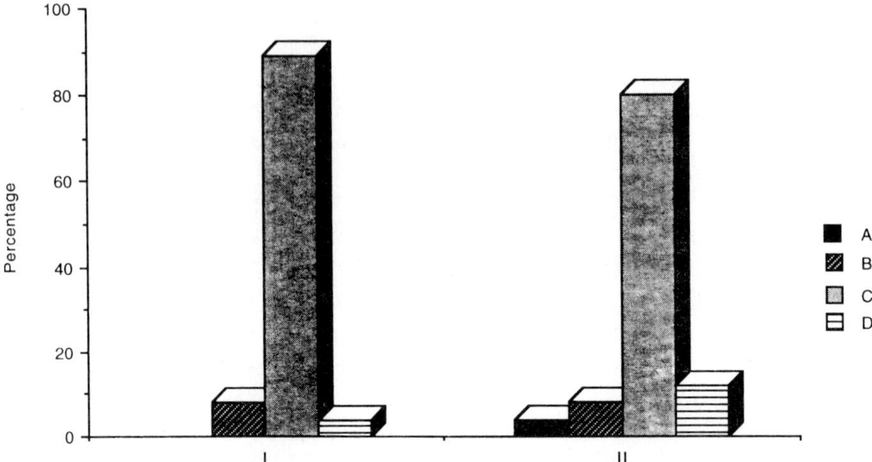

Figure 30. General composition of gastrointestinal contents of the fossil horse: I) macroremains;
II) pollen and spores of: (a) tree species, (b) shrubs and low shrubs, (c) grasses, (d)
mosses and ferns.

particles in the cavities (Vereshchagin, 1977). This fact proved to be of great importance for paleobotanists, namely anatomo-morphologists, a carpologist and particularly, palynologists, since it ruled out the possibility of foreign plants, their pollen and spores into the gastrointestinal tract thus suggesting that the investigators dealt with plant remains eaten by the animal just before a fatal accident. The fact that forage (grass, hay) remains in a horse's stomach for no more that 6-8 (10) hours (Azimov et al., 1954) indicates that the horse could not have gone far away from the site of its pasture and **hence, the composition of plants and pollen and spores produced by them and found in the horse's stomach, indicates the floristic composition and the character of vegetation in the area under study during its lifetime.**

Solonevich and Vikhireva-Vasilkova (1977) determined that remains of herbaceous plants account for more than 90 % of vegetational mass, found in colon of the horse; remains of young shoots of the birch (*Betula nana*) and willow (*Salix* sp.) and fine fragments of bark of these trees as cork total 5-7 %; remains of various mosses come to 1-2 % (Fig. 30). Remains of herbaceous plants were dominated by grasses. Most remains of epiderm ranked among different species of the genus *Festuca*. Rare are remains of epidermis typical of species of the genus *Calemegrostis,* as well as fragments of upper and lower epidermis having long hair, which resemble modern *Helictotrichon krylovii* (N. Pavl.) Henrard in structure. Remains of sedges, primarily represented by epidermis of their leaves, proved to be a much smaller amount than remains of grasses. No remains of sedge stems have been discovered. Remains of epidermis of dicotyledonous plants were noted but they remain unidentified.

Mosses, though extremely insignificant in amount, are diverse in species composition; they are represented by the following species:[8] *Bryum* sp.,

Distihium capillaceum (Hedw.) B.S.C., *Drepanocladus* sp., *Polytrichum stric-tum* Sm., *Polytrichum* sp.1, *Polytrichum* sp.2, *Rhutidium rugosum* (Hedw.) Kindb., *Thuidium abietinum* (Hedw.) B.S.C., *Thuideum* sp., *Tortula rugalis* (Hedw.) Crome, *Sphagnum* sp. They are dominated by fragments, complete leaves and even twigs of *Polytrichlum* sp.; fragments of leaves of the Calliergon sp. type and some other true mosses are rather common. Remains of bog mosses were very rare.

The great majority of fruits, found in food remains of the horse, belong to specimens of the sedge family (Egorova, 1977); an insignificant amount of fruits belongs to the families Rosaceae, Plantaginaceae, Polygonaceae. Egorova (1977) found 880 fruits of sedges belonging to two genera, namely, *Kobresia* Wild. and *Carex* L., and to six species with three species belonging to the genus *Kobresia* and three to the genus *Carex*. Of 880 fruits of sedges, 800 belong to *Kobresia capilliformis* Ivanova; eleven to *Kobresia filifolia* (Turcz.) Clarke; six to *K. simpliciuscula* (Wahl.) Mack; one fetal sac almost intact is assigned to *Carex pediformis* C.A. Mey, and one fruit to *Carex* sp. (a more accurate iden-tification could not be ascribed). All fruits present in the gastrointestinal tract of the horse are representatives of mountain types of vegetation. Pollen 2nd spores allow identification of 86 taxa of plants with 45 of them being ascribed a specific identification (Tikhomirov and Kultina, 1973; Ukraintseva (Kultina), 1977; Table 20 , present work).

The total composition of palynological spectrum is dominated by pollen of herbs (83 %); spores of *Bryales, Sphagnum,* and *Pteridophyta* account for 11 %; pollen of shrubs and low shrubs, such as *Pinus pumila, Salix* spp., *Juniperus* sp., *Betula fruticosa, B. exilis,* and *Alnus fruticosa* totals 7 %; pollen of trees constitutes 2 % only; it is represented by rare pollen grains of *Larix* sp., *Picea obovata, Picea* cf. ajanensis, *Populus suaveolens, Betula* sp., (sect. *Costatae), Betula platyphylla, Alnus hirsuta* and others.

The composition of pollen and macroremains in the gastrointestinal tract of the horse points to a leading part of herbs in its forage, dominated by grasses, sedges, Rosaceae, compositae, pinks; besides, the horse ate such plants as *Allium strictum, A. shoenoprasum, Juncus* sp. It also ate stems of mosses, primarily, true mosses, and lower young twigs of trees (*Betula, Alnus, Populus*), and shrubs and low shrubs (*Salix* sp., *Betula exilis, Alnus fruticosa*) (see Plates I-VIII).

A diversity of ecological groups of plants whose macroremains, fruits and pollen were found in the gastrointestinal tract of the horse suggests that prior to the fatal accident the horse grazed on: (i) steppe-like pastures, composed of *Helictotrichon krylovii, Kobregia filifolia, Carex pediformis, C. bigelowii* ssp. *rigidioides, Thalictrum foetidum, Potentilla stepularis, Allium strictum, Selaginella rupestris* and others; (ii) mesophytic meadows of forest clearings where *Lathyrus pilosus, Festuca* sp., *Hedysarum* sp., *Astragalus* sp. grew; and (iii) sites, sufficiently moistened, such as wet and swampy meadows on river-sides with abundance of *Kobresia capiliformis, K. simpliciuscule, Epilobium* cf. *palustre, Juncus* sp., *Potentilla multifida* and other specimens of forbs whose

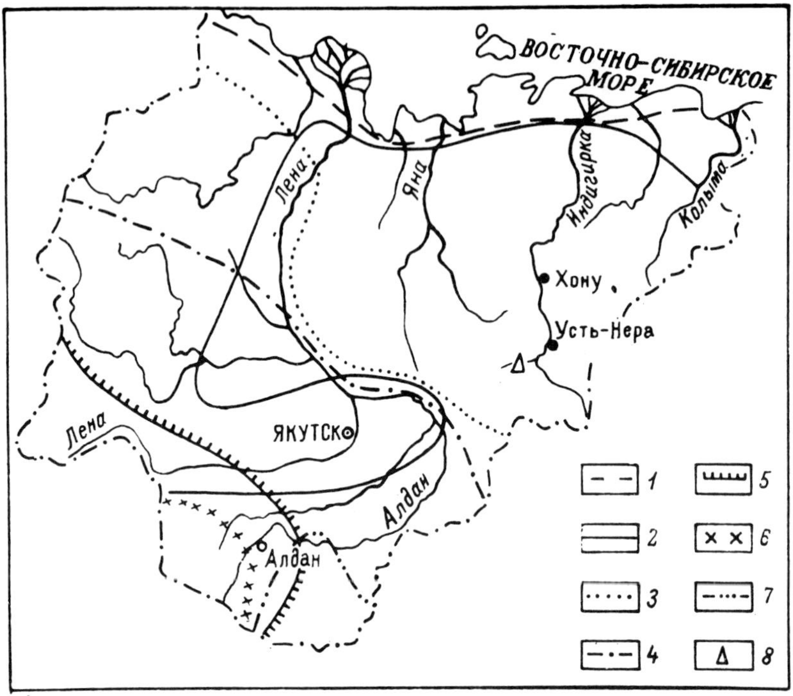

Figure 31.   Ranges of coniferous species in the territory of Yakutia. After Karavaev and Skryabin, 1971: 1) Dahurian larch, 2) mountain pine, 3) Siberian spruce, 4) Scotch pine, 5) cedar pine, 6) silver fir, 7) Jeddo spruce, and 8) the horse's burial site.

pollen is scant, and with some amounts of true mosses in the ground layer.

The horse perished in the period when most species of herbaceous plants were in blossom, probably in late July or early August, as suggested by: (i) pollen of such plants as *Lathy ruspilosus, Artemisia vulgaris, Oxytropis* sp., *Equisetum* sp. which are eaten by present Yakut horses in the second half of summer (Andreev, ed., 1974); (ii) the bulk of fruits of *Kobresia* cf. *capilliformis* which is in blossom in late May - early June and bear fruits in July - August; (iii) and the lack of cyryopses of grasses as they had not ripened by the time.

Almost all the plants whose pollen, macroremains end fruits were found in the gastrointestinal tract of the horse, grow today in the Indigirka River Basin and in other areas of northern Yakutia. *Picea obovata, Pinus sylvestris, Alnus hirsuta, Betula* sp. ex sect. *Costatae* are exception to this rule. Their current ranges - are removed from the burial site of the horse at more by than 1000 km (Figs 31, 32). Pollen of the above-mentioned tree species was found in the preparations in minor, but constant amounts. Small amounts of pollen of trees and shrubs in the content of the horse's gastrointestinal tract are no doubt explained by the fact that by the time of the fatal accident the trees and shrubs had already finished blossoming (most trees and bushes blossom in April - early May), therefore pollen of trees and bushes, accumulated on other plants and in

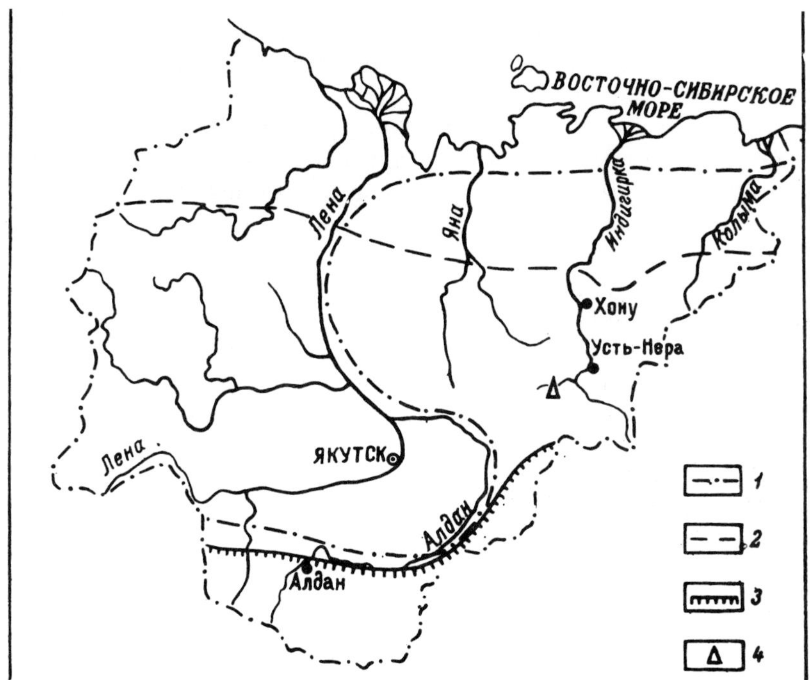

Figure 32. Ranges of foliage tree species in the territory of Yakutia. After Karavaev and Skryabin, 1971: 1) chozenia, 2) *Betula pendula*, 3) *Betula Ermanii*, and 4) the horse's burial site.

water basins, such as creeks, rivers or lakes, got in the horse's gastrointestinal tract together with herbs eaten by the horse.

In this connection, even such low amounts of pollen of tree species (*Betula platyphylla, Betula* sp. ex sect. *Costatae, Picea obovata, Picea* cf. *ajanensis, Pinus sylvestris, Alnus hirsuta, Larix gmelinii*) point to their presence in the vegetational cover of the area and adjacent territories during horse's lifetime. Of particular interest is pollen of the elm (*Ulmus pumila, Ulmus* cf. *japonica*[9]) and pollen of the beaked hazel (*Corylus* cf. *cornuta*), a bush species which occurs now in Canada and in the northern portions of the United States. Even if pollen of these plants is deliberately regarded as transported to the burial site from more southerly areas, then, taking into account that pollen of larch species is not brought in mass for very large distance (Tikhomirov, 1950b; Fedorova, 1952, 1956), it may be assumed that in the past, about 40,000 yr B.P. their ranges extended much farther north and north-east than at present.

Broad-leaved species occur nowhere in the territory of Yakutia despite the fact that its southern regions lie at the latitude of oak forests of the Moscow Region. However, along the river valleys, essentially those of the Amur River tributaries, representatives of mixed such coniferous-broad-leaved forests such as oak, maple, elm, lime, hazel and some others are spread rather far north, but only in a few places reach the boundaries of Yakutia. It should be noted that in the region the current boundary of mixed coniferous-broad-leaved forests lies near the southern limit of permafrost.

Of herbs such plants as *Kobresia* cf. *capilliformis, Nuphar pumilum, Allium shoenopragum, Stellaria jacutica* and some others are vital for reconstruction of the character of vegetation and natural environment. *Kobresia* cf. *capilliformis* is a typical plant of highlands of Middle and Central Asia, Mongolia which is rare in the upper part of forest zone. The plant dwells on wet and swamp meadows, along rivers and creeks, in cryophytic steppes. It grows at sites which have a sufficiently thick snow cover. During the horse' s lifetime it may have been abundant in the upper Indigirka River Basin, as suggested by a large amount of its fruits in the food remains. Egorova (1977) believes that the extinction of *Kobregia* cf. *capilliformis* in the area was associated with a progressive development of continental climate of Eastern Siberia in the succeeding period, manifested, particularly, in a sharp change of snow cover.

The pit in which the horse was found was partly caved in and flooded in the process of further mining works and thus proved to be inaccessible for studying the deposits which directly enclosed the animal's corpse and for sampling for palynological analysis.

Two sections were studied near the burial site. The first was made by O.V. Grinenko and P.A. Lazarev in the summer of 1968 at 30 m from the burial site; according to Lazarev (1982), it exposed the following deposits (in descending order):

Deposits	**Thickness (m)**	**Depth (m)**
1. Soil horizon of the recent valley	0.2	0-0.2
2. Loam dark-grey with inclusions of sandy loam, gravel, poorly rounded pebbles, up to 3 cm in diameter	1.6	0.2-1.8
3. Loam dark-grey, impregnated with fine plant debris	0.6	1.8-2.4
4. Moderately and poorly rounded pebbles and gravels with inclusions of rock debris (the bed is not persistent in thickness and strike)	0.4	2.4-2.8
5. Loam with numerous inclusions of twigs, roots, plant debris, including a birch twig with well preserved bark	0.7	2.8-3.5
6. Sand inequigranular, fine pebbles, sandy loam, loam, rock debris; material is unsorted, poorly rounded. Pebbles and rock debris 7-10 cm in diameter with the predominance of 3-5 cm	1.1	3.5-4.6
7. Sand inequigranular, containing pebbles,		

sand loam, loam. Pebbles and gravels derived from sandstones and shales, moderately or poorly rounded, 3-5-7 cm in diameter. The bed contains thick (up to 1.5 m) lenses of cloudy ice showing bubbles and vertical banding	2.3	4.6-6.9
8. Loam and sandy loam with inclusions of scattered pebbles and unrounded material, up to 2-3 cm in size, and plant remains. Horizontal bedding, cut by ice lenses, is traced. The bed is not persistent in strike; it is a lens. It is in this bed that the horse was buried	1.1	6.9-8.0
9. Sand, sandy loam and loam with inclusions of rare pebbles and rock debris, thick ice lenses	0.9	8.0-8.9

Results of palynological analysis allow subdivision of the sequence, made up of alluvial-diluvial deposits into two beds. The lower bed (depth 6.8-8.9 m) which enclosed the horse's corpse contains a steppe-like sporo-pollen complex, dominated by pollen of low shrubs and herbs (up to 73%). The herbs contains abundant grasses (51%), wormwood (44.1%), sedges (35.5%), pinks (18%), *Ranunculus* (16.5%), Ericaceae (16.5%), forbs (12%), buckwheat (8.5%), cruciferae (5.3%). Pollen of wood group (24%) is represented by pollen of willow (up to 82%), birch (83.5%), dwarf alder (26%), alder (up to 11.2%). Spores average 28% with the predominance of *Selaginella rupestris* (up to 74%), liverwort (up to 46.2%), true moss (up to 30.6%), and Polypodiaceae (up to 19.4%).

The upper part of sequence (from surface to a depth of 6.0 m) includes forest-type complexes in which pollen of wood plants (up to 82%) prevails over that of herbs (up to 42%) and spores (up to 18%). The wood vegetational group is composed of birch (up to 74%), alder thicket (up to 39%), larch (up to 19%). Mountain pine (*Pinus* sub./g. *Diploxylon* and *Pinus* sub./g. *Haploxylon*) is subordinate. Herbs are represented by grasses (up to 58%), ericaceae (up to 22%), wormwood (up to 14.5%), pinks (up to 13.6%), sedge family (up to 9.2%), whereas spores encompass those of bog mosses (up to 44.5%), polypodiaceae (up to 44.5%, *Selaginella rupestris* (38.4%), true moss (up to 33.9%), liverworts (up to 2%), and pteridophytes (up to 7.9%). The above authors believe that spectra of samples, taken around the corpse, and that of one sample from the corpse-enclosing monolith, are similar in composition to the above-characterized spectra of the lower bed.

The sporo-pollen complex of the lower bed which reflects the steppe-like vegetation, dominated by grass-wormwood-sedge associations, is dated "as the end of Zyrianka Ice Age or as the end of Karginsky Interglacial and late

Pleistocene" (Lazarev, 1982).

The second section of loose deposits resulted from thermokarst subsidence in the summer of 1969 at 150 m from the site of discovery; the section was studied and described by Belorusova (1977). Deposits, essentially composed of debris of clay shale and sandstone lie at the base of section. As a whole, the outcrop is made up of organo-mineral rocks with inclusions of subsurface ice. The mineral part is built up of silt and sandy loam containing moderately rounded pebbles, abundant debris, 0.5-2.5 cm in diameter, and horizontal intercalations of gravel and scree, mixed up with coarse-grained sand. Silt and sandy loam alternate with horizontal beds of organic remains, composed of unripe peat, roots, bark and trunks of wood.

Vegetational remains form four main beds which lie at depths of 0.3-0.5 m, 1.8-2.5 m, 3.5-3.9 m, and 5.1-6.5 m. Fragments of larch trunks up to 2.15 m in length and 16 cm in diameter were found in the bed occurring at a depth of 3.5-3.9 m; a larch trunk 30 cm in diameter was extracted from the bed at depth interval 1.8-2.5 m. The bed (5.1-6.5 m) which is in contact with auriferous "sand" is replaced along the strike by oozy sandy loam which contains not only floristic, but also faunistic remains. In the process of mining, this bed may have yielded, to date, surface remains (bones, horns, tusks) of mammal fauna, i.e. of mammoth, bison, musk ox, woolly rhinoceros and others, found by workers of the Selerikhan mine. "The horse's corpse was found at contact of the bed (5.1-6.5 m) with the underlying beds, in loam containing debris, at depth of 8-9 m from the surface. The corpse lay between "peats" (according to miners' terminology), 8-9 m thick, and Mesozoic rocks occurring in section at a depth of 10.8 m at the top of productive series, i.e. auriferous 'sand'" (Belorusova, 1977, p. 70).

The following is a complete description of the section, performed by Belorusova (in descending order).

Outcrop 3a, August 23, 1969	**Thickness, m**
1. Sandy loam whitish-yellow, with debris and scree of clay shale and sandstone.	0.30
2. Sandy loam yellow-grey with few acute-angles debris and frequent thin (1-3 cm) seams of broken peat and abundant vegetational remains, represented by branches and trunks of trees, up to 5-7 cm in diameter	0.20
3. Sandy loam yellow-grey with debris (up to 30%) and rare thin (1-2 cm) lenses and seams of peat	1.30
4. Sandy loam yellow-grey with rare clastic products (debris or, less commonly, pebbles up.to 2 cm); in places sandy loam is oozy, peaty, with abundant vegetational remains, represented by flexuous roots, stems, birch bark, trunks of trees up to 30 cm in diameter	0.70
5. Sandy loam oozy yellow-grey with clastics up to 10% and peat bands	1.00

6. Sandy loam oozy dark-grey with sparse clastics and numerous wood remains. Unusual flattened saber-end disk-shaped larch trunks, 1.9 m and 2.5 m in length, 4 X 16 cm in size, were extracted from the lower part of bed   0.40
7. Sandy loam dark-grey with debris and scattered vegetational remains   1.20
8. Peat with seams (1-2 cm) of sandy loam and wood remains (branches, bark, trunks of trees 10 cm in diameter   0.60

Radiocarbon analysis of two specimens of part of the larch trunks, collected from beds 6 and 8, has given a Holocene age of the deposits. The lower sequence of a section (Bed 8) was formed in the early Holocene, 9,480±300 yr B.P. Ancient Holocene and Sartan deposits "cannot be clearly recognized in alluvium of Balkhan Creek. This can be explained by either their small thickness or the fact that main mine workings embrace not only productive strata, but also these horizons. Part of deposits may have been washed out since it should be borne in mind that accumulations of bone remains in the Elga River Basin occur at the contact of Sartan and Karginsky beds" (Belorusova et al., 1977, p. 274).

Results of palynological analysis (Fig. 33 , bed 8) suggest that in early Holocene time (9,480±300 yr B.P.) the vegetational cover of the area was dominated by shrub and undershrub associations, composed of Betula sp. ex sect. Fruticosae, B. exilis, Betula sp. (sect. Nanae), Alnus fruticosa, Salix sp. sp. A sporo-pollen spectrum of sample 17 from bed 8 is characterized by the predominance of pollen of shrubs and low shrubs (66%); pollen of herbs account for 11.3%; spores of mosses come to 20.2%. Pollen of tree species is represented by rare grains of Larix sp., Pinus sylvestris, Pinus sp.,and totals only 2.5%.

Larch pollen is known to be poorly preserved in fossil state (Vaskovsky, 1959; Mironenko and Savina, 1975; Karevskaya, 1979. This results in skewness of true percentages of pollen of trees and bushes, as well as other plant groups in the composition of palynological spectra. This suggests larch forests and open woodlands in the area in early Holocene time. The climate may have been no warmer than that of today, nevertheless, frost embraced deeper horizons. This is suggested by the larch trunk, 10 cm in diameter, 90 years old, with the tap root exceeding 60 cm in length, which was found in bed 8, whereas present trees have sprawling root systems even in the Selerikhan area (Belorlsova et al., 1977). It is interesting that in more northerly areas of the Indigirka River Basin the early Holocene (10,500-9,000 yr B.P.) witnessed a partial degradation of permafrost and the formation of alases within grass-sedge tundras, composed of large woody birch and relatively abundant grass flora (Lavrushin et al., 1963).

In ensuing time (Fig. 33, bed 7), the vegetational cover underwent no substantial variations. From time to time only the areas under alder and birch forests reduced due to the development of herbaceous vegetation (Fig. 33, sample

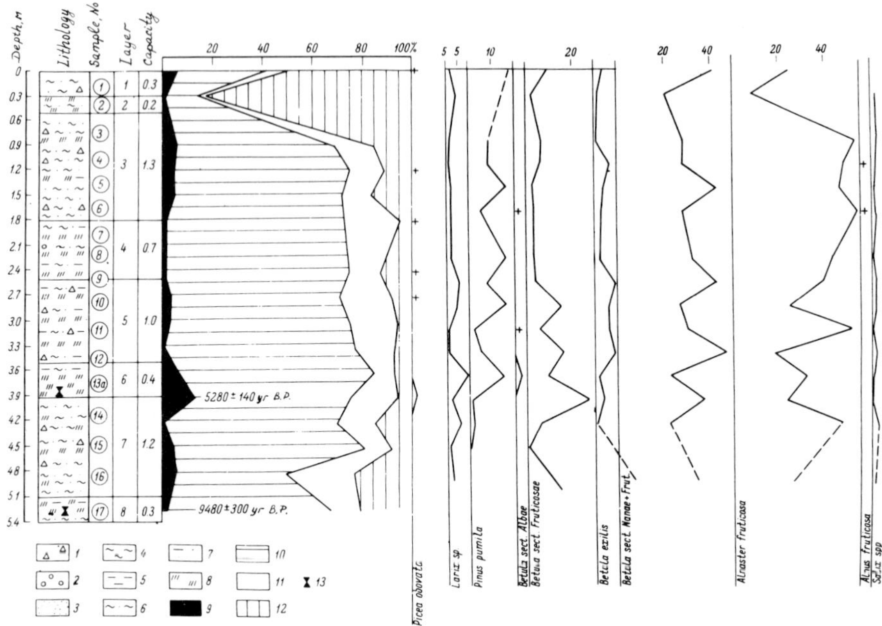

Figure 33. Spore-pollen and sediment diagram of outcrop No. 3a. After Belorusova et al., 1977:
1) detritus, 2) pebble, 3) sand, 4) clay, 5) ooze, 6) sandy loam, 7) silt, 8) peat with

16). A low content of pollen of *Pinus pumila* is noted in all the samples from beds 7 and 8; this points to its insignificant part in phytocoenoses of the terrain during the formation of the beds.

In middle Holocene time, 5,280±140 yr B.P. (Fig. 33, bed 6), birch-larch forests, composed of *Larix* sp. with minor *Picea obovata* and *Betula* sp. (sect. *Betula*) already existed there. The grass cover, made up of grasses, sedges and forbs, was well developed. True and bog mosses play a minor part in phytocoenoses. Xerophytic aggregations including *Selaglenella rupestris* were common on southern aspects, stony and debris slopes. All the above facts suggest a warmer climate during the formation of bed 6. The radiocarbon date permits synchronizing this warming with the time of climatic optimum middle Holocene. According to the current notions, in middle Holocene time the summer temperatures were higher than present ones as suggested by not only the composition of sporo-pollen spectra, but the abundance of trunks of fossil woody birch beyond its current range. According to Vaskovsky (1959), summer warming was sufficient to completely thaw glaciers in the mountains of Northeast Siberia, nevertheless, it exerted a lesser influence on change in frost soils regime. Permafrost did not disappear completely during this period, but its top greatly lowered thus causing an increase in thickness of active layer and, hence, intensification of slope processes, instability of soils of and their mass downslope movement. This is suggested by disk- and saber-shaped trunks of the two fossil

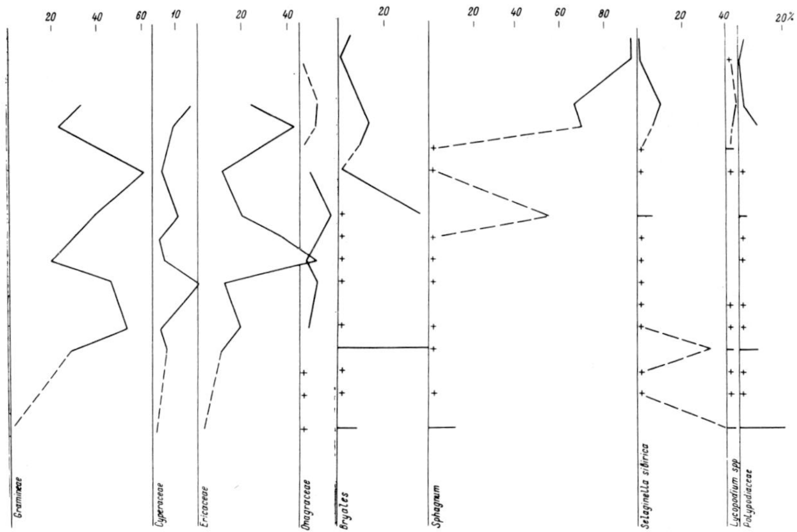

woody remains; pollen of: 9) tree species, 10) shrubs and undershrubs, 11) grasses; 12) spores of Pteridophyta and Bryophyta; and 13) site where wood was taken for ^{14}C analysis.

larches, extracted from bed 6, depth 3.9 m (Belorusova et al., 1977, Fig.s 3, 4). The flattened form of the trunks cannot be explained by secondary roundness because of good preservation of annual rings which allowed them to be dated at 141 and 211 yr, respectively.[10] Analysis of their growth and the rate of growth permitted N.V. Lovelius to reveal secular and minor variations of natural environments during the lifetime of trees. Extremely low growths in separate years and decades, as well as a smooth curve of growth in these models indicate that they grew under unfavorable conditions. The upper limit of forest was higher than that of today during the climatic optimum of Holocene hence solifluction and soil drifting, inevitable on slopes during warmings, should be classified as unfavorable conditions of forest growing. Intensification of solifluction was caused by a change in permafrost regime and, first of all, a greater thickness of active layer. Degradation of permafrost resulted in lowering its top and local development of thermokarst, but analysis of pitdrilling lines (Belorusova et al., 1977) showed that permafrost had never disappeared completely in alluvium of Badran and Balkhan creeks.

Starting from a depth of 3.6 m up section a decrease of pollen of tree species down to 6.6-0.5% is observed in spectra. In specimens 1 and 2, the amount of spores (50.0-84.6%) increases 2 to 10 times, as compared to the fourteen specimens described above, whereas the amount of forb pollen (7.5-2.6%) correspondingly decreases 2 to 9 times. Minimal amounts of pollen of trees

(0.5%), herbs (7.0%), shrubs and undershrubs (12.3%) were fixed in specimen 2 with pollen of mountain pine (62%) dominating in the latter group of spectrum. Spores of true mosses account for 84.6% of this specimen. As can be seen from the diagram (Fig. 33), its upper part suggests an evident cooling. Deterioration of climatic conditions in the subboreal and subatlantic of the Holocene caused: (i) a change in coenotic composition of the forests, resulted in the disappearance, at first, of firs, pines and, then, woody birches, and (ii) lowering of the upper boundary of forests in mountains.

The above-mentioned data indicate that two sequences of diachronous deposits were studied near the burial site of the horse. The section of late Pleistocene - Zyrianka - early Karginsky deposits was exposed and studied at 30 m from the burial site by Lazarev (1982), and the sequence of Holocene deposits at 150 m from the site was studied by Belorusova and coworkers (1977). After Belorusova (1977), the horse's corpse was buried 1-2 m below the Holocene deposits at a depth of 8-9 m from surface in the bed, composed of loam with scattered debris. Examination of the frozen monolith showed that the corpse of Selerikhan horse was pressed in bluish loam which contains poorly rounded pebbles and clasts 50-60 mm in diameter (8-12 vol. %), both ferruginated on surface and along joints and made up of dark-bluish relatively loose micaceous sandstone" (Vereshchagin, 1977, p. 79). The bed of loam in the outcrop, studied by O.V. Grinenko and P.A. Lazarev, is not traced far along the strike as it is a lens (Lazarev, 1982). Spectra of the "steppe-like type" are typical of the specimens, taken from the bed of loam (depth interval 8.9-6.0 m, sp. 9-22) and specimen 33, taken from the monolith. Spectra of the gastrointestinal contents of the horse are similar to those of the loam in percentage of pollen of trees, shrubs and low shrubs (totalling 9%), forbs (83%), spores of sporophyres (11%), but not identical to them in taxonomic composition. They also differ from spectra of the "forest type" of the overlying sequence. This suggests that both sequences, exposed in the outcrop studied, are not synchronous to the period of fossil horse life.

Taphonomic conditions of horse's burial, investigated by N.K. Vereshchagin, suggest two alternatives for its death, namely, either in a thermokarst scour or a solifluction stream. Collation of geological-geomorphological and paleobotanical data, based on studies of the previously mentioned outcrops and the examination of horse's gastrointestinal contents, combined with the study of taphonomic conditions of the horse burial in the deposits (Vereshchagin, 1977), provide evidence for the first alternative, i.e. the sudden death of the animal in a thermokarst scour. The scour, formed 38,530±1,120 yr B.P., exposed strata of Karginsky age. The horse fell in it by accident and was buried in the lower part of exposed sequence of deposits which may have been accumulated in early Karginsky time. This may have been simultaneous with the deposits exposed and studied in the outcrop at 30 m from the burial site (depth interval 8.9-6.0 m from surface). During the formation of the latter and during the accumulation of loams in which the horse was pressed the area was dominated by open treeless landscapes as indicated by spectra of the "steppe-like type". The overlaying strata (depth interval 6.0-0.0 m) was formed in late stages of the Karginsky Interglacial when forest landscapes had already been developed, as shown by spectra of the "forest type". As suggested by the composition of these spectra, during the accumulation of deposits enclosing them, monodominant larch forests with undergrowth composed of shrubby and undershrubby birches, willows and alders, were distributed in the area.

The lifetime of the horse falls in the optimal phases of the Karginsky Interglacial, as reliably established by ^{14}C analysis of the remains of the plants, eaten by the horse and preserved in its gastrointestinal tract. This is also confirmed by geological-geomorphological analysis of the area where the horse was found (Belorusova, 1977). However, palynological data suggest that no deposits, associated with the optimal phases of the Karginsky Interglacial, have been represented in the outcrop exposed and studied at 30 m from the site of horse's death. Optimal phases in the history of flora end vegetation of the area, fixed by results of paleobotanical analysis of the food remains of the horse, suggest the existence of forest coenoses, composed of spruces (*Picea ajanensis, P. obovata*), woody birches (*Betula platyphylla, Betula* sp.) and other tree species, shrubs and low shrubs in the area at the time of its death [Ukraintseva (Kultina), 1977; Ukraintseva, 1979]. The forests were similar in composition to the present middle taiga forests in the basins of Lena and Aldan rivers and on the Stanovoi Highland. The ranges of some broad-leaved species, the elm can be cites as an example, advanced much farther to the north then, than today; therefore their long-distance transported pollen occurred as far as the latitude of the Elga River and possibly farther north. Various meadows occupied the river valley, and meadow-steppe communities occured on southern hillside exposures; reconstructions of some elements of paleoclimate showed that this was associated with a higher thermal provision of the area in question (Table 12 Fig. 34) and adjacent territories. Summer which was somewhat warmer, as compared to the present, was conductive to lowering levels of permafrost ground thawing and the development of thermokarst processes and solifluction, as proved, with assurance, by the well preserved corpse of fossil horse.

Two subsequent coolings (Konoshel and Sartan) resulted in degradation of the forests, with an initial disappearance of spruce. Jeddo disappeared first, followed by Siberian spruce, there was a reduction of the ranges of mountain pine and alder. The development of shrub and undershrub associations in the valleys was promoted, as well as various tundras in the mountains which, as stated above, dominated the area in early Holocene time (9,480$\pm$300 yr B.P.).

Conclusions

1. The fossil Chersky horse (*Equus lenensis* Russanov), found in the Balkhan Creek Basin, upper Indigirka River, was short; as suggested by some characters in the structure of skull and skeleton, the Chersky horse resembles the present Yakut horse which grazes on grass all the year round.

2. The geological age of the horse is dated as late Pleistocene - Q3 - Würm I-II.

Table 12

Values of heat- and moisture provision

Site, area	Temperature, C			Σt° for months (V)VI-IX	Σ Annual precipitation	Σ Precipitation for months (V)VI-IX (mm/% for months I-XII)
	July	January	mean annual			
1. Elga, 596 m	14.7	-48.5	-22.0	1148	233	126/54%
2. Marshalsky Mine, 707 m	7.8	-46.0	-20.5	520	223	134/60%
3. Selerikhan, reconstructed for 38,500 ±1,120 yrs B.P.	17.0	-30.0	-7.0	1680	377	255/68%
Deviation	2.3	-18.5	-15.0	+532	+144	+129/90%

1) Calculated by interpolation of data of six meteorological stations, namely, Lensk, Nyuya, Tanya, Dzhikimsa, Sion-Tit, Aldan.

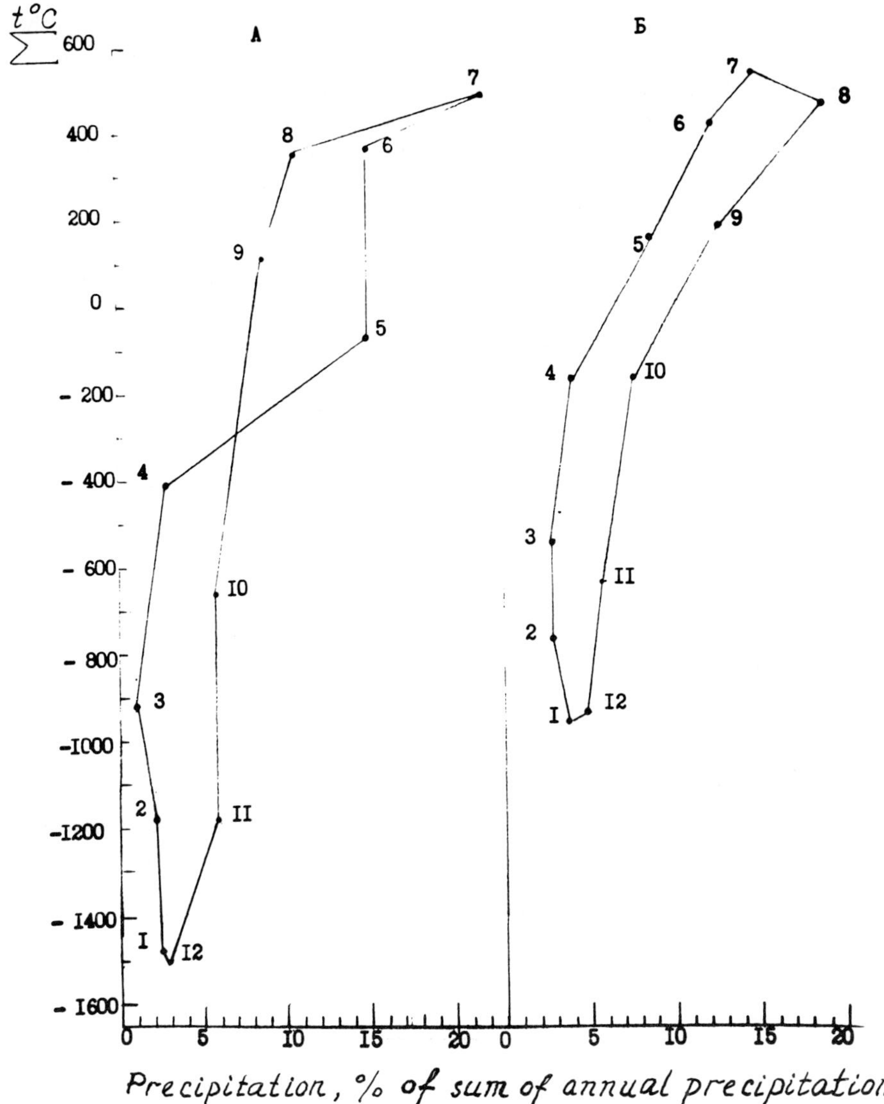

Figure 34. A) climagraph characterizing present climatic conditions of the area of fossil horse's burial, Elga River Basin; B) paleoclimagraph characterizing temperature and moisture conditions of the Elga River Basin 38, 590±1120 yr. B.P.

3. The geographic range of the species was confined to northeastern regions of Siberia, Beringia land and, probably, Alaska in late Pleistocene time.

4. The horse perished during the optimal phases of the Karginsky Interglacial; it is believed to have fallen in the thermokarst scour and struggled in vain to get out of it, as suggested by corpse's attitude. The animal proved to be pressed on the bottom of the scour which exposed the sequence of deposits,

accumulated in earlier stages of Karginsky time. This is supported by: (i) the ^{14}C date of horse's death; (ii) the composition of paleoflora; and (iii) geological-geomorphological analysis of the environs of the burial site. The animal perished in late July or early August when most herbs were in blossom; never-the-less, fruits of some sedges (*Kobresia, Carex*) had already ripened. However, caryopses of grasses had not ripened by the time.

5. The composition of macroremains, pollen and spores of plants and their percentages, determined in analysis of the food remains of the horse, suggests herbs, grasses, sedges, and forbs were major components of its forage; it also ate leaves and thin twigs of shrubs and low shrubs, plus the lower branches of some trees.

6. The variety of ecological groups of plants which were determined from their remains, preserved in the gastrointestinal tract of the animal showed that shortly before its death the animal grazed on habitats of different types such as dry steppe communities, mesophytic meadows, and riparian aquatic aggregations. However, forest coenoses played an important part in the vegetational cover, as suggested by (i) the forest habit of paleoflora, shown by findings of arboreal pollen and shrubs; (ii) high systematic diversity of herbaceous flora; (iii) the presence of some aquatic plants, ferns and mosses. Although the paleoflora synchronous to the horse's life was revealed in detail, nevertheless it displays richer vegetation than that of today in the area. The recent Lena-Aldan forests, mountain forests of southern Yakutia and the Stanovoi Highland can be regarded as modern equivalents of the forests which existed in the middle Indigirka River Basin in the optimal phases of Karginsky Interglacial.

7. The presence of some species of steppe communities, meadow-steppes and dry meadows in paleoflora suggests that the communities had been developed in the area since the optimal phases of Karginsky Interglacial; this was related to greater warmth in the area and adjacent territories.

4.5. Mylakhchin bison

In 1971 A.M. Struchkov, the hunter, found a fossil animal on the middle Indigirka River, 50 km upstream from the settlement of Belaya Gora (Fig.35). The animal proved to be a bison. It was buried in loess-like loam at the base of the right Indigirka loess-ice terrace, 35-40 m high, which is being intensely destroyed at present (Fig.s 36, 37).

B.S. Rusanov and P.A. Lazarev, workers of the Institute of Geology, Yakutsk, extracted the corpse from permafrost and transported the monolith to Yakutsk where it was placed into a freezing pit of the Institute of Permafrost. In January, 1973, the monolith was transported to the Novosibirsk, Institute of Biology which played host to the meeting at which the participants examined two finds, namely, the bison and the mammoth, discovered on the Shandrin River. K.K. Flerov examined the animal and stated that its fore-part was badly injured by an ice wedge, whereas its rear part was well preserved. The hair coat was entirely preserved; skin, some muscles and the skeleton were more or less

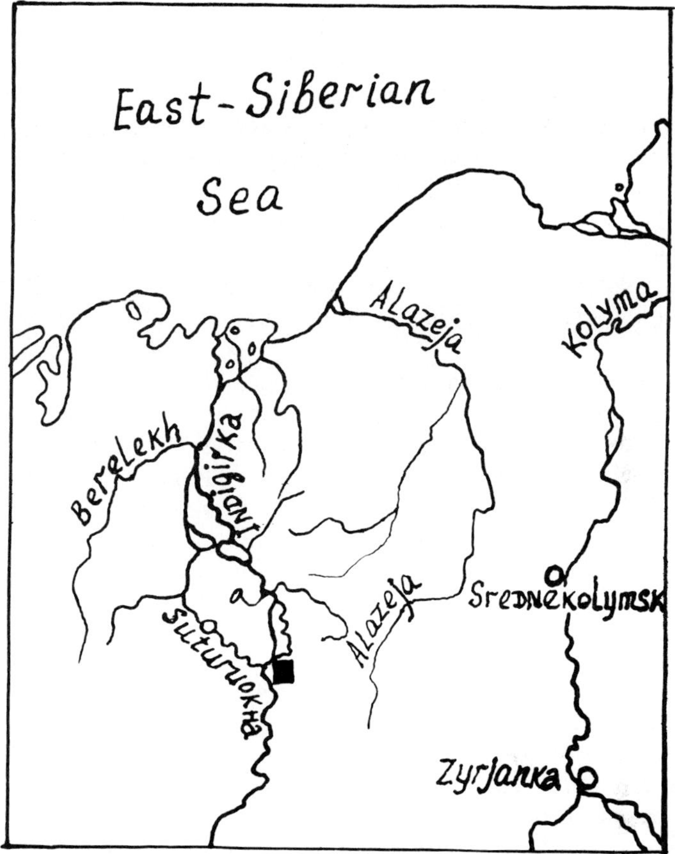

Figure 35. Index map showing the area of fossil bison's burial site (marked black square).

well preserved. The skull was badly broken, but its lower part proved to be quite intact. Some internal organs, including the genitals and the gastrointestinal tract, were also preserved; the intestinal tract contained some food remains.

Exterior features of the animal led K.K. Flerov to the conclusion that it was the female *Bison priscus occidentalis* (Lucas), 2.5 years old (Ukraintseva *et al.*, 1978; Flerov, 1979). Histological and cytological examinations of the preserved reproductive organs suggest that the animal was no younger than a year and eight months when it died (Korobko and Kursanov, 1979). Despite the fact that the mammal died 29,500±1000 yr B.P.,[11] as suggested by ^{14}C analysis, the hair coat and its color proved to be well preserved and showed that it was identical to that of the recent wood bison *Bison priscus athabascae* (Rhoads) (Flerov, 1979), (Fig. 38).

The site where the bison was buried is located on the Abyi Lowland. The elevation of the lowland is 50 m in the central part and 200-300 m around the periphery. The lowland is bounded by The Moma Range on the south; the

Figure 36. Indigirka River loess-ice terrace in the area of fossil bison's burial. Photo by V.V. Ukraintseva.

Polousny Range on the north; the Saltaga-Tag and the Andrei-Tag ranges on the west. The Indigirka River Valley is forested, but contains many swamps; plus multiple lakes; forests alternate with dwarf birch communities and bogs. The larch [*Larix gmelinii* ssp. *cajanderi* (Mayr) J. Kozhevn]. is the only forest-forming species. The area is mainly occupied by sedge-cotton grass back bogs and cotton grass-sedge hillock bogs, combined with light larch forests, dwarf birch communities and, in places, polygonal sphagnum bogs (Karpenko, 1958). Topographic highs are dominated by lichen-moss larch forests with shrubs and moss-undershrub larch forests and open woodlands, generally combined with bogs (Fig.s 39, 40).

As a whole, flora is poor and contains no more than 200 species of plants (Kozhevhikov, 1981). According to Kozhevnikov, the paucity is caused by a small number of species in any complex of habitats of Abyi Lowland and a small number of species of the complexes proper; this is related to little-contrasting relief, swamping of depressions, afforestation of hills, and severe climate. Therefore no mountain pine, large woody birch, poplar, and chozenia grow there, although their north limits are shown downstream the Indigirka River (Komarov, 1926).

The food remains were found out to be digested to such a degree that identification of plant macroremains involved difficulties. Larger plant sections which showed the best preserved structures, do not exceed 0.7-1.0 cm in length and 1.5 mm and below in width. Herbs were primarily represented by fragments of epidermis of sedge and grass leaves. A few relatively large remains belong-

Figure 37.   Heavy destruction of the loess-ice terrace in the area of bison's burial. July, 1977. Photo by V.V. Ukraintseva.

ed, according to V.V. Vikhireva-Vasilkova, to some large forbs. In addition, few fragments of epidermis of leaves with the cell structure typical of leaves of lilies were found - which were relatively large, elongate along the leaf axis. Bits of cork and bark of the low birch (*Betula nana*), alder (*Alnus* sp.), willow (*Salix* sp.), remains of venation of fine leaves of *Ericaceae* were also found. According to N.G. Solonevich, a sample weighing 50 grams was dominated by remains of plant tissues which were primarily different parts of sedges and grasses. In addition, a small amount of well preserved sedge fruits and sacs and a few fruits of *Kobresia* sp. of rather poor quality were discovered. No grass caryopses were found. Moss remains were relatively abundant. They included: a ramal leaf of

Figure 38. Late Pleistocene bison (*Bison priscus occidentalis*). Two years old female bison. Reconstruction by K.K. Flevov after a specimen found in permafrost, middle Indigirka River.

Sphagnum sp., some leaves of *Polytrichum* sp., leaves of *Tomenthypnum nitens*, leaf fragments of some species of *Bryum* and some other true mosses.

Palynological analysis of the bison's food remains revealed a more complete composition of plants which grew in the vicinity of its habitat (Table 13). Palynological spectrum showed almost equal proportions of pollen of herbs and spores of mosses and ferns (Bryophyta and Pteridophyta) which account for 49.7% and 43.4%, respectively (Table 13 , Fig. 41). Tree pollen is represented by rare grains; it comes to only 1% of the sum total of all the forms calculated; pollen of shrubs and undershrubs amount to 5.9%.

These results from the examination of the bison's food remains suggest that the mammal was basically fed on sedges (*Carex* spp., *Kobresia* sp.), grasses (*Poa* spp., *Poaceae*) and various plants of forbs. As suggested by abundant macroremains of mosses and their spores in the forage mass, the bison also ate mosses which were common in the ground cover. However, like today, mosses were subordinate in the nutrition of the bison (Soper, 1941). The presence of moss remains and their taxonomic composition suggest numerous swamps in the localities where the bison grazed shortly before its death and its range as a whole. A low amount of fruits of *Carex* spp., *Kobresia* sp., found in the forage mass of the bison, suggests that it died in early summer, i.e. in late June or at the very beginning of July when fruits of these plants had just started ripening. Seeding had not started yet, that is why no caryopses were found. At present, mass ripening of fruits of sedges and cotton grasses falls on mid-July in the In-

Figure 39. Wild rosemary-mossy larch forest with *Alnus fruticosa* and *Rosa accicularis* in the vicinity of bison's burial site. Photo by V.V. Ukraintseva.

Figure 40. Paludal sedge-cotton grass larch forest in the vicinity of bison's burial site. Photo by V.V. Ukraintseva.

digirka River Basin. At this time, caryopses of some grasses only, namely, *Poa botryoides, Festuca kolymensis,* and *Agrostis trinii* start ripening. In August, caryopses of *Arctophila fulva, Calamogrostis langadorfii, Bromus pumpellianus* and others ripen (Andreev *et al.*, 1974).

Table 13

Composition of plants, identified from pollen and spores, in the large intestine contents of the bison.

Ord. No.	Plant	The amount of pollen and spores	
		abs.	%
1	2	3	
Pollen of the trees, shrubs and low shrubs			
1.	*Larix sp.*	1	0.11
2.	*Pinus sibirica*	2	0.2
3.	*P. pumila*	30	3.0
4.	*Salix spp.*	5	0.5
5.	*Betula platyphylla*	3	0.3
6.	*B. exilis*	1	0.1
7.	*Betula sp. (ex sect. Nanae)*	18	1.8
8.	*Alnus cf. hirsuta*	3	0.3
10.	*Ulmus sp.*	1	0.1
Pollen of herbs			
11.	*Typha sp.*	1	0.1
12.	*Arctophylla fulva*	1	0.1
13.	*Poa sp.*	1	0.1
14.	Poaceae	91	9.1
15.	Cyperaceae	146	14.7
16.	*Polygonum viviparum*	4	0.4
17.	Chenopodiaceae	17	1.7
18.	*Stellaria jacutica*	4	0.4
19.	*Stellaria sp.*	7	0.7
20.	Caryophllaceae	100	11.0
21.	Ranunculaceae	1	0.1
22.	*Papavar sp.*	1	0.1
23.	Brasicaceae (Cruciferae)	2	0.2
*24.	*Saxifraga sp.*	1	0.1
*25.	*Potentilla spp.*	2	0.2
*26.	Rosaceae	1	0.1
*27.	*Lathyrus pilosus*	2	0.2

Table 13 (cont.)

*28.	Fabaceae	1	0.1
29.	Apiaceae (Umbelliferae)	2	0.2
30.	*Gentiana sp.*	1	0.1
31.	Ericaceae	1	0.1
32.	*Phlox sibirica*	10	1.0
33.	*Polemonium boreale*	19	1.9
*34.	*Valeriana capitata*	8	0.8
*35.	*Artemisia spp.*	14	1.4
*36.	*Aster sibiricus*	2	0.2
37.	*Centaurea sp.*	1	0.1
*38.	Asteraceae	33	3.3
*39.	Cichoriaceae (Liguliflorae)	9	0.9
40.	Monocotyledoneae indeter.	1	0.1
41.	Dicotyledoneae indeter.	7	0.7

Spores of Bryophyta, Pteridopyta

42.	Hepaticae	4	0.4
*43.	*Bryales sp. -sp.*	314	31.4
*44.	*Dicranum spp.*	30	3.0
45.	*Pottia sp. (Pottiaceae)*	5	0.5
*46.	*Sphagnum sp.*	1	0.1
47.	*Dryopteris sp.*	1	0.1
48.	Polypodiaceae sp.	52	5.2
*49.	*Equisetum sp.*	8	0.8
50.	*Lycopodium sp.*	4	0.4
51.	*Selaginella sibirica*	1	0.1
52.	*Selaginella sp.*	1	0.1
53.	*Sporites indeter.*	9	0.9

Total amount of pollen and spores including those of	1000	
trees	10	1.0
shrubs and low shrubs	59	5.9
herbs and dwarf shrubs	497	49.7
spores	434	43.4

*The taxon is present in the recent flora

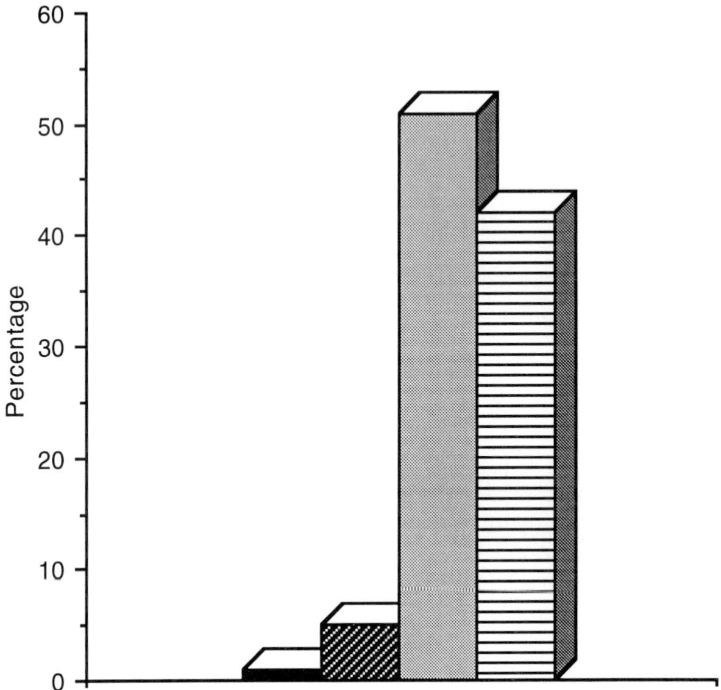

Figure 41. Palynological spectrum (general composition) of gastrointestinal contents of the bison. (For legend see Figure 18).

Not long before its death, the mammal may have pastured at some body of water, for example, swampy grass-sedge meadows with some forbs. According to Soper (1941), it is in early summer that bison of Buffalo Park (Canada) eat early species such as *Equisetum pratense, Calamogrostis inexpansa, Juncus balticus* and other plants which grow around some waterbodies or wet depressions. The young female bison is likely to have been stuck in a quagmire of a river or lake bank, made up of very fine loess-like loams, and could not get out. The bison could also come down into a trembling bog, similar to those which are now common in the area (Kozhevnikov, 1981) and in other areas of northern Siberia. A point worthy to mention here is that there is a very good preservation of pollen and spores, found in the intestinal tract of the bison, a somewhat poorer preservation of pollen and spores, discovered in the mummy-enclosing sediments, although their almost identical taxonomy suggests an in-situ burial of the animal.

The taxonomic composition of plants, determined from their remains in the intestinal tract of the bison, is poor (Tables 13, 20, Chapter 5). The paucity is no doubt caused by a paucity, first, of local flora and, second, of a species set of vegetation of the habitats which served as pastures of the animal shortly before its death, as well as by the season when it perished. These data suggest that the paleoflora synchronous to the bison's life has been revealed in insufficient detail. Nevertheless, comparison of its composition with that of recent flora (Kozhevnikov, 1981) shows that many of plants, represented in recent flora, grew in the area as early as 29,500±1,000 yr B.P. At that time, the larch, pollen of which was found in both food remains and corpse-enclosing loams, was probably the only tree species in the area.[12] It cannot be ruled out that mountain pine was then represented in local flora as undergrowth of the valley larch forests together with dwarf alders, shrubby and low shrubby birches, and various species of willows. Reed mats probably grew on shores of lakes and ponds. Pollen of this plant is not transported for long distances by air currents (Krattinger, 1975, cited from Ritchie *et al.*, 1983); this suggests that about 30,000 yr B.P. reeds entered the flora of the area, but it may have been already relict at that time. *Aster sibiricus, Phlox sibirica, Stellaria jacutica, Selaginella rupestris* that were also relict grew at drier sites. At present, they do not grow at the locality. Plants of overmoistened and swampy habitats (sedges, cotton grasses, grasses, valerian, jointweed and others) grew in topographic lows.

The bison died during a transitional period between Konoshel cooling (33,000-30,000 yr B.P.) and Lipovsko-Novosyelovskoe warming (30,000-22,000 yr B.P.) within the Karginsky Interglacial. At that time the area under study was still dominated by open treeless landscapes - tundras and forest-steppe meadows. However, the Konoshel cooling probably was not very intense and forests did not disappear completely during that time interval. Warming which started about 30,000-22,000 yr B.P. led to an increase in afforestation and swamp development; a decrease in meadow, steppe meadow and steppe-like communities which were widely distributed there during the climatic optimum of Karginsky Interglacial. This caused deterioration of living conditions of the large herbivorous mammals - mammoths, bison, woolly rhinoceros which lived there at that time (Flerov, 1979).

Conclusions

1. The complete corpse of the female bison *Bison priscus occidentalis* (Lucas) was the first discovered in the territory of Siberia; the corpse showed an excellent preservation which made it specifically determinable.

2. The geologic age of this species is dated as late Pleistocene (middle Wurm, middle Wisconsin).

3. The geographic distribution of the species was confined to Eastern Siberia. On the west, the range reached the Yenisei River. The species dwelt on Alaska and in Canada as well.

4. The animal in question died at the age of a year and eight months 29,500±1,000 yr B.P. The date falls on the transitional period between Konoghel cooling (33,000-30,000 yr B.P.) and Lipovsko-Novosyelovskoe warming (30,000-22,000 yr B.P.) within the Kargingky Interglacial (middle Wisconsin).

5. The gross composition of plants, their pollen and spores, percentages, established in examining food remains, suggest a leading role of sedges and grasses in the bison's nutrition; plants of the herbaceous group, represented by

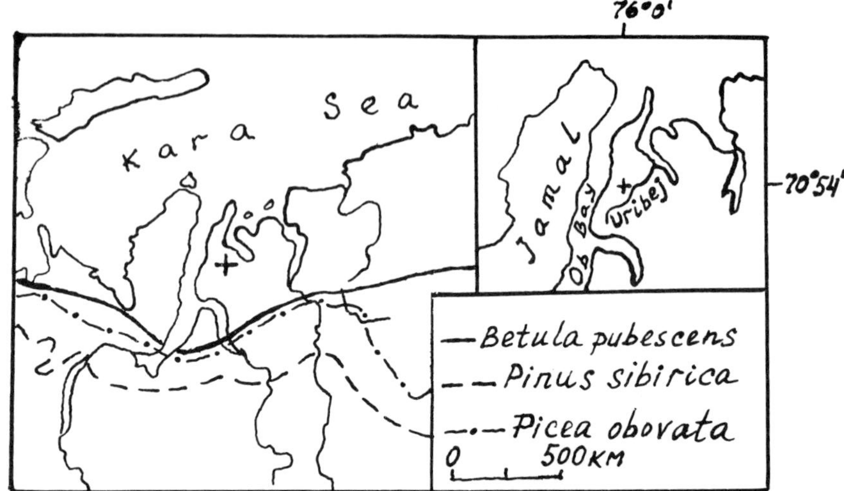

Figure 42. Index map showing the area of mammoth's burial (marked "+") and ranges of tree species in the north of West Siberia. After Sokolov et al., 1977.

a narrow species composition, included leaves and thin sprigs of low birch, willow, alder, heather and other plants, as well as various true mosses, abundant in the ground cover, which were also eaten by the animal.

6. The animal perished in early summer, therefore the composition of flora synchronous to its life, was insufficiently revealed.

7. In general, the floristic composition is poor and hence no zonal type of vegetational cover is reliably established. The paucity is a function of both paucity of local flora and species composition of plants of the habitats which served as pastures for the bison shortly before its death; another factor is the season when it perished. The presence of rare pollen grains of larch in the food remains and the corpse-enclosing deposits suggests that it grew at that time in the middle Indigirka River Basin. However, the landscape was probably dominated by open treeless communities such ag tundra and forest-steppe meadow ones.

4.6. Yuribei mammoth

The Yuribei mammoth is the most westerly and the "youngest" find of all known specimens of the "mammoth" faunal complex, discovered in Siberia (Fig. 1.8, 42).

The animal remains lay in a hill on the right bands of the Yuribei River, 1.5-1.6 m from the slope edge near the very surface of slope (Figs. 43, 44). The animal died 10,000±70 yr B.P. (Arslanov et al., 1980). Excavations undertaken showed that its wool coat, skeleton, gastrointestinal tract, tightly filled with plant remains, as well as other internal organs, were entirely preserved. Muscle tissues and hypoderm were almost wholly decomposed and their fragments showed up

Figure 43. General view of mammoth's burial site, before excavation. Mammoth corpse remains in the foreground. Photo by V.V. Ukraintseva.

Figure 44. General view of the right bank of Yuribei River near mammoth's burial site. Photo by V.V. Ukraintseva.

white on the terrace slope surface. The fore-part of stomach proved to be also slightly injured and its contents were exposed and stood out as a dark distinct spot against a background of light-colored slope deposits (Fig. 43). Thorough

Figure 45. Outcrop No. 1, exposing mammoth-enclosing deposits on the right bank of Yuribei River. Sites where samples were taken for palynological analysis are marked black rectangles. Photo by V.V. Ukraintseva.

examination showed that other sections of the intestinal tract had not been injured. This fact was vital to the examination of forage mass which allowed determination of the composition of plants which served as food for the animal and, hence, to paleobiogeographical reconstructions.

Dwarf willow and birch tundras with a shrub layer made up of *Betula nana* and *Salix* sp. are widely developed near the burial site, on flat interfluves of the right bands of the Yuribei River; undershrubs include the marsh tea (*Ledum palustre*), bog bilberry (*Vaccinium uliginosum*), and cowberry (*Vaccinium vitig-idaea*). The schinlead (*Pyrola grandiflora* and *Arctols alpine*) were encountered as well. The sedge (*Carex* sp.) is abundant. At lower poorly drained sites, the bushy tundra gave way to moss tundra passing into moss-sedge fens and lake-filled bogs in depression. The left bands of the Yuribei River is occupied by bushy willow tundras with *Betula nana* growing to 60-80 cm in height. Grass meadows on sand and clay banks of the river and lake, overgrown with grass, are of limited extension.

According to Gorodkov (1944) and Aleksandrova (1977), the area of the discovery belongs, respectively, to the subzone of typical tundras and the southern belt of subarctic tundras of the Gydan Region, the Yamal-Gydan-West Taimyr subprovince where vast areas on hillsides are occupied by shrubs with a closing canopy, composed of *Betula nana, Salix lanata, S. pulchra* growing to 50-80 cm in height. The medium belt is characterized by a reduction of areas, occupied by shrub formations and by the development of hillocky and "spotty" tundras on flat interfluves. According to Aleksandrova, the amount of shrubs drastically decreases in the northern belt. This is particularly typical of *Betula nana* which completely disappears in Arctic tundras, whereas the part of the willow (*Salix reptans*) increases.

Knowledge of the paleobotany of the area is inadequate. Only one peat-bog is believed to be studied in the Yuribei River Basin (Zubkov, 1931), situated 80 km from the mouth and a few kilometers north-east of the area under consideration. Therefore, integrated studies undertaken in the middle Yuribei River Basin are essential for both paleobiogeographical reconstructions and stratigraphic correlations.

Twelve samples were taken by the author for palynological analysis from deposits of terrace I above the flood plain which enclosed the mammoth and from mammoth-underlying and overlying deposits which were in intimate contact with the mammoth's wool in different parts of the body. Stripping was made in deposits occurring on terrace I slope, 2.5 m from the burial site (Fig. 45).

The following beds were exposed (in descending order):[13]

Outcrop No. 1 (September 19-22, 1979)   **Thickness, m**

1. Soil-cover layer 0.05
2. Light sandy loam, yellow-buff 0.30
3. Loam yellow-buff, in places sandy, with signs of
   ferrugination 0.35
4. Sandy loam, buff, interbedded with fine-grained sand,
   greenish-grey 0.25
5. Loam, greenish-grey, intercalated with buff sandy loam and
   vegetational detritus 0.35

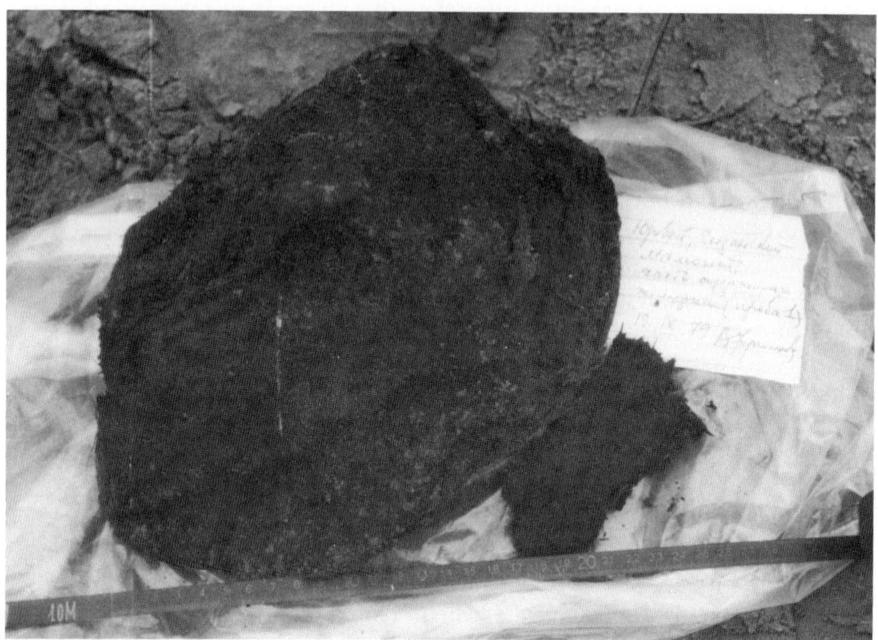

Figure 46. General view of part of stomach contents (when wet) of the Yuribei mammoth. Photo by V.V. Ukraintseva just after excavation from ground at excavation site.

6. Sandy loam, greenish-grey, interbedded with loam and fine-grained sand, in places ferruginated 0.60
7. Sand, yellow-grey, obliquely laminated, inter-bedded with vegetational detritus 0.60
8. Loam, dark-grey, interbedded with very gine vegetational detritus 0.40
9. Sands, light yellow, fine-grained, with signs of strong fer-rugination; laminated in the upper part and massive in the lower part 0.60
10. Clay, dark-grey . . . .

The author collected 18 samples at 20-25 cm interval for paleobotanical analysis from this described section, exposing practically all the deposits of terrace I above the flood plain (Fig. 45).[14]

Samples were taken for botanical, carpological, palynological and radiocarbon analyses from the mammoth's stomach, (Fig. 46) the middle part of intestine, colon and rectum. The stomach was thoroughly cleaned where ruptured.

Botanical analysis of the sample taken from the stomach showed that vegetational mass under examination was mainly composed of remains of herbaceous plants, namely, ground tissues, the vascular-fibrous bundles of stems

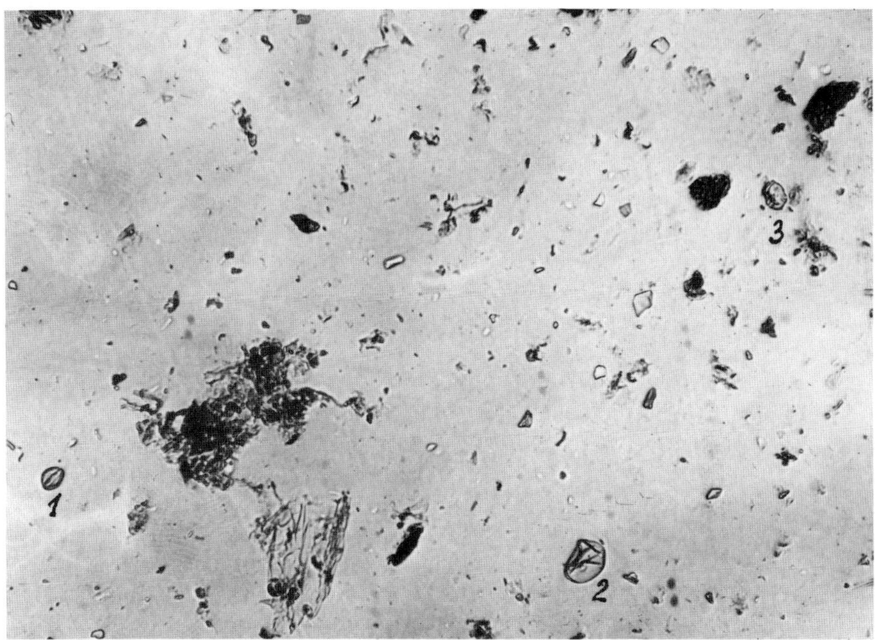

Figure 47. Pollen and fine plant fragments, extracted by centrifuging from the colon contents (X 100): 1) *Salix* sp., 2) *Poaceae* gen., 3) *Betula nana* Photo by V.V. Ukraintseva.

and leaves (95%); remains of trees, namely, sprigs crushed and cut, came to only 1%; leaves of true and bog mosses totalled 0.5%; plant remains badly decomposed and hence unidentifiable accounted for 3.5% of the total (Gorlova, 1982). An overall list of plants, identified by R.N. Gorlova, contains 27 taxa of flowering plants and cryptogsmous plants with species, genus and family ascribed (see composite list, Table 20 , Chapter 5).

The list of plants, determined by Stanishcheva (1982) from plant fruits and macroremains, includes nine names only. These are springs, caps and leaves of Bryales; fruits and fruit fragments of the sedge (*Carex* spp.), fruits of two species of cotton grasses (*Eriophorum* sp. *brachyanthemum* and *Eriophorum* sp.); seeds of the rush (*Juncus* sp. and *Luzula* sp.), one seed of the currant (*Ribes* sp.); one seed of *Saxifraga* sp; one intact leaf and 20 more or less broken leaves of *Dryas* sp. The examination of gastrointestinal contents, carried out by the author, showed that palynological spectra of the stomach and colon contents are dominated by pollen of herbs (Fig. 47); pollen of shrubs and low shrubs constitutes 8.7-9.5%; spores of mosses range from 24.0% to 44.0%; pollen of tree species is represented by sporadic grains of the cedar pine (*Pinus sibirica*), the Siberian spruce (*Picea obovata*), woody birch [(*Betula* sp. (sect. *Betula*)]. The group of pollen of herbaceous plants is dominated by pollen of the grass (*Poaceae*) (67.9-87.3%); pollen of the sedge (*Cypersceae*) accounts for 5.8-27.0%; pollen of forbs is represented by rare grains of the pinks

Table 14

Percentages of main plant groups in the gastrointestinal contents of the Yuribei mammoth from palynological data.

Plants	Part of gastrointestinal tract							
	Stomach		Middle part of intestine		Colon		Rectum	
	abs.	%	Abs.	%	Abs.	%	abs.	%
Trees	3	2.0	12	6.4	5	3.8	12	5.6
Shrubs and undershrubs	18	8.7	33	17.5	13	9.5	31	14.6
Herbs	68	45.3	37	20.1	86	62.9	34	16.0
Mosses	66	44.0	106	56.0	33	24.0	135	63.8
Total amount of pollen and spores	150		188		137		212	

(*Caryophyllaceae, ranunculi Ranunculaceae*), dryad (*Dryas octopetalla*), valerian (*Valeriana capitata*), and wormwood (*Artemisia* spp.), totalling 5.6%. The group of spores is dominated by spores of no less than 10 species of true mosses (cf. *Calliergon* sp., *Dicranum* sp., aff. *Drepanocladus* sp.1-3). The sample taken from the stomach and that from the colon contain one spore of *Sphagnum* sp. and one underdeveloped spore of *Lycopodium* sp., respectively.

The total composition of sporo-pollen spectra of contents of the middle intestine and rectum is dominated by spores of mosses, accounting for 56.0-63.8%; pollen of shrubs and undershrubs and herbs comes to 14.6-17.5% and 16.0-20.1%, respectively; pollen of trees in spectra of two sections of the intestine ranges from 5.6% to 6.4%; pollen of trees is represented by few grains of cedar pine and large woody birch. The group of pollen of shrubs and low shrubs is absolutely dominated by pollen of the low birch and (*Betula nana*); rare grains of at least two species of willow, and alder were noted (Table 14).

Attention is drawn to the following facts: (i) poor saturation of plant remains of the intestinal tract with pollen and spores; (ii) poor taxonomic composition of palynoflora; (iii) the presence of a relatively large amount of damaged (broken, torn and slightly torn, cracked) pollen (*Poaceae, Valeriana capitata, Artemisia* spp., *Picea* sp.) and spores (*Bryales*); (iv) the abundance of mineral particles in the sample from the mid-stomach and their low amount in other sections of the gastrointestinal tract.

A taxonomic paucity of palynofloras and poor saturation of vegetational remains of the intestinal tract with pollen and spores can be explained, on the one hand, by a paucity of flora itself in the vicinity of the burial site, and, on the other hand, by the fact that the animal died in either early spring when vegetation and flowering had not begun yet or in late autumn when they had ceased. This is suggested by an extremely low quantity of fruits and seeds in plant remains of the intestinal contents.

The presence of damaged pollen and spores suggests that the animal swallowed pollen and spores of plants which had flowered and borne spores a few months earlier, the pollen settled on other plants and ground and started decomposing. The presence of a relatively large amount of injured pollen and spores in spectra of the intestinal tract cannot be explained by the action of gastric juice, since analogous phenomena have not been observed when examining pollen and spores from the gastrointestinal tracts of the Selerikhan horse (Ukraintseva, 1977), Shandrin mammoth (Solonevich et al., 1977), or the Mylakhchin bison (Ukraintseva et al., 1978). This suggests that before they got to the mammoth's intestinal tract, pollen and spores had already been broken, i.e. they had long been exposed to aerobic conditions which first caused damage and then complete destruction. The composition of palynological spectra (Table 15, Fig. 48) and the gross composition (Gorlova, 1982; Stanishcheva, 1982) of the mammoth's intestinal tract suggest that it fed mainly on herbs. Shortly before its death, the animal grazed on bottomland tundra meadows, dominated by no less than four species of grasses with sedges, some specimens of forbs (*Ranun-*

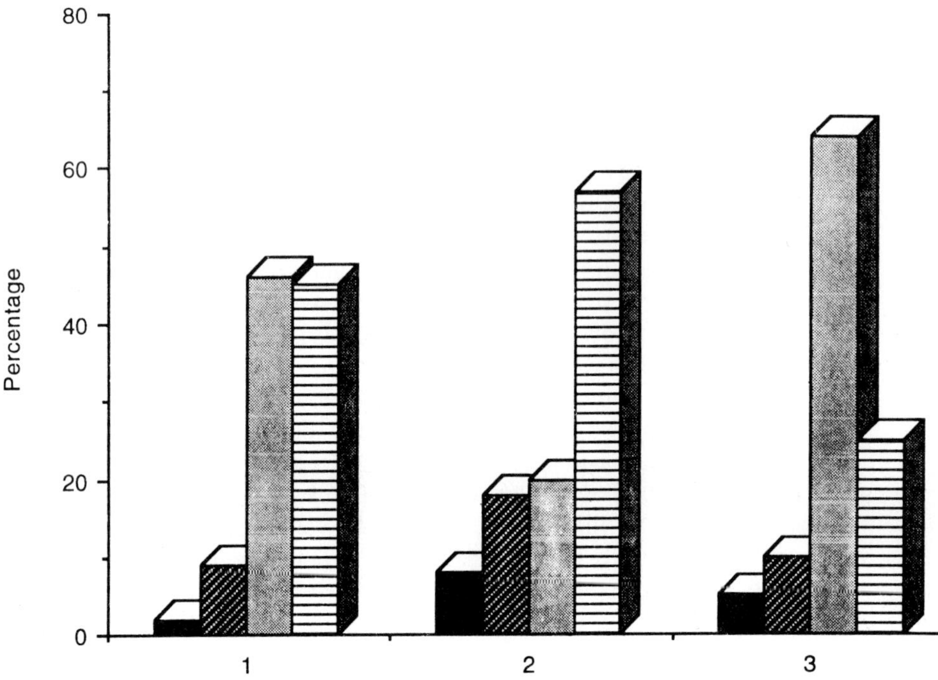

Figure 48. Palynological spectra (general composition) of gastrointestinal contents of the Yuribei
mammoth and enclosing deposits: 1) stomach, 2) middle part of intestines, 3) colon,

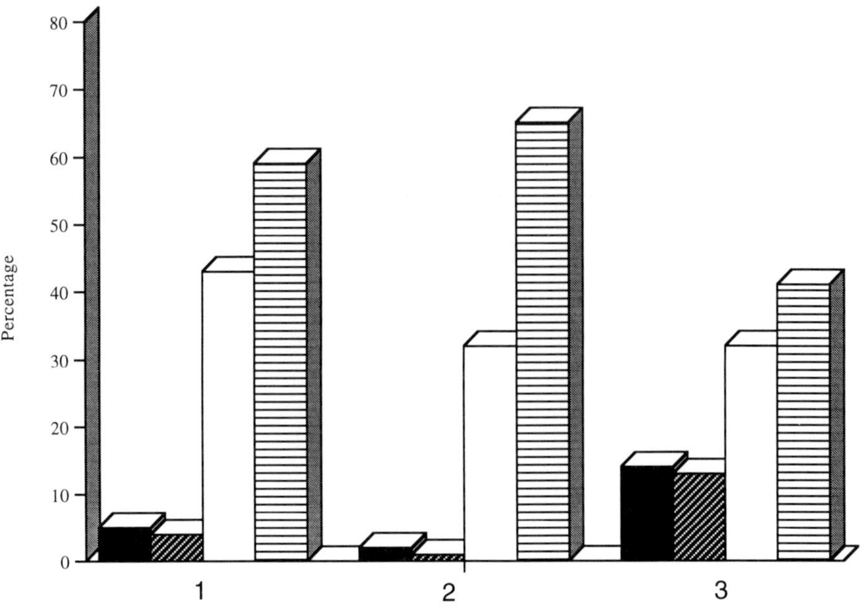

Figure 49. Palynological spectra (general composition) of deposits which enclosed the Yuribei
mammoth: 1) sample No. 8, 2) sample No. 20, 3) sample No. 16, 4) sample No. 14;

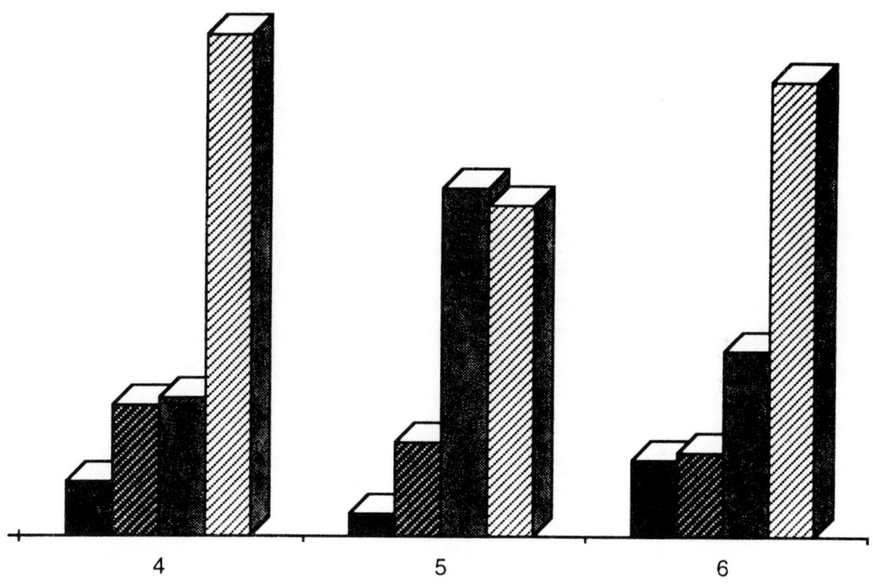

4) rectum; 5) outcrop No. 1, sample No. 15; 6) sample No. 12, vegetational detritus under the mammoth. (For legend see Figure 30).

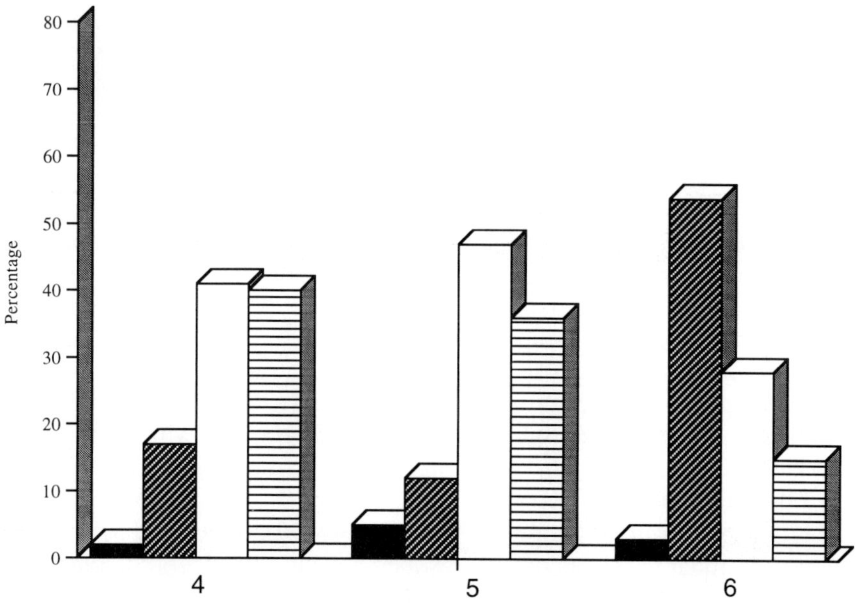

5) outcrop No. 1, sample No. 10; the same place, sample No. 6 (For legend see Figure 30).

culus sp., *Saxifraga* sp., *Valeriana capitata, Artemisia* spp. and others) and mosses of the ground layer being minor; shrub-moss bottomland meadows and grass-sedge (*Arctophila fulva, Carex stans* s.l.) associations of lake depressions.

Results of palynological analysis of the mammoth-enclosing deposits are given in Table 15 and sporo-pollen diagram (Fig.49)[15]. Spectra of the samples, taken from the bed of sand interlayered with loam, occurring above the bed of laminated clay (depth interval 3.5-1.9 m), are characterized by the predominance of redeposited pollen and spores accounting for 54-89%)[16]. Quaternary pollen and spores come to only 11-45% in the spectra of the samples; they are represented by rare pollen grains of the fir (*Abies sibirica*), Siberian spruce (*Picea* obovata), cedar pine (*Pinus sibirica*), large woody birch [*Betula* sp. (sect. *Betula*)], low birch (*Betula nana*) and willow (*Salix* sp.). Pollen grains of herbaceous plants, namely, grasses (*Poaceae*), sedges (*Cyperaceae*), pinks (*Caryophyllaceae*), wormwood (*Artemisia* sp.) are also few in number. Both quantitative and qualitative composition of spores is poor; these are mainly spores of true mosses (no more than four species) and rare spores of *Sphagnum* sp. and *Equisetum* sp.

The spectra characterized above suggest that the enclosing deposits (depth interval 3.5-1.9 m) were formed under severe climatic conditions. It was just this bed of sand, interlayered with loam, in which the mammoth has been buried.

Starting from a depth of 1.9 m (bed 1, sandy loam), the composition of sporo-pollen spectra varies markedly. At first the proportions of redeposited and Quaternary pollen and spores are almost equal and account for 53% and 47%, respectively (sample 10, depth interval 1.85-1.80 m). Upsection, Quaternary pollen and spores start dominating; this suggests the stabilization of denudation which prevailed over accumulation at the time of deposition of the sand bed. Spectra in depth interval 1.9-0.5 m are alternatively dominated by either pollen of herbs (*Cyperaceae, Poaceae, Caryophylaceae, Saxifraga* sp., *Artemisia* sp., *Ericaceae*) and spores of mosses (*Bryales* sp.1-7, *Sphagnum* sp.) or pollen of herbs and shrubs (*Betula nana, Alnus fruticosa, Salix* sp.1, sp.2). Pollen of trees of obvious adventitious character is represented by rare grains of the spruce (*Picea obovata*), cedar pine (*Pinus sibirica*), woody birch (*Betula alba* s.l.). Only starting from a depth of 0.50 m, pollen of the low birch (*Betula nana*) and alder (*Alnus fruticosa*) starts playing a leading part in sporo-pollen spectra.

Comparison of the composition of spectra of: (i) the mammoth's intestinal tract; (ii) samples, taken close to its body; and (iii) samples of outcrop No. 1 shows that spectra of three samples (15, 18, 13), taken near the corpse, are similar to those of samples, taken from the bed of sand interlayered with loam, in outcrop No. 1 (depth interval 3.6-1.9 m), but they are markedly different from spectra of the gastrointestinal contents.

Only spectrum of sample 14 (Table 15, Fig. 49), dominated by pollen of grasses, is similar to spectra of the intestines contents. Spectra of sample 8 (bed underlying the mammoth), sample 12 (vegetational detritus under the mammoth), and sample 16, with predominant pollen of sedges and minor pollen of

Table 15

Results of palynological analysis of gastrointestinal contents of the Yuribei mammoth and enclosing deposits.

Samples studied Plants	Stomach contents. Sample 1.	Contents of the middle part of intestinal. Sample 6.	Colon contents	Rectum contents	Mammoth - overlying sandy loam. Sample 7.
1	2	3	4	5	6
Larix sp.	-	2	-	2	-
Abies sibirica	-	-	-	-	-
Picea obovata	-	1	1	1	-
Pinus sibirica	2	2	1	1	-
Betula sp. (sect. *Betula)*	1	7	3	8	1
Betula sp. (sect. *Fruticosae)*	1	3	-	2	-
Betula exilis	-	-	-	-	-
Betula nana s.l.+ *Betula* sp.	11	27	12	22	-
Alnus incana	-	-	-	3	-
A. fruticosa	-	-	-	3	-

To be continued

Table 15 (cont.)

1	2		3		4		5	6
Salix spp.	1	2	3	2	1		4	-
Poaceae	40	67.9	18		75	87.3	9	2
Cyperaceae	16	27.0	9		5	5.8	17	4
Polygonum bistorta		-		-		-	-	-
Rumex artica		-		-		-	-	-
Polygonaceae		-		-		-	-	-
Stellaria sp.		-		-		-	-	-
Caryophyllaceae	1	1.7		-		-	-	-
Chenopodiaceae		-		-		-	1	-
Ranunculus sp.		-		-		-	-	-
Thalictrum alpinum		-		-		-	-	-
Ranunculaceae		-	1		1	1.2	-	-
Draba spp.		-		-		-	-	-
Saxifraga spp.		-		-		-	1	-
Dryas sp.	-	2	1		1	1.2	-	-

To be continued

Table 15 (cont.)

1	2		3	4		5	6
Rubus chamaemorus	-		-	-		-	-
Fabaceae (Leguminosae)	-		-	-		-	-
Umbelliferae	-		-	-		-	-
Ericacae	-		-	-		-	-
Armeria arctica	-		-	-		-	-
Polemonium sp.	-		-	-		-	-
Valeriana capitata	1	1.7	-	1		-	-
Pedicularis sp.	-		-	-		-	-
Parnasis sp.	-		-	-		-	-
Artemisia spp.	1	1.7	7	3	3.5	6	4
Nardosima sp.	-		1	-		-	-
Asteraceae	-		-	-		-	1
Dycotyledonae indeter.	-	2	-	-		-	-
Sphagnum spp.	-		-	-		-	-
Bryales spp.	66		104	32		135	1

To be continued

Table 15 (cont.)

1	2		3		4		5		6
Polypodiaceae	-		-		-		-		-
Equisetum sp.	-		2		-		-		1
Lycopodium alpimun	-		-		1		-		-
Huperzia selago	-		-		-		-		-
Hepaticae	-		-		-		-		-
Total amount of pollen and spores	150		188		137		212		15
Including:									
trees	3	2.0	12	6.4	5	3.6	12	5.6	1
shrubs	13	8.7	33	17.5	13	9.5	31	14.6	-
herbs	68	45.3	37	20.1	86	62.9	34	16.0	11
spores of sporophytes	66	44.0	106	56.0	33	24.0	135	63.8	3
	150		188		137		212		31
Including those:									
of Quaternary age	150		188		137		212		15
redeposited (Mz-N)	-	-	-		-		-		16

To be continued

Table 15 (cont.)

Samples studied Plants	Mammoth-underlying sand. Sample 8.	Vegetational detrius below the mammoth. Sample 12.	Mammoth underlying sand. Sample 14.	Sand in contact with underwool. Sample 16.	Inequigranual sand, containing organic matter from under the pelvis. Sample 20.	Mammoth - overlying sandy loam. Sample 7.
1	7	8	9	10	11	12
Larix sp.	3	5	-	-	-	2
Abies sibirica	-	4	-	1	1	-
Picea obovata	-	-	-	6	4	6
Pinus sibirica	-	-	-	3	3	5
Betula sp. (sect. *Betula*)	2	-	1	3	3	4
Betula sp. (sect. *Fruticosae*)	-	1	4	-	2	-
Betula exilis	-	7	-	8	-	-
Betula nana s.l.	3	4	9	3	3	9
Alnus incana	-	1	-	-	-	-
A. fruticosa	1	-	-	-	-	2

To be continued

Table 15 (cont.)

1	7	8	9	10	11		12
Salix spp.	2	2	-	1	1		1
Poaceae	8	7	19	13	5	3.6	12
Cyperaceae	18	15	9	12	101	61.1	24
Polygonum bistorta	-	-	-	-	2	1.5	-
Rumex artica	-	1	-	-	-		-
Polygonaceae	-	-	-	-	-		-
Stellaria sp.	-	-	-	-	-		1
Caryophyllaceae	-	-	-	-	38	2.2	1
Chenopodiaceae	-	-	-	1	1	0.7	-
Ranunculus sp.	1	-	-	-	-		-
Thalictrum alpinum	-	-	-	-	1	0.7	-
Ranunculaceae	-	-	-	-	7	5.1	-
Draba sp.	-	-	-	-	1	0.7	-

Saxifraga sp.	-	-	-	-	1	0.7	-
Dryas sp.	2	2	-	-	1	0.7	-
Rubus chamaemorus	-	-	-	-	1	0.7	-
Fabaceae	-	-	-	-	1	0.7	-
Umbelliferae	-	-	1	-	1	0.7	-
Fricaceae	-	-	-	-	1	0.7	-
Armeria arctica	-	-	-	-	1	0.7	-
Polemonium sp.	-	-	-	-	2	-1.5	3
Valeriana capitata	1	-	-	1	3	2.2	-
Pedicularis sp.	-	-	-	-	-		1
Parnasia sp.	-	1	-	-	-		1
Artemisia spp.	4	3	3	2	7	5.1	7
Nardosmia sp.	-	1	-	1	1	0.7	-

	1	%	2	%	3	%	4	%	5	%	6	%
Asteraceae	-		-		-		-		1	0.7	5	
Dycotyledoneae indeter.	1	2	-	2	-		-		-	-	-	
Sphagnum spp.	3		8		1		3		1	0.3	3	
Bryales spp.	52		56		29		28		327	91.6	27	
Polypodiaceae	-		-		-		-		2	0.6	-	
Equisetum sp.	2		1		-		2		21	5.9	10	
Lycopodium alpinum	-		-		-		3		1	0.3	-	
Huperzia selago	-		-		-		-		-		-	
Hepaticae	1		-		-		2		4	1.3	-	
Total amount of pollen and spores	100		115		76		93		510		124	
Including:												
trees	5	5.0	10	8.6	1	1.3	13	14.0	11	2.2	10	8.0

shrubs	4	4.0	12	10.4	13	17.1	12	13.0	6	1.2	12	9.7
herbs	33	33.0	28	24.0	32	42.1	30	32.0	137	27.0	62	50.3
spores of sporophytes	58	58.0	65	56.7	30	39.5	38	41.0	356	69.6	40	32.0
Total number of forms calculated	200		260		128		130		540		149	
Including those:												
of Quaternary age	100	50.0	115	44.2	76	59.4	93	71.5	510	94.5	124	83.2
redeposited	100	50.0	145	55.8	52	40.6	37	28.5	30	5.5	25	16.8

grasses or with equal proportions of the two groups of pollen, are similar to spectra of the middle intestine and rectum; pollen of the forb group is represented by sporadic grains of *Ranunculus* sp., *Rumex arcticus, Valeriana capitata, Nardosmia* sp., and *Artemisia* sp.

Spectra of samples 8, 12, 16 are similar to those of samples 10, 11 from outcrop No. 1, which were collected from the sandy loam (depth interval 1.85-1.55 m); they are also dominated by pollen of herbs (*Cyperaceae, Poaceae, Caryophyllaceae, Saxifraga* sp., *Artemisia* sp., *Taraxacum* sp., *Ericaceae*) and spores of mosses (*Bryales* spp., *Sphagnum* sp.).

Palynological spectra of samples: (i) from the gastrointestinal tract; (ii) collected near the mammoth's body (overlying and underlying deposits); (iii) from outcrop No. 1 suggest that the mammoth died in the period of deposition of the lower sandy loam (depth interval 1.90-1.50 m). It may have drowned in the Yuribei River or in a rather large lake. Under gravity the body gradually sank into the underlying sand, interbedded with loam; then it was buried under later deposits. It is noteworthy here that the beds of sand, sandy loam and even the upper clays were heavily deformed under the mammoth, whereas no deformation was noted in stripping of outcrop No. 1, 2.5 m right of the burial site (Fig. 50).

The spectrum of sample 20 reflects the most complete floristic composition and the character of vegetation growing in the area in the period of mammoth's death (Table 15, Fig. 49). The spectra is dominated by spores of mosses, accounting for 59.6%; pollen of herbs constitutes 27%; pollen of shrubs and undershrubs comes to only 1.2%; pollen of trees accounts for 2.2%. The group of spores is dominated by spores of no less than 7-10 species of true mosses (91.6%); spores of horse-tail account for 5.9%; spores of *Lycopodium alpinum, Polypodiaceae, Sphagnum* sp., and *Hepaticeae* are rare. The group of pollen of herbs is seminated by pollen of the sedges (*Cyperaceae*) (61.0%); pollen of the grasses (*Poaceae*) comes to only 3.6%; pollen of the ranunculi (*Ranunculaceae*) and wormwood (*Artemisia* sp.1-sp.3) are equal in proportion (5.1% each); rare grains represent pollen of *Dryas octopetala, Polygonum viviparum, Valeriana capitata, Artemisia arctica, Polemonium* sp., *Nardosmia* sp., *Agtragalus* sp., *Saxifraga* sp., *Ericaceae* sp. and others (see Table 15). This kind of spectrum suggests the stage of shrub-moss bottomland meadows in the development of vegetation in the area under study in the period of mammoth's life and death. According to Gorodkov (1944), at present the shrubmoss meadows occupy up to 40% of the whole area of flood plain terrace of Yuribei River; moss bogs, mainly made up of true mosses, account for 30% of the area; waterbodies which are not yet bogged and their littoral successional stages account for the balance, i.e. 30% as well.

The vegetation of shrub-moss bottomland meadows of Gydan tundra is characterized as follows: "At the next stage (shrub-moss bottomland meadows), vegetation forms a solid, although thin, ooze-impregnated cover, made up of *Drepanocladus uncinatus, Aulacomnium turgidum, Hylocomnium alascanum, Camptothecium trichoides, Polytrichum alpinum* and other mosses, totalling 15 in number. Minor lichens, particularly Peltigera aphtosa, appear as well. The grass cover is rather thick; many grasses such as *Alopecurus borealis, Calamagrostis neglecta* have survived from the meadow stage, but the rhizomatous sedges (*Carex stans* and *Eriophorum angustifolium*) are dominants which approach the association to tundras. *Equisetum arvense, Nardosmia frigida, Pedicularis sudetica, Saxifraga punctata, Valeriana capitata*, and *Luzula nivalis* are also very typical. The grass layer contains about 50 species, including a few low shrubs (*Salix polaria*)" (Gorodkov, 1944, p. 5).

Results of palynological analysis of deposits of the Yuribei River terrace, 6-8 m, suggest a wider distribution of valley shrub-moss meadows and, probably, their equivalents around lake depressions in the divides at certain stages in the past (Ukraintseva, 1982b). Thus, the above data point to two main stages in the development of vegetation and environment in the middle Yuribei River Basin.

First stage coincides in time with the formation of the bed of sand interlayered with loam and the upper part of bed of laminated clay in depth interval 3.6-1.9 m (Fig. 50). The date of mammoth's death, $10,000\pm70$ yr B.P., suggests that the stage corresponds to the uppermost Paleoholocene (HL1, QIV). At that time, the terrain under consideration may have been occupied by polar deserts and/or Arctic tundras, characterized by an extremely sparse vegetational cover. This was no doubt caused by severe climatic conditions. Glaciation of the polar basin is estimated at 7 to 8 for that time (Borisov, 1975). At the stage, the northern boundaries of tree species were displaced far south, as suggested by rare pollen grains of the fir (*Abies sibirica*), spruce (*Picea obovata*), cedar pine (*Pinus sibirica*), woody birch (*Betula tortuosa*), transported to the area from afar.

Second stage coincides in time with the formation of the beds in depth interval 1.95-0.0 m (beds 1-6) and corresponds to the Eoholocene (HL2, Q2IV). It marks the emplacement of stable vegetational cover, related to a global improvement of climatic conditions (Khotinsky, 1977). The time of mammoth's death falls on the early subperiod of Boreal when grass-sedge-forb, grass-forb and essential grass associations were widely developed on fresh alluvial deposits of the river valley of the terrain studied, under-shrub-moss meadows of terraces and lake depressions, moss tundras appeared and developed later. Bush and low bush tundras bogged in the valley and drier in flat interfluves - were formed. The latter were common at the end of second stage (spectra in depth interval 0.30-0.00 m) when climatic conditions may have been close or analogous to those of the present. At that time, meadow bogs as brakes of the sedge (*Carex aquatilis* ssp. *stans*), grass (*Arctophila fulva*), horse-tails (*Equisetum* spp.) and other plants were formed on shores of lakes.

The present climate of Gydan Peninsula is very severe: mean January temperature is -26^0, -30^0, the hardest frosts reach -57^0, -60^0; summer is sunny and relatively warm, mean July temperature ranges from 4^0 to 11.5^0; the bulk of precipitation, 175-200 mm, falls in summer, and 30-120 mm falls during a

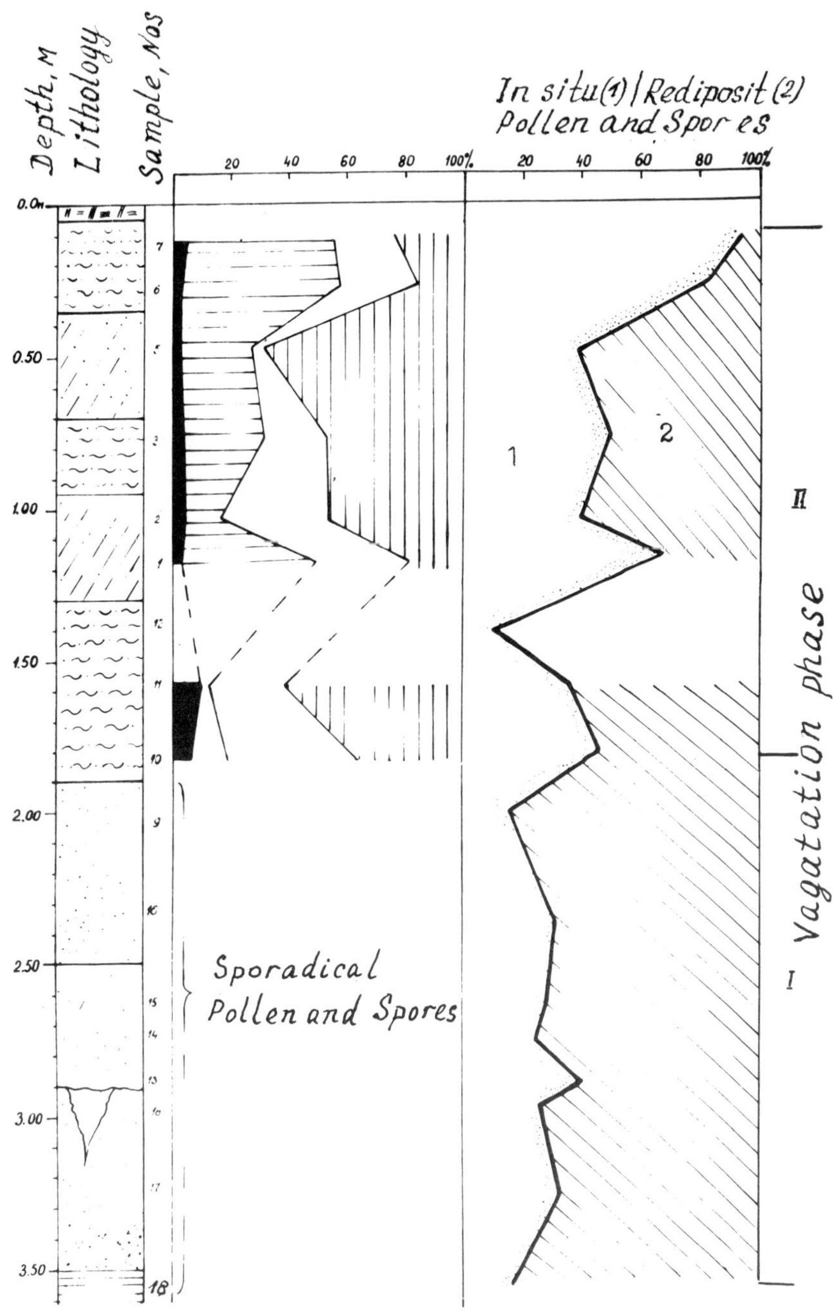

Figure 50.   Spore-pollen diagram of deposits of Yuribei River terrace I above the flood plain (outcrop No. 1), 2.5 m from the mammoth's burial site.

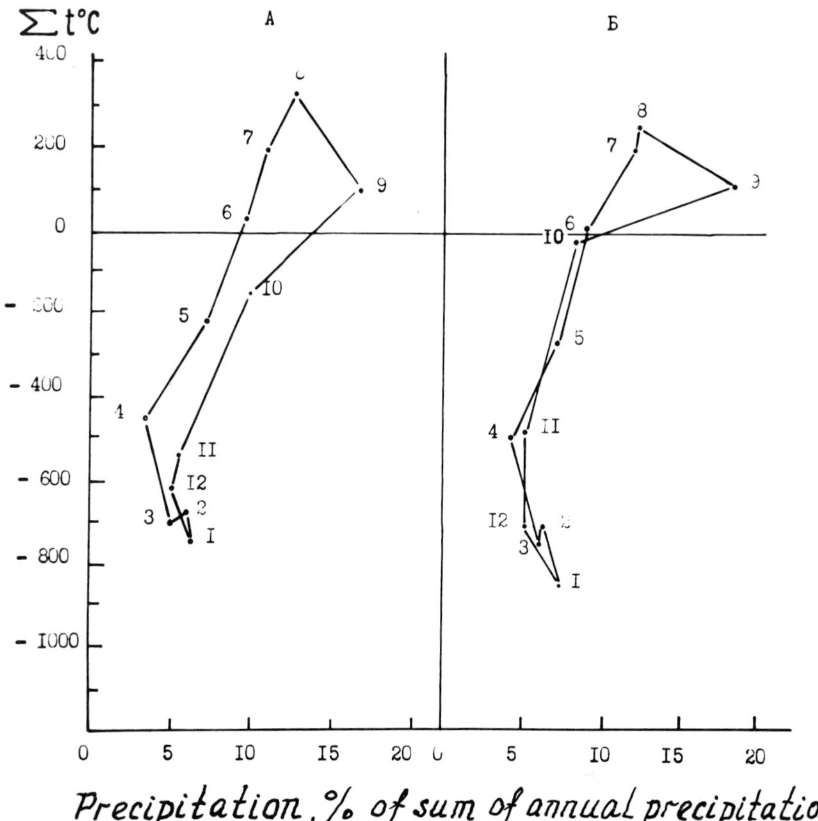

Figure 51. Climagraph characterizing the annual course of temperature and moisture conditions in the boreal hypoarctic tundra subzone. Gydan Peninsula and Ymal (A) and in the middle Yuribei River Basin, Gydan Peninsula (B).

longer cold period.

Climatic conditions of environment of the mammoth's burial site can be characterized by observational data, obtained at the nearest Tadibyayakha meteorological station, 70 km west of the site (Table 16). A climagraph, constructed on the basis of multi-year data of the station, illustrates an annual trend of warmth and moisture provision of the area (Fig. 51). A climagraph, based on data obtained by interpolating mean multi-year characteristics of four meteorological stations in this subzone, gives an indication of the general character of warmth and moisture provision in the subzone of southern hypoarctic tundras of Gydan Peninsula and Yamal (Fig. 51).

Paleobotanical data (Gorlova, 1982; Stanishcheva, 1982; Ukraintseva, 1982) indicate a climate like that of the present or somewhat warmer during the mammoth's lifetime, i.e. 10,000 yr B.P.; this led to an advancement of larch and some large bushes (*Ribes* sp., *Salix* spp.) along the river valley and its tributaries some what north of their present ranges.

Table 16

Main values of temperature and moisture conditions in the mid-Yuribei River Basin, Gydan Peninsula.

Area	Temperature, °C			+ °C months VI-IX	Annuall precipitation	Precipitation during months VI-IX: mm/% of annual sum
	August	January	mean annual			
Mid Yuribei River Basin	7.7	-27.6	-10.8	551	425	220
						51.8%

Based on mean multi-year observational data of Tadibyayakh meterorological station.

On the basis of study of the lithofacies composition of the mammoth-enclosing deposits, Evseev and coworkers (1982) arrived at the conclusion that during the accumulation of sedimentary strata of terrace I above the flood plain, in the Yuribei River Basin, the climate was within the confines of northern humid zone. As a whole, it wag humid and cool, as suggested by the following features of sediments (i) dominantly grey color; (ii) a complete absence of chemical and even clastic calcium carbonate throughout the strata; (iii) the polymictic composition of clastic material; (iv) a relatively high (3-4%) content of heavy minerals (and, particularly, high contents of fresh hornblende, epidote, and zeolite) in silty fraction; (v) a very poor diagenetic reworking of a clay component of the sediments. A qualitative estimation of the character of paleoclimate for the middle Yuribei River Basin in the period of mammoth's life and death (Evseev et al., 1982) is quite consistent with that based on paleobotanical data. Hence, quantitative characteristics of main elements of present climate can provide a correct idea of the character of warmth- and moisture provision of the terrain about 10,000 yr B.P. In any case, the fact that the quantitative characteristics of main elements of paleoclimate and climate of the present in the area under study are of the same order is unquestionable. Like today, August may have been the warmest month in a year at that time; its mean month temperature did not exceed 8^0, and mean January temperature may have been no lower than -28^0; mean annual air temperature did not rise in excess of -11^0; the sum of air temperatures above 0^0C constituted no lower than 600^0. No less than 50% of annual precipitation fell on the period when air temperatures were positive. Such a relation between warmth and moisture gave impetus to the development of tundra landscape at the latitude of mammoth's discovery and the distribution of flora very similar compositon to that of today.

Conclusions

1. Palynological examinations of deposits of terrace I above the flood plain (6-8 m) in the middle Yuribei River Basin suggest two main stages in the development of vegetation and environment. The first stage corresponds to the uppermost Paleoholocene or the late subperiod of subarctic period when the territory studied was occupied by polar deserts and/or Arctic tundras; this was related to severe climatic conditions. The second stage corresponds to the Eoholocene or Boreal period. It marks the emplacement of stable vegetational cover in the area and adjacent territories which was caused by a global improvement of climatic conditions at that time.

2. The mammoth's death falls on the beginning of second stage, namely, the early subperiod of Boreal period when the Yuribei River Valley was occupied by forb-grass-sedge, in places essentially sedge-grass communities, undershrub-moss meadows; meadow bogs, formed by the sedges (*Carex* spp.), grass (*Arctophila fulva*), horse-tail (*Equisetum* sp.) and other plants, were developed on lake shores. Bushy tundras may have been poorly developed at that time. They became more widely developed at later stages.

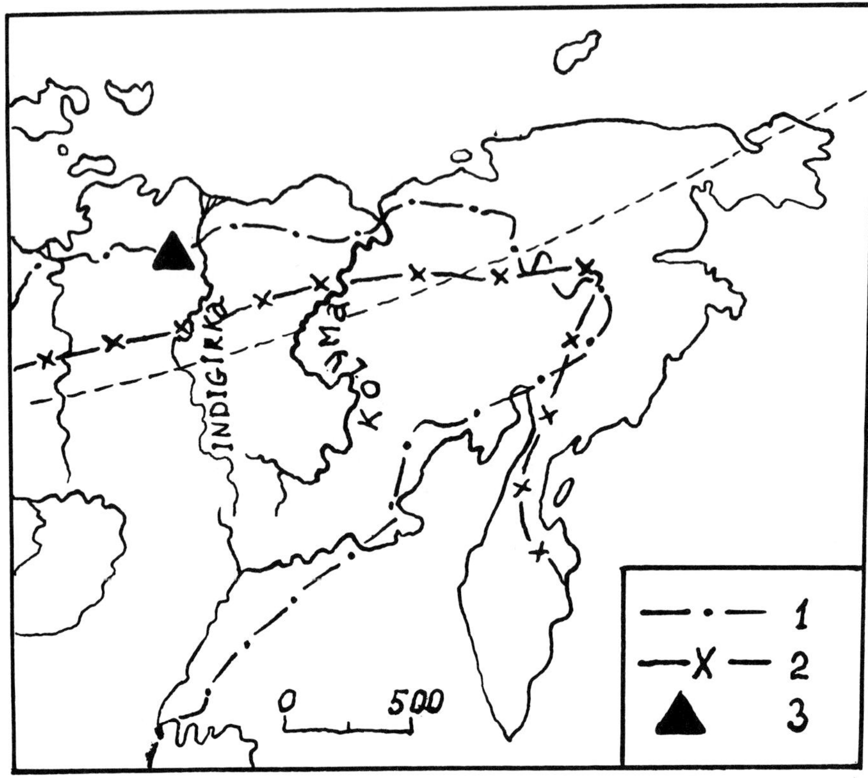

Figure 52. Index map showing the location of Berelekh mammoth cemetery. Northern limit of: 1) larch, 2) *Betula pendula* (after Sokolov et al., 1977; and 3) the location of the cemetery.

3. Paleobotanical and lithofacies data indicate a climate like that of the present or somewhat warmer during the mammoth's lifetime, i.e. 10,000 yr B.P.; this caused an advancement of the larch (*Larix* sp.) and some large bushes (*Salix* spp., *Ribes* sp.) along the river valley and its tributaries north of the present ranges of above-mentioned plants.

4. The animal drowned in the river or a lake, gradually sank into the underlying sands (beds 7-9). Then the body was buried under later deposits.

5. A poor preservation of mammoth's muscle tissues can be explained by weakening and, possibly, complete ceasing of the action of permafrost - the main factor of preservation of dead mammals - at the Holocene optimum, as the level of its occurrence strongly lowered because of a more favorable general paleogeographic situation at that period.

4.7. Berelekh mammoth population

A cemetery of mammoths, the largest burial site in the North-East of Siberia (Vereshchagin, 1977), is situated at 71°N and 145°E, among marginal "islands" of larch forests on the middle Berelekh River, the left tributary of the

Figure 53a). Berelekh River terrace II above the flood plain, general view (a); upper part of the terrace exposed by stripping (b). Photo by N.K. Vereshchagin, 1980.

Indigirka River (Fig. 52). The height of Berelekh River terrace reaches there 10-12 m above low water level (Fig. 53) and 15-20 m at localities of gently sloping ground-ice mounds. A bone bed is traced on the riverside for a distance of 180 m due to dark coloring of rocks. In 1970, the bed occurred at a depth of 3.5-4.0 m from the terrace edge and varied within 0.50-0.60 m to 1-2 m in thickness at different sites. The bone bed overlain by a bed which contained branches and roots. The bed formed in time interval 11,870$\pm$60 - 10,260$\pm$150 yr B.P. (Lozhkin, 1977). Radiocarbon analysis of branches and roots, collected by N.K. Vereshchagin in 1970 at a depth of 2.5 m from the terrace edge yielded a date of 11,830$\pm$110 yr B.P. The mammoth remains from the bone bed yielded two values, namely, ages of 13,700$\pm$400 yr B.P. and 12,240$\pm$160 yr B.P. were obtained on skin and ligament fragments (Lozhkin, 1977) and a tusk fragment from a depth of 3.5 m from the terrace edge, respectively. Thus, the dates of death of only two of 140 specimens which lived and died in the area were determined (Baryshnikov *et al.*, 1977) Nevertheless, these values give, on the one hand, an indication of the lifetime of Berelekh mammoth population and, on the other hand, date the bone marker bed.

A late Palaeolithic site yielding man's artifacts, made of mudstone and mammoth's tusk fragments, and bones of wolves, hares, and birds, was discovered at 200 m from a "head" site of cemetery (downstream) at a depth of 2.5 m from the terrace edge. This is the most northerly site of all the primeval

Figure 53b).

man's sites known in the North-East of the Asian continent. According to Mochanov (1977), this Palaeolithic site belongs to "Dyuktai" culture; its ^{14}C ages fall in the range of 25,000 to 10,700 yr B.P.

Species composition and the amount of bones of the Berelekh cemetery

Table 17
Bone frequency by species for the Berelekh cemetery.

Species	The amount of bones	The amount of specimen	%of the total amount of bones
Mammoth	8,431	140	99.3
Glutton	7+1 corpse	4	0.08
Cave lion	1	1	0.01
Woolly rhinoceros	1	1	0.01
Chersky horse	12	3	0.14
Reindeer	32	4	0.37
Primeval bison	8	2	0.09
Willow grouse	8,491	1 corpse	100

Calculations were made by G.F. Baryshnikov (cited from Vereshchagin, 1977).

(Table 17) show that the paleopopulation was dominated by mammoths, essentially young and semi-grown up specimens. According to Zherekhova (1977), females accounted for 60% of the population.

The geological setting of cemetery, the position of bone bed at three sites of the terrace, and taphonomic conditions suggest a long-term period and complex conditions of formation thereof (Vereshchagin, 1977). The development of landslides, solifluction and thaw depressions makes the situation even more complicated. for instance, in July, 1971, a massive block of landslide measuring 12 m in length and 4.5 m in height was found by N.K. Vereshchagin at 100 m from the "head" site; the block slid downslope to 4.5 m above the low water level of the Berelekh River.

Taking into account the extremely complex geological setting of cemetery formation, in 1980 Vereshchagin stripped the upper part of terrace scarp in the middle of the bone site (Fig. 53). He took seven samples for palynological analysis from an exposed 5 m bed of loam including thin (5-10 mm) intercalations of sandy loam and badly decomposed peat. Another sample was collected at the "head" site of bone bed at a depth of 3.9 m from the terrace edge.

Results of palynological analysis made by the author are given in Table 18 and sporo-pollen diagram (Fig. 54). Two samples taken from a depth interval 4-5 m, contained: (i) scanty pollen grains of *Betula exilis* and *Alnus fruticosa*; (ii) those of herbs such as *Poaceae, Valeriana capitata, Chamerion* sp.; and (iii) scarce spores of true and bog mosses. Redeposited minor pollen and spores of *Picea* sp., *Betula* sp., *Pinus* sp. subgen. *Haploxylon, Alnus* sp., *Polypodiaceae,* and *Zonotriletes* were encountered.

Species composition of palynological spectra and quantitative percentages of their main components indicate four sporo-pollen complexes equivalent to four phases in the vegetational history of the area (Fig. 54). The following are characteristics of the complexes recognized (in ascending order).

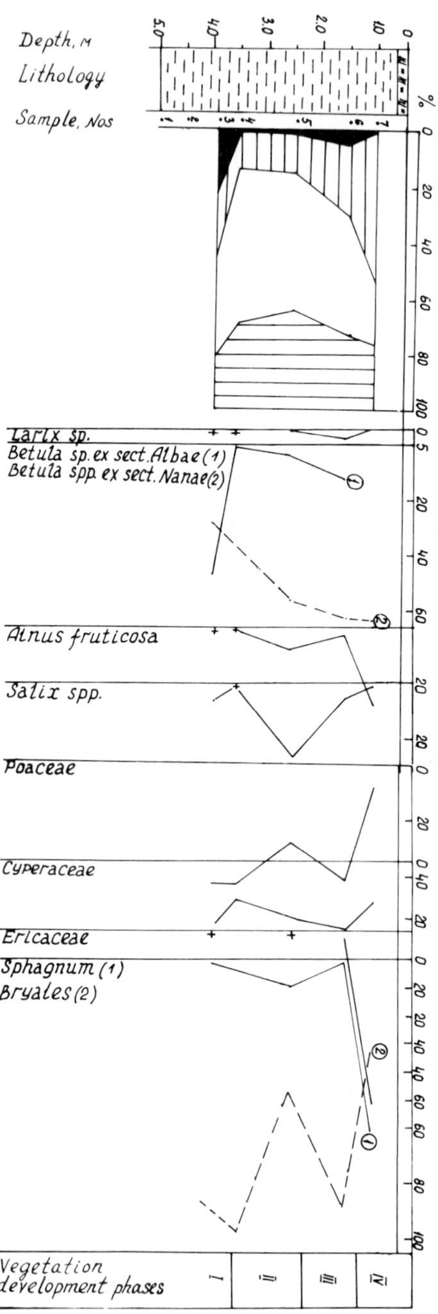

Figure 54. Spore-pollen diagram of Berelekh River terrace II above the flood plain, lower Indigirka River. (For legend see Figure 33 and the end of the book).

Complex I is characterized by spectra of two samples: sample No. 3 taken at a depth of 3.9 m from the stripping and sample No. 8 collected at the "head" site of bone bed, also at a depth of 3.9 m from the terrace edge. Pollen is represented by (i) about equal proportions of pollen of tree species, and shrub and undershrubs, 21% and 23% respectively; (ii) pollen of herbaceous plants, 36%. Spores of mosses (*Bryales* and *Sphagnum*) account for 20%. The group of trees and shrubs is dominated by pollen of woody birch (45.6%). Scarce pollen grains of the larch (*Larix* sp.), cedar pine (*Pinus sibirica*) and alder (*Alnus fruticosa*) were noted. Pollen of the shrubby birch [*Betula* sp. (sect. I)] and low shrubby birch [(*Betula exilis, Betula* sp. ex sect. *Nanae*)] accounts for 12.0% and 33.7% respectively. Pollen of willows comes to 6%. The group of herbs is dominated by pollen of the grasses (*Poaceae*) (43.0%) and sedges (*Cyperaceae*) (22.7). Pollen of the rockfoil (*Saxifraga* spp.) and wormwood (*Artemisia* spp.) constitutes 9.3% and 8.0%, respectively. There were observed rare pollen grains of the following plants *Sparagnium* sp., *Sanguisorba officinalis*, *Comastoma tenellum* (=*Gentiana tennela*) and others. The group of cryptogamous plants is dominated by spores of no less than four species of true mosses accounting for 89.4%; there were found scanty spores of *Sphagnum* sp. and *Selaginella rupestris*. With regard for a general poor preservation of fossil pollen of larch, accumulated in the groundmass under trees, the fact which was previously established (Dylis, 1948) and indicated by its low percentage in subrecent sporo-pollen spectra (Vaskovsky, 1957; Piavchenko, 1968; Ukraintseva, 1981 and others), there is good reason to believe that birch-larch forests with undergrowth of shrub and low shrub birch were distributed in the area during the formation of strata in a depth interval 4.0-3.6 m. Alder and willow were minor as suggested by low amounts of pollen of these plants in the spectra. The present northern limit of the woody birch (*Betula pendula*) lies no less than 500 km south of the Berelekh cemetery of mammoths (Fig. 52). This suggests that during the formation of lower bed of deposits contemporary to Berelekh terrace II above the flood plain in depth interval 4.0-3.6 m, the climate of the area was warmer than that of the present and auspicious for growth of woody birch.

Complex II is characterized by spectra of two samples taken in depth interval 3.5-2.5 m. It shows a sharp decrease in pollen of woody birch and a simultaneous increase in that of shrub birch and willow. The complex is dominated by pollen of herbs (*Poaceae, Cyperaceae*) and spores of true mosses (*Bryales*), accounting for 50.5-54.5% and 31.5-36.6%, respectively. Pollen of trees is represented by rare grains of the larch *Larix* sp. only. The composition of above-characterized spectra (Table 18) and curves of main components (Fig.54) indicate that the bed of deposits formed in depth interval 3.50-2.50 m under conditions of a sharp cold spell which resulted in a reduction of forest formations and disappearance of woody birch. During this period, the vegetational cover may have been dominated by sedge-grass communities, developed on banks of rivers and lakes with subordinate moss-low bush tundras, composed of *Betula exilis, Salix* sp. (two species *Alnus fruticosa* and abundant true mosses in the ground layer.

Complex III is characterized by results of analysis of a sample, taken at a depth of 1.5 m. It is dominated by pollen of herbs (44.5%); pollen of shrubs and undershrubs (*Betula exilis, Betula* sp. ex sect. *Nanae, Alnus fruticosa, Salix* spp.) constitutes 24.9%; spores of true mosses, horsetails and ferns total 27.0%. Sporadic pollen of trees (*Larix* sp., *Betula* sp. ex sect. *Betula, Pinus sibirica, P. sylvestris*) totals 4.6% only. Percentages of main components of the complex indicate that during the formation of complex-enclosing deposits, the area was mainly occupied by low bush tundras. Larch forests and open woodlands, if any, may have been concentrated in the river valley.

Complex IV is also characterized by results of analysis of a sample taken at a depth of 1 m. The total composition of this palynological complex is dominated by pollen of shrubs and undershrubs totalling 52%; pollen of herbs and spores of true and bog mosses is represented by about equal proportions thereof, 22.0% and 24.0% respectively. Pollen of trees is practically absent. Only one pollen grain of *Pinus sibirica* (0.2%), undoubtedly adventitious, was observed. The group of shrubs and undershrubs is dominated by pollen of dwarf birch (*Betula exilis, Betula* sp. ex sect. *Nanae*), accounting for 62.0%. Pollen of the alder (*Alnus fruticosa*) comes to 28.0%. Rare pollen grains of *Pinus pumila* (3.4%), *Betula* sp. ex sect. *Fruticosae* (4.8%), *Salix* sp. (0.5%) were encountered. The group of herbs is dominated by pollen of *Ericaceae,* accounting for 67.0%. The amount of pollen of grasses and sedges drops to 8.0% and 14.0%, respectively. Pollen of forbs is represented by scanty grains of *Ranunculus* sp., *Valeriana capitata,* two species of wormwood. The group of spores is dominated by spores of peat mosses (61.4%) and true mosses (33.0%).

The composition of complex IV suggests that during the formation of complex-enclosing strata, the area was mainly occupied by both low bush tundras and moss-bush (*Cassiope, Ledum, Vaccinium*) marshy tundras with abundant peat mosses. Subarctic tundras of the Yana-Indigirka subprovince can be regarded as their present equivalents. They show an extremely high abundance of peat mosses (Aleksandrova, 1977).

The four above-characterized sporo-pollen complexes fix four phases in the vegetational history of the middle Berelekh River Basin. They are: I. Phase of larch-birch and birch forests. II. Phase of herb-moss tundras and polygonal bogs. III. Phase of forest-tundra and low bush tundras. IV. Phase of moss-low bush (*Cassiope, Ledum, Vaccinium*) tundra with abundant peat mosses, and low bush tundras.

Hence, results of palynological analysis of the upper strata of deposits making up Berelekh terrace II above the flood plain suggest repeated successions of vegetation in the area, first associated with marked warming (phase I) and then prolonged cooling (phases II and IV). Slight warming occurring in phase III resulted in some change in the character of vegetation; during this period, tundra gave way to forest-tundra.

Table 18

Results of palynological analysis of deposits of Berelekh River Terrace II above the flood plain in the area of Berelekh cemetery of mammoths (lower Indigirka River).

Plants	Sample Nos, depth in meters (from terrace edge), lithology											
	3 3.90 grey loam		4 3.50 bluish grey loam		5 2.50 peat		6 1.50 buff loam		7 1.70 buff loam		8 3.90 grey loam	
1	2		3		4		5		6		7	
	abs.	%	abs.	%	abs.	%	abs.	%	abs.	%	abs.	%
Larix cf. gmelinii	1	0.5	2		1	2.0	3	2.9	-		3	3.8
Pinus sibirica	3	1.6	-		-		1	1.0	1	0.5	-	
P. sylvestris	-		-		1	2.0	-		-		-	
P. pumila	-		-		5	2.0	1	1.0	7	3.4	-	
Betula sp. (sect. *Betula*)	84	45.6	-		2	4.0	12	12.0	-		24	30.0
Betula sp. (sect. *Fruticosae*)	22	12.0	1		-		14	14.0	10	4.8	12	15.0
Betula exilis	5	2.8	-		-		-		1	0.5	7	8.8
Betula sp. ex (sect. *Nanae*)	24	13.0	22		25	50.0	52	51.0	127	61.0	13	16.2
Betula spp.	33	18.0	-		3	6.0	11	11.0	2	1.0	15	18.7

To be continued

Table 18 (cont.)

1	2		3		4		5		6		7	
	abs.	%	abs.	%	abs.	%	abs.	%	abs.	%	abs.	%
Alnus fruticosa	1	0.5	2	-	4	8.0	3	2.9	59	28.0	1	1.2
Salix spp.	11	6.0	4	-	13	26.0	5	4.9	1	0.5	5	6.3
Sparganium sp.	1	0.7	-	-	1	0.6	-	-	-	-	-	-
Rumex sp.	-	-	-	-	-	-	-	-	1	1.0	-	-
Festuca sp.	-	-	-	-	1	0.6	4	2.7	-	-	-	-
Poa sp.	-	-	-	-	1	0.6	6	4.0	-	-	-	-
Poaceae	65	43.0	49	41.0	47	28.0	50	33.3	8	8.0	40	50.7
Eriophorum sp.	-	-	-	-	1	0.6	-	-	-	-	-	-
Cyperaceceae	34	22.7	15	12.5	33	19.6	34	22.7	14	14.0	15	17.7
Allium sp.	-	-	-	-	1	0.6	1	0.7	-	-	-	-
Chenopodiaceae	-	-	-	-	-	-	1	0.7	-	-	-	-
Minuartia sp.	-	-	-	-	-	-	2	1.3	-	-	-	-
Silene sp.	-	-	-	-	-	-	3	2.0	-	-	-	-
Stellaria jacutica	4	2.7	1	0.8	-	-	-	-	-	-	1	1.2
Caryophyllaceae	14	11.7	-	-	-	-	-	-	-	-	1	1.2

To be continued

Table 18 (cont.)

1	2		3		4		5		6		7	
	abs.	%	abs.	%	abs.	%	abs.	%	abs.	%	abs.	%
Ranunculus sp.	1	0.7	-		-		-		-		3	3.5
Ranunculaceae	-		-		2	1.2	1	0.7	4	4.0	-	
Papaver sp.	-		1	0.8	-		2	1.3	-		-	
Cardamine sp.	3	2.0	-		-							
Cruciferae	-		-		-		1	0.7	2	2.0	-	
Saxifraga emarginata		-		-	18	10.7		-		-		-
Saxifraga spp.	14	9.3	2	1.7	10	6.0	4	2.7		-	7	8.0
Parnasia palustris	2	1.4		-		-	2	1.3		-		-
Potentilla sp.	1	0.7		-		-	1	0.7		-		-
Sanquisorba officinalis	2	1.3	1	0.8	1	0.6		-		-		-
Apiaceae (Umbellifere)		-	6	5.0	1	0.6	3	2.0		-		-
Chamaenerion dahuricum		-		-	6	3.6		-		-		-
Chamaenerion sp.	1	0.7	1	0.8		-	2	1.3		-		-
Hedysarum hedysarbides		-		-		-	1	0.7		-		-
Cassiope tetragona		-		-	5	2.8	3	2.0		-		-
Fricaceae	1	0.7	2	1.7	2	1.2		-	67	67.0		-

To be continued

Table 18 (cont.)

1	2		3		4		5		6		7	
	abs.	%	abs.	%	abs.	%	abs.	%	abs.	%	abs.	%
Gentiana tenella	1	0.7	-		-		-		-		-	
Phlox grumondii	-		-		-		2	1.3	-		-	
Polemonium acutifoliom	-		-		-		2	1.3	-		-	
Polemonium sp.	-		-		1	0.6	-		-		-	
Lagotis minor	-		-		-		-		-		1	1.2
Galium verum	-		-		-		1	0.7	-		-	
Valeriana captiata	-		4	3.5	2	1.2	4	2.7	2	2.0	1	1.2
Aster sp.	3	2.0	-		-		-		-		2	2.4
Tanacetum sp.	-		-		-		-		-		3	3.5
Artemisia vulgaris	-		-		-		5	5.3	-		-	
Artemisia sp.	13	8.0	17	14.0	-		7	4.7	2	2.0	6	7.0
Saussurea sp.	-		1	0.8	17	10.0	-		-		-	
Taraxacum sp.	1	0.7	-		-		-		-		-	
Asteraceae	-		4	3.5	13	7.7	4	2..7	-		-	
Dicotyledon indeter.	4	2.7	2	1.7	-		-		-		2	2.4
Hepaticae	1	1.2			3	0.7	1	1.1			2	4.0

To be continued

144

Table 18 (cont.)

1	2		3		4		5		6		7	
	abs.	%	abs.	%	abs.	%	abs.	%	abs.	%	abs.	%
Sphagnum sp.	1	1.2	2	3.0	11	9.0		-	54	61.4	1	2.0
Bryales sp.	74	89.2	66	97.0	57	46.7	86	94.0	29	33.0	40	30.0
Polypodiaceae		-		-	1	0.2	3	3.2	1	1.0	2	4.0
Equisetum sp.	6	7.2		-	53	43.4	2	2.0	1	1.0	5	10.0
Lycopodium sp.		-		-		-	-	1	1	1.0		-
Lycopodium selago		-		-		-		-	1	1.0		-
Selaginella sibirica	1	1.2		-		-	2	2.0	1	1.0		-
Total amount of pollen and spores calculated	417		222		340		356		396		215	
Including those of:												
trees	88	21.1	2	0.5	4	1.2	16	4.6	1	0.2	27	12.5
shrubs and low-shrubs	96	23.0	30	13.5	46	13.8	86	24.9	207	52.3	53	24.5
herbaceous plants	150	36.0	120	54.5	168	50.4	150	44.5	100	25.2	85	40.0
sporophytes	83	19.9	70	31.5	122	36.6	94	27.0	88	22.3	50	23.0

Sample was taken at the head site of bone bed at a depth of 3.9 m from terrace edge

Three radiocarbon values are crucial for the determination of chronology of environmental variations and dating of the sequence of sediments under study. These values are as follows:

1) 13,700±400 yr B.P. obtained on the mammoth's skin and ligament fragments from the bone bed (Lozhkin, 1977);

2) 12,240±160 yr B.P. obtained on the mammoth's tusk fragments, taken at a depth of 3.5 m from the terrace edge (Vereshchagin and Ukraintseva, 1985);

3) 11,830±110 yr B.P. obtained on branches and roots, taken in 1970 by N.K. Vereshchagin at a depth of 2.5 m from the terrace edge (Vereshchagin and Ukraintseva, 1985).

The lower sequence of sediments[17], composing Berelekh terrace II above the flood plain, started forming in Sartan time; this is supported by the date of 13,700±400 yr B.P. The date 12,240±160 yr B.P., obtained on the mammoth's tusk fragments taken at a depth of 3.5 m from the terrace edge (8.5 m from the water edge) indicates on the existence of the mammoth and its associates at the period in the pre-Berelekh River area. The species composition of the Berelekh mammal population which practically contained no forest species (Vereshchagin, 1977), palynological spectra of the bone bed, dominated by pollen of herbs (Lozhkin, 1977), as well as a fauna of coleopterous beetles in which tundra, steppe-tundra and steppe species obviously prevailed over forest species (Medvedev and Voronova, 1977) all suggest the long-term existence of treeless landscapes in the area and adjacent territories.

Spectra of two samples, one taken from stripping at a depth of 3.9 m from the terrace edge (8.1 m from the water edge) and the other from the "head" site of bone bed, also taken at a depth of 3.9 m from the terrace edge, mark a fundamental change in natural environment, caused by a first wave of warming at the end of late Pleistocene. The date of 12,240±160 yr B.P., obtained on the mammoth's tusk from the overlying bed (depth 3.5 m from the terrace edge), indicates that the phase of larch-birch forests (phase I) corresponds to the first post-Sartan warming, probably equivalent to Kokorevo warming of Siberia or Bölling warming of Europe (Kind, 1973, 1974) . This warming is also fixed in spectra of deposits of the right Khroma River terrace at a depth of 9.1 m from the terrace edge (Rybakova, 1979[18]), as well as in spectra of sediments of Mus-Khaya exposure on the Yana River (Kondratieva et al., 1976). Warming was short and soon it gave way to a sharp cooling, fixed by the composition of complex II in the stripping on the Berelekh River and by spectra of samples, taken from deposits on the Chroma River ln depth interval 8.5-7.5 m.

Phase III in the vegetational history of the Berelekh River Basin marks a second wave of upper Pleistocene warming which may be considered as equivalent to Taimyr warming of Siberia and Alleroed warming of Europe.

Phase IV marks a new stage of cooling which can be correlated with Norilsk cooling, late Wisconsin of North America and upper Dryas of Europe.

It should be noted that the evolutionary phases of vegetation and climate, established for the Berelekh River Basin, are in good agreement with the trend of paleotemperature curve, constructed for the time interval 15,000±10,000 yr B.P. by Dansgard and coworkers (1970, cited from Khotinsky, 1977).

Conclusions

Analysis of results of the integrated study of the Berelekh cemetery of mammoths leads us to the following conclusions.

1. In the area of mammoth cemetery, the Berelekh River exposed late Pleistocene (Sartan) and late-, post-glacial deposits which make up present terrace II above the flood plain, 10-12 m high.

2. The cemetery began to form at the close of Sartan time when the area under study and adjacent territories were mainly occupied by treeless landscapes, namely, grass and low bush-grass tundras on highs and sedge-grass in lows. This is confirmed by the lack of forest species in the Berelekh mammal population and the predominance of pollen of herbs in spectra of the bone bed, as well as the predominance of tundra, tundra-steppe and steppe species in fauna of coleopterous beetles.

3. The upper part of terrace, exposed by stripping (depth interval 8-12 m from the water edge), was formed by deposition under lacustrine-alluvial conditions, accompanied by alluvium accumulation at lower levels.

4. Palynological analysis of the upper strata of deposits which compose the terrace allows tracing of repeated changes in vegetational cover in late-postglacial time; the changes are fixed by four phases of vegetational evolution such as I. Phase of larch-birch and birch forests; II. Phase of moss-low bushgrass tundras and polygonai bogs; III. Phase of forest-tundra and low bushy tundras; IV. Phase of low bushy tundras (*Cassiope, Ledum, Vaccinium*), with abundant peat mosses,and bushy tundras.

5. A first fundamental change in the character of natural landscapes in the Berelekh River Basin, caused by a first wave of warming at the end of late Pleistocene, is fixed by phase I - the phase of larch-birch and birch forests; the phase is most likely to be equivalent to a first post-Sartan, so-called Kokorevo warming of Siberia, Bolling and Fjeres Interstadials of Europe about 12,800 yr B.P.

6. Phases II and IV which fix cooling periods can be correlated with those occurring during the ancient and young Dryas, whereas phase III may be equivalent to Taimyr warming of Siberia and Alleroed of Europe.

7. Thus, the evolutionary phases of vegetation and climate, recognized in the middle Berelekh River Basin, indicate that the period of Sartan cooling gave way to a period of unstable natural environment in late- and post-glacial time when relatively warm rhythms (phases I, III) were replaced by cold rhythms (phases II, IV). The period can be regarded as an interval of unstable development of natural environment. Naturally, the period was unfavorable for the mammoth and its contemporaries such as musk ox, saiga, bison and others, adjusted to life under severe climatic conditions in treeless or poorly forested landscapes.

8. In the author's opinion, primeval hunters who settled along paths of

ancient migration of large herbivorous mammals, in the Berelekh River Valley, lower Indigirka River, contributed to a reduction of their populations which drastically dropped in warm periods, unpropitious for the animals (Ukraintseva, 1979).

Data obtained on the material discussed above and previously, from studied finds are summarized in Tables 19, 20. Data shown in Table 20 (Chapter 5) provide an idea of the composition of plants growing in the habitats of the fossil animals discovered, whereas data of Table 19 give an indication of the character of vegetation dominating in each of the areas studied in the past. From Table 19 and results of the above paleoclimatic reconstruction it is inferred that most of the fossil mammals which lived in various areas of northern Siberia and which were dug up in the period 1900-1979 existed in either somewhat more favorable environment (Berezovka mammoth, Selerikhan horse, Shandrin mammoth, bison, Taimyr mammoth) or an environment similar, or analogous to, that of today (Yuribei mammoth). In any case, radiocarbon analysis suggests that five of the nine mammals discovered lived in various periods of Karginsky Interglacial which embraced the time period, according to the current notion, from 50,000 to 25,000 yr B.P. (middle Wisconsin of North America). For instance, the Selerikhan horse and the Shandrin mammoth lived in its optimal phases,whereas the bison in the final phase. Two of the nine fossil mammals only lived in cold periods of late Pleistocene, namely, the Vereshchagin mammoth lived during Zyrianka cooling (glaciation), and the Kirgilyakh mammoth, Dima, during early Karginsky or Kirgilyakh cooling. The Berelekh paleopopulation of mammoths and their associated existed in the lower Indigirka River Basin for a relatively prolonged period. The formation of mammoth cemetery began there at the end of Sartan time when the area under study and adjacent territories were dominated by open landscapes, and completed in late- and post-glacial time when the terrain was mainly covered by forest-tundra and light larch forests.

At present we are adequately aware of living conditions of some specimens of the "mammoth" faunal complex, including some mammoths, horse, bison, even a rather large population of mammoths and their associates. In order to gain a more complete understanding of living conditions of the "mammoth" fauna as a whole, we should consider the character of variations in vegetation and environment of norther Siberia in late Anthropogene (Quaternary) time, first of all, in the areas with which the finds of specimens of the fauna, studied in detail, are associated. The problem is discussed in Chapter 6.

Table 19
Data on the vegetation of the region of the animal carcasses (from V. Ukraintseva, 1986).

Fossil animal	14C-dated time of perish: yr B.P.	Pattern of reconstructed vegetation	Modern vegetation pattern of the find district
Mammoth Large Forested Rassokha River, 1978	>53,170	grass-forb and sedge-grass communities	bush and low bush tundras, polygonal mires
Mammoth Terekhitkh River, 1971	44,540±1870	tundras, forest tundras (?)	larch taiga
Mammoth Beresovka River, 1990	44,000±3500	larch taiga, occassionally with *Betula platphylla, Betula pendula, Alnus hirsuta, Pinus sibirica.* Meadows, dry meadows, steppe like meadows	larch taiga
Mammoth ("Dima") Kirgilyach River, 1977	41,000±1100 39,590±870	hypnum and cotton-grass swamps, bush and low bush tundras, larch taiga, occassionally with *Betula platyphylla,* grass forb communities , mountain (dry) tundras , alpine bush and low bush tundras, larch forests along river valleys, mountain steppe like communities	larch taiga, *Betula platyphylla* only on south slopes. Bushy tundras with *Betula exilis*
Mammoth Shandrin River, 1977	40,350±880	larch taiga, bush and low bush tundras, sedge and hypnum bogs	bushy tundras, heavily paludified

To be continued

Mammoth East Taimyr, 1948	11,450±450	bushy tundras forest tundras (?)	cotton-grass tundras, sedge-grass tundras, polygonal tundras, "spotty" tundras
Mammoth Yuribei River, 1979	9730±100	flood plain meadows, shrub moss meadows, sedge-cotton-grass tundras with *Betula exilis*	bushy tundras with *Betula exillis*, Salix ssp., moss tundras, polygonal bogs
Horse Selerikhan River, 1968	38,590±1120	larch taiga with spruce and wooky birch, grass-forbs meadows in the valleys and on the slopes	larch taiga with *Betula platyphylla* on the south slopes, flood plain meadows, mesoxerophytic conditions
Bison Indigirka River, 1971	29,500±1000	larch taiga, heavily paludified, grass-forb and sedge-grass communities	larch taiga

Abramov and Abramova 1981; Belaya and Kislerova 1978; Egerova 1977; Kupriyanova 1957; Medvedev and Forenova 1977; Solonevich et al. 1977, 1978; Sukaczev 1914; Tikhomirov and Kultina 1973; Tikhomirov and Kupryanova 1954; Ukraintseva (Kul'tina) 1977; Ukraintseva 1981; Ukraintseva 1982; Ukraintseva et al. 1978.

Chapter 5

Food Remains of Fossil Herbivorous Mammals as an Indicator of Late Anthropogenic: Floras in the North of Siberia

When the contents of gastrointestinal tracts of all the animals, found in various areas of northern Siberia (during the last 85 years), were studied in detail through the efforts of many researchers, one can say with assurance that vegetational remains of their food should be considered, to some extent, as "hay-crops" or unique "surficial samples", taken by the animals before their death .

Sukachev (1914) stated that the Berezovka mammoth perished in late July-early August when heads of sedges had ripened but had not shed their grains yet, and ripened fruits of *Beckmannia erlciformis* and *Hordeum violaceum* had not fallen, but many plants; were still in blossom (Kupriyanova, 1957). Just before its death, the mammoth had eaten plants growing within reach of its trunk [as stated by Gerz (1902); the animal could not move as it had a bad fall from a bluff and had broken its hipbone]. It could not swallow them and they remained tightly clenched in its teeth. Remains of these plants with teeth imprinted on them are displayed in an unique photo taken by V.N. Sukachev.

As shown in Chapter 4, the horse also perished in late July or early August when most herbaceous plants were in bloom in its habitat and adjacent areas; nevertheless, heads of sedges had already ripened, but caryopses of grasses had not ripened yet. Fruits of such families as Polygonaceae, Plantaginaceae, Rosaceae may have started ripening only, therefore they were found in food remains of the horse in small amounts. The horse perished suddenly (Tikhomirov and Kultina., 1973; Vereshchagin, 1977; Ukraintseva (Kultina), 1977) and had had no time to completely digest the plants eaten, therefore their remains allowed determination of the composition of plants which grew in the vicinity of horse's habitat.

The bison perished in the mid-Indigirka River Basin in late June or early July when only a few plants flowered, however, fruits of sedges had begun to ripen, but caryopses of grasses had not ripened yet (Ukraintseva *et al.*, 1978). In the mid-Shandrin River Basin the mammoth died in early spring because of asphyxia of the gastrointestinal tract (Yudichev and Averikhinov, 1982, Ukraintseva, this work, Chapter 3). Early spring at latitude 72° is the time when plants rest under a dense snow cover. Therefore the animal had to feed on remains of the last year's vegetation, spurs of low shrubs, shrubs and even young seed-bearing twigs of larch. It is precisely this fact that explains a paucity of taxonomic composition of flowering plants, and abundance of macroremains of mosses and their spores in the mammoth's gastrointestinal tract. And, finally, 7-8 months old juvenile mammoth that perished in the mid-Kirgilyakh Creek Basin 41,900±1,000 yr B.P. was a suckling and therefore no remains of vegetational food have been found in its gastrointestinal tract. If the mammoth had been older it could eat the sedges (*Carex concolor, C. tripartita,* and *C. vaginata),* stems of grasses, leaflets of willows and tufts of bog mosses, i.e. the plants which, as shown above, were collected by A.V. Logachev directly from under the little mammoth's body, and many other plants which grew at that time at the site (Belaya and Kisterova, 1978; Ukraintseva, 1981a; Shilo *et al.,* 1983).

The data given above suggests that the taxonomic composition of all the plants which can be identified in examination of food remains of the fossil animals' is dependent on many factors such as the season and the foraging area of the animal, the degree of digestion of plants and even the age of the animals as is the case with the baby mammoth. Nevertheless, a complex approach to the examination of food remains allows us to reveal a sufficiently complete composite of plants which grew in the habitats of the fossil animals. An illustrative example is furnished by Table 20 which is a composite list of all the plants determined at the current level of study from their remains (ground tissues, fibrovascular bundles, seeds, fruits, pollen, spores) in examination of the gastrointestinal contents of the fossil animals, discovered in various areas of northern Siberia in the period 1900 to 1979.

The lists of plants, recognized for each one of the concrete finds, can be considered as relatively complete paleofloras.This is supported by comparing their taxonomic composition (Table 20) and that of the recent local floras of each one of the areas where the fossil animals were found (Sokolova, 1977; Ukraintseva and Kozhevnikov, 1979, 1981; Kozhevnikov, 1981; and others). However, it should be noted that the composition of these paleofloras could be expanded due to the identification of remains of grasses and sedges, both their macroremains and pollen. Additional work needs to be done in this direction. Two examples can be cited: *First example:* 264 species of plants were determined in the recent flora of the area where the fossil horse had been discovered (Sokolova, 1977). The grasses are represented by 33 species, assigned to 16 genera; the sedges include 26 species belonging to two genera, thus accounting, respectively, for 12.6% and 9.8% of all the taxa, recognized by M.V. Sokolova. The grasses and sedges contain 59 species of plants or account for 22% of the total floristic composition. Ninety six taxa of flowering plants which were ascribed specific, generic and family identification were determined in the paleoflora synchronous to the horse's life period and dated at 38,590±1,120 yr B.P. At the present stage, only 11 taxa of grasses with determinable species, genus, subfamily and family were identified from macroremains and pollen. It is better with sedges: their

Table 20

Composition of plants, found in the content of gastrointestinal tracts of the fossil animals of Siberia.

Sample Number	Site, animal, time of death					
	Berezovka River, 68°30' mammoth 44,000±3,500 yr B.P.	Shandrin River, 71° mammoth 40,350±880 yr B.P.	Elga River, 64°30' horse 38,590±1,120 yr B.P.	Indigirka River, 68° bison 29,500±1,000 yr B.P.	Kirgilyakh River, 63°30' mammoth 41,000±900 yr B.P.	Yuribei River, 70°54' mammoth 9,730±300 yr B.P.
1	2	3	4	5	6	7
Larix gmelinii (Rupr.) Rupr.		x +	+	+		x +
Larix sp.	x +				+	x +
Picea obovata Ledeb.		+				+
Picea of. ajanensis Fisch.			+			
Picea sp.					+	
Pinus pumila (Pall.) Rege		+	+	+	+	
P. sibirica Du Tour	+			+	+	+
P. sylvestris L.			+		+	

To be continued

Table 20 (cont)

1	2	3	4	5	6	7
Pinus sp.			+		+	
Juniperus sp.			+			
Salix of. caprea L.			+			
S. cf. hastata L.			+			
S. glauca L.						x +
S. nummularia Anderss.			+			
S. polaris Wahlenb.						x
S. pulchra Cham.			+			x
Salix sp.	x +	x +	x +	x +	x +	x +
Populus suaveolens Fisch.			+			
Populus sp.			+			
Betula sp. (sect. Costatae)			+			
Betula platyphylla Sukacz.			+	+	+	
B. pubescens Ehrh.	+					
Betula sp. (sect. Betula)	x		+		+	+

To be continued

154

Table 20 (cont)

1	2	3	4	5	6	7
B. fruticosa Pall			+			+
Betula sp. (sect. Fruticosae)					+	
Betula exilis Sukacz.		+	+	+	+	+
B. nana L.			x	x		x +
Betula sp. (sect Nanae)		x +	+	+	+	+
Betula sp., sp.		+	x +		x +	x +
Alnus hirsuta (Spach) Turoz. ex Rupr.	x	x +	x +	+	+	+
Alnus sp.	+	+	x(?)	x		
Corylus cf. cornuta Marsh			+			
Ulmus japonica (Rehd.) Sarg.			+")			
U. pumila L.			+")			
Ulmus sp.			+")	+")		

To be continued

Table 20 (cont)

1	2	3	4	5	6	7
Caragana jubata (Pall.) Poir.	+					
Typha latyfolia L.			+	+		
Potamogeton sp.			+			
Agropyron cristatum (L.) Beauv.	x +					
Agropyron sp.	+					
Agrostis sp.						
Alopecurus alpinus Smith	x +					
Arctophila fulva (Trin.) Anderss.					+	+
Beckmenia eruciformis (L.) Host	x +					
Bromus sibiricus Drob.	+					
Calamagrostis sp.	x	x				
Deschampsia sp.			x			
Elymus sp.	x					
Helictotrichon krylovii (Pavl.) Henrard			x			
Festuca sp.		x +	x +		x	

To be continued

Table 20 (cont)

1	2	3	4	5	6	7
Festucoideae sp.			x			
Glyceria sp.			x			
Hordeum violaceum Biss. et Huet	x					
Phalaris sp.			x			
Phragmites communis Trin.	+					
Phragmites sp.			x			
Poa arctica R.Br.		+	+		x	
Poa sp.		x +	+		x	+
Poaceae	x +	x +	x +	x +	x +	x +
Carex begelowwi Torr. ex Schwein. subsp. rigidioides (Gorodk.) Egor.			x			
C. concolor Br.					x	x
C. pediformis C.A. Mey			x			

To be continued

Table 20 (cont)

1	2	3	4	5	6	7
C. tenuiflora Wehlenb.					x	
C. tripartita All.					x	
C. vaginata Tausch					x	
Carex sp.	x +		x	x +	x +	x +
Eriophorum cf. brachyantherum Trautv. et Mey.						x
E. polystachion L.					x	x +
E. scheuchzeri Hoppe						x
E. vaginatum L.						x
Eriophorum sp.		x +				x +
Kobresia cf. capilliformis Ivanova			x			
K. filifolia (Turcz.) Clarke			x			
K. simpliciuscula (Wahlenb.) Mackenz.			x			x
Kobresia sp.			+	x		
Cyperaceae	x +	x +	x +	x +	x +	x +
Juncus castaneus Smith						+

To be continued

158

Table 20 (cont)

1	2	3	4	5	6	7
Juncus sp.			x +			
Luzula sp.						x
Allium schoenoprasum L.			+			
A. strictum Schrad.			+			
Allium sp.					+	
Liliaceae				x		
Oxyria digina (L.) Hill	+					
Polygonatum sp.				+		
Polygonum aviculare L.			+		+	
P. bistorta L.				+		+
P. foliosum Lindb. fil.				+		
P. scabrum Moench			+			
Polygonum sp.			+		+	x
Rumex acetosa L.	+		+		+	
R. acetosella L.	+					

To be continued

Table 20 (cont)

1	2	3	4	5	6	7
R. arcticus Trautv.	+		+		+	
Polygonaceae			x			
Atriplex sp.	+					
Chenopodium sp.			+			
Cerastium maximum L.						x
Cerastium sp.	+					
Dianthus sp.	+					
Lychnis sibirica L.			+			
Minuatria arctica (Stev. ex Ser.) Graebn.			+			
M. macrocarpa (Pursh) Ostenf.			+			
M. rubella (Wahlenb.) Hiern						x
Minuartia sp.		+				
Gastrolychnis affinis (L. Vahl ex Fries) Tolm. et Kozh.						x
Sagina sp.	+		+			
Silene sp.			+			

To be continued

Table 20 (cont)

1	2	3	4	5	6	7
Stellaria jacutica Schischk.			+	+		
Stellaria sp.				+	+	+
Caryophyllaceae		+	+	+	+	+
Nyuphar pumila (Timm) DC.			+			
Hymphaea tetragona Georgi			+			
Caltha sibirica (Regel) Makino						
C. palustris L.	+					
Ranunculus acris L.	x					
R. affinis R. Br.						
R. gmelinii DC.						
Ranunculus sp.		+	+	+		+
Thalictrum cf. alpinum L.			+		+	
Th. foetidum L.			+		+	
Ranunculaceae			+		+	
Papaver pulvinatum Tolm.				+	+	+

To be continued

Table 20 (cont)

1	2	3	4	5	6	7
Rubus arcticus L.			+			
R. chamaemorus L.						x +
Sanguisorba officinalis L.	+		+		+	
Rosaceae	+	+	x +	+		
Astragalus sp.			+			
Hedysarum hedysaroides (L.) Schinz et Thell.			+	+		
Lathyrus pilosus Cham.			+	+		
Oxytropis sordida (Willd.) Pers.	x +					
Fabaceae			+			
Epilobium sp.			+			
Chamerion sp.						+
Aegopodium podagraria L. (?)	+					
Angelica dahurica (Fisch. ex Hoffm.) Benth. et Hook. fil. ex Franch. et Savat.			+			
Angelica sp.	+					

To be continued

Table 20 (cont)

1	2	3	4	5	6	7	
Papaver sp.			+	+	+		+
Draba sp.							+
Brasicaceae	+		+	+	+		
Sedum telephium L.			+			x	
Ribes sp.						x	
Saxifraga hirculus L.						X	
Saxifraga sp., sp.		+	+	+	+	x	+
Comarum palustre L.						x	+
Dryas punctata Juz.		+				x	+
Dryas sp.						x	
Potentilla hyparctica Malte			+				
P. multifida L.							
P. stepularis L.			+			x	
Potentilla sp., sp.	+		+	+	+		
Rosa sp.	+						

To be continued

Table 20 (cont)

1	2	3	4	5	6	7
Apiaceae			+	+	+	+
Cassiope tetragona (Pall.) D.don		+				
Vaccinium vitis-ideae L.		x +				
Ericaceae		x +	+	x +	+	x +
Gentiana sp.	+			+	+	
Phlox sibirica L.				+		
P. boreale Adam				+		
Polemonium sp.					+	+
Lamiaceae		+	+			
Pedicularis sp.		+				
Plantago media L.	+					
Plantago sp.	+					
Plantaginaceae			x			
Valeriana capitata Pall.		+	+	+	+	+
Valeriana sp.						+

To be continued

Table 20 (cont)

1	2	3	4	5	6	7
Aster alpinus L.	+		+	+	+	
Artemisia borealis Pall.			+			
A. dracunculus L.	+					
A. furcata Bieb.						+
A. tilesii Ledeb.						x
A. vulgaris L.	+	+	+			
Artemisia sp.	+	+	+	+	+	+
Cirsium sp.			+			
Lactuca sibirica (L.) Maxim.	+		+			
Nardosmia sp.						+
Saussurea sp.			+			
Tanacetum vulgare L.	+					
Asteraceae		+	+	+	+	+
Cichoriaceae				+		
Dicotyledoneae indeterm.	x +	x +	x +	x +	x +	x +

To be continued

Table 20 (cont)

1	2	3	4	5	6	7
Monocotyledoneae indeterm.	x +	x	x +	x +	x	x
Hepaticae				+		
Sphagnum angustifolium C. Tens		x				
S. girgensohnii Russ.		x				
Sphagnum sp. (sect. Subsecunda)		x				
Sphagnum sp. (sect. Palustria)		x				
Huperizia selago (L.) Bernh.			+		+	
Lycopodium alpinum L.			+			
L. annotinum					+	
Lycopodium sp.				+		+
Selaginella rupestris (L.) Spring	+		+	+	+	
S. selaginoides (L.) Link					x	
S. pleistocenica Ukraints.			+			
Selaginella sp.				+	+	
Dryopteris sp.			+	+		

To be continued

Table 20 (cont)

1	2	3	4	5	6	7
Polypodiaceae	+	+		+	+	
Sporites inteterm.	x +		+	+	+	

1) Sukachev, 1914; Tikhomirov and Kupriyanova, 1954; Kupriyanova, 1957;   2) Solonevich et al., 2977; Gorlova, 1982; Ukraintseva, this work; 3)Tikhomirov and Kultina, 1973; Ukraintseva (Kultina), 1977; Meteltseva, 1979; 4)Ukraintseva et al., 1978;   5) Belaya and Kisterova, 1978; Nikitin, 1981; Ukraintseva, 1981; 6) Gorlova, 1982; Stanishcheva, 1982; Ukraintseva, 1982; 7) wood in the enclosing deposits.

x -- macroremains (ground tissues fibrovascular bundles, needles, seeds, fruits)

+ -- pollen, spores

") -- adventitious pollen [=foreign]

fruits, preserved in the horse's stomach, enable even species of the plants to which they belonged to be reliably determined (Egorova, 1977). *Second example:* 183 species of flowering plants were recognized in the recent flora of the area where the baby mammoth was found, on the mid-Kirgilyakh River (Ukraintseva and Kozhevnikov, 1979). The list gives a sufficiently complete flora of the area, though some species may have been omitted; for instance, only one species of cotton grass was collected. As a whole, an average of 200-220 local floristic species are typical of the southern Magadan Region (Kozhevnikov and Khokhryakov, 1976). The flora, studied in the mid-Kirgilyakh River Basin, contains: 21 species of grasses, assigned to 11 genera, and 12 species of sedges belonging to two genera, i.e. they account, respectively, for 11.5% and 6.5% of the total local flora. Grasses and sedges, taken together, account for 18% of the list of plants, identified in the area. The palynoflora coeval to the lifetime of the baby mammoth and dated at 41,000±900 yr B.P. contains 52 taxa of flowering plants with identifiable species, genus and family. The grasses and sedges are represented by five and six taxa, respectively, with species, genus and family identified (Table 20). The data presented above suggests that the lists of paleofloras, ascertained from plant remains in the gastrointestinal tracts of the fossil animals, could be enlarged mainly due to a more precise identification of remains of grasses, sedges and, possibly, willows.

The lifetime of revealed paleofloras is fixed by the date of animal's death, obtained by the method of radiocarbon analysis of remains of vegetation once eaten by the animal and/or analysis of its skeletal bones. The facts that: (i) the paleofloristic composition is established from results of complex investigations, and (ii) the paleofloras established are precisely dated by the method of ^{14}C analysis and geological-geomorphological data allow us to consider them as standards. Five of six revealed paleofloras, associated with the Kolyrma and Indigirka Basins, are dated by the Karginsky Interglacial which embraced, according to the current notions, a time span 25,000±50,000 yr B.P. Only one of the revealed paleofloras, associated with the middle Yuribei River Basin (Gydan Peninsula) is dated as early Holocene (or Praeboreal) (Table 20).

The flora synchronous to the life of the mammoth in the Berezovka River area, lower Kolyma River, as dated at 44,000±3,500 yr B.P., characterizes the floristic type of the early Karginsky stage of warming. The flora coeval to the life of baby mammoth on the mid-Kirgilyakh River, upper Kolyma River, and dated at 41,000±900 yr B.P., characterizes the floralistic type of the first cold interval within the Karginsky Interglacial which is called the Kirgilyakh cooling (Shilo et al., 1983).

Two floras, namely, the flora contemporaneous to the life of the horse in the middle Elga River Basin, dated at 38,500±1,120 yr B.P., and that contemporary to the life of the mammoth in the middle Shandrin River Basin, dated at 40,350±880 yr B.P., characterize the types of floras of the optimal phases of the Karginsky Interglacial the Indigirka River Basin.

The flora synchronous to the life of the bison and dated as 29,000±1,000 yr B.P., characterizes the type of floras of the initial phase of the final stage of Karginsky Interglacial; the phase coincided with the initial phase of Lipovsko-Novosyelovskoe warming of Siberia (Kind, 1973).

The paleoflora of the optimal phases of Karginsky Interglacial in the middle Elga River Basin is the richest one as far as the taxonomy is concerned; this paleoflora reflects, more or less completely, a rich flora of the area. Comparison of its taxonomic composition with that of recent flora in the vicinity of the horse's burial site (Sokolova, 1977) shows that:

1. Some species of vegetation, primarily, tree species, represented in the paleoflora (*Picea ajanensis, P. obovata, Pinus, Betula* sp. ex sect. *Costatae*), are absent in the recent flora and the nearest limits of their ranges lie at a distance exceeding 1000 km from the area in question.

2. The paleoflora contained aquatic and riverside aquatic plants, such as *Nuphar pumila* and *Typha latifolia* which are absent in the recent flora.

3. The recent flora contains no genus *Kobresia* whereas the paleoflora included three species of *Kobresia*; at present, one of them, *Kobresia* cf. *capiliformis*, is a character plant of the upper part of the forest belt of highlands of Middle and Central Asia and Mongolia; the other species, *K. filifolia* is absent in the Indigirka River Basin but present in the Altai, Eastern Siberia, Mongolia, northwestern China and less common in the Asian Arctic; only the third species, *K. simpliciuscula*, is present in the Indigirka River Basin but absent in the middle Elga River Basin.

4. Most species of shrubs, low shrubs and herbs which today grow in the middle Elga River Basin were present in the flora of the area as early as 38,590±1,120 yr B.P.

5. Some species of the recent steppe communities proper, for instance, *Allium strictum*, of meadow end less dry steppe varieties (*Helictotrlchon krylovii, Thalictrum foetidum, Aster alpinus*), as well as species entering into the communities of dry meadows and meadow steppes (*Sangui sorba officinelis, Polemonium boreale, Kobresia filifolia*) had been already present in the composition of paleoflora.

6. However, the forest habit of paleoflora revealed is determined by finding pollen of tree species, bushes, some aquatic and riverside aquatic plants and remains of cryptogamous plants. Although the flora has been revealed rather completely, nevertheless it renders a richer vegetation than that of today, the latter being similar in character to the recent forest vegetation of southern Yakutia. As a whole, this conclusion is in harmony with that arrived at by Grichuk (1973, 1976) who showed that the species and genera of vegetation, identified from pollen in the upper Pleistocene deposits of the Indigirka-Kolyma mountainous region, currently live together in amounts of 99-100% in the Stanovoy Highland area.

The paleoflora of the middle Shandrin River Basin, lower Indigirka River, which also characterizes the floras of optimal phases of the Karginsky Interglacial is the poorest one of all the paleofloras revealed; this is associated, first, with its zonal position and, second, as shown above, with the season of the animal's death. The paleoflora renders vegetation of the forest and forest-tundra type, which is very poor in composition and similar in character to the vegetation of recent monodominant larch forests and light forests at the northern limit of their current distribution. No larch grows now in the area where the mammoth was found. The nearest area whose flora contains larch is the Yercha River Basin, 200-250 km south of the area under study.

Food remains of *Mammuthus columbi* in the form of dung boluses have been found recently in dry caves of the Colorado Plateau, Utah, Southwestern USA. The finds and their study (Jennings, 1980; Agenbroad *et al.*, 1984a, b Davis *et al.*, 1984; Mead *et al.*, 1986), enable the composition of plants, identified from food remains of the mammoths of Eurasia and North America to be compared. Comparison, undertaken by Agenbroad and coworkers (1984b)[19], showed that 42 % of plants of family identification proved to be common thus confirming, first of all, a common character of the floras. Study of *Ovibos'* excrement, performed by the author for comparison, supports the conclusions, obtained from fossil material, and suggests that food remains of herbivores can provide beneficial information for reconstruction of floras, vegetation and natural environments of the past.

Conclusions

1. Integrated investigations of the gastrointestinal contents of six fossil animals, namely, the horse, bison, and four mammoths, found in various areas of Siberia, indicate that plant remains of their food can be regarded as unique samples, "taken" by the animals in their habitats shortly before their death.

2. The taxonomic composition of plants, ascertained in examination of the samples, reliably (and more or less completely) reflects the floristic composition of areas where the animals lived.

3. Since the paleofloras, determined in the Kolyma and Indigirka basins (Yakutia) and in the Yuribei Basin (Gydan Peninsula), were accurately dated by ^{14}C analysis and geological-geomorphological data, they can be treated with good reason as standard for the above areas, characterizing the types of paleofloras of different stages of the Karginsky Interglacial and the early Holocene.

4. The taxonomic composition of the paleofloras of optimal phases of the Karginsky Interglacial in the middle Elga River Basin, upper Indigirka River area, and in the middle Shandrin River Basin, lower Indigirka River area, points to a well defined floristic differentiation and zoning of vegetational cover at that time.

5. Each of the paleofloras revealed is peculiar and distinct which is caused by some factor, mainly, their zonal position. The core of the floras was formed by species which grow in the areas under question at present.

6. The most completely revealed flora is that of optimal phases of the Karginsky Interglacial in the Elga River Basin, upper Indigirka River area; 118

taxa were identified in its composition; 98 taxa belong to higher flowering plants and 20 to higher cryptogamous plants. Twelve taxa, specifically and generically determined, fall within trees, accounting for about 12 % of the flowering plants; shrubs and undershrubs are represented by 14 taxa which are specifically and generically identified; they account for 15 % of the flowering plants. The flora is dominated by forbs and low shrubs, represented by 72 taxa with identifiable species, genus, subfamily and family; they account for 73 % of the flora. Boreal and hypoarctic species predominate.

7. The presence of some species of steppe communities proper, meadow steppes' and dry meadows in the paleoflora of the middle Elga River Basin suggests that the above species and their communities had been present in the area under study since the optimal phases of the Karginsky Interglacial.

Chapter 6

Vegetation and Environment of Siberia in the Late Anthropogene (Quaternary)

6.1. Regional features of vegetation and environment of the north of Siberia

The specimens of the "mammoth" faunal complex, found during the last 85 years (from 1900 to 1985) in the north of Siberia, are concentrated in three areas, namely, in the Gydan, Taimyr and Indigirka-Kolyma areas. In order to gain some insight into living conditions of not concrete mammal specimens found in the region, but the "mammoth" fauna as a whole, it is necessary to consider the character of vegetational and climatic variations in the above-enumerated areas during the last 50,000-55,000 year, i.e. in the time interval which corresponds, in the sense of Kind (1973), to the Late Anthropogene (Quaternary) and which have yielded most of the finds of "mammoth" faunal remains, dated by the method of radiocarbon analysis.

6.1.1. Gydan area

The area includes the Gydan Peninsula territory which formed as early as the middle Pleistocene Yamal transgression and later transgressions, penetrated to the territory along major V-shaped depressions which formed the surface of the fourth marine and third lagoon-laida terraces with elevations of 50-30 m and 30-40 m, respectively. Middle and late Pleistocene rocks crop out in scarps of riversides and in the basements of erosion terraces. Outcrops of Karginsky-Zyrianka laida and lagoon-laida deposits are common, composed of clay, loam, sand, intercalated with plant remains and peats (Lazukov, 1973). However, they remain practically unknown in respect to paleobotanical.

The present topography of Gydan Peninsula is rather uniform and monotonous. In general, the peninsula represents flat, paludal plains with many lakes; as a rule, elevations are under 100 m, locally 150-160 m. The flatness of the country is locally broken by knobs and trails, as well as ridges which formed due to erosion, frost action and tectonic movements; previously they were taken for glacial forms (Lazukov, 1970). During the last (Sartan) ice age, the territory of Gydan Peninsula and most of the neighboring territory of northern and central Yamal and the Tazovsky Peninsula were not ice-covered (Belorusova, 1964;

Lazukov, 1972; Arkhipov *et al.*, 1977; Velichko and Faustova, 1986). It must have been during this epoch when representatives of the "mammoth" faunal complex such as woolly mammoth, reindeer, the horse (*Equus caballus* s.l.) penetrated there. Their remains are not so common in the deposits of the area as in coeval deposits of other areas of the north of Siberia, in particular, Taimyr, Yakutia, the North East. From the last century to the present, remains of the fauna have been found at 14 only localities, including the site where the corpse of female mammoth has been found in the middle Yuribei River Basin (Lazukov, 1970; Evseev *et al.*, 1982; Chapter 4.6). The last find radiocarbon dated suggests that mammoths lived in the area as late as about 10,000 yr B.P. The area is entirely in the tundra zone. The vegetational cover includes three subzones, namely, those of: (i) arctic tundras; (ii) boreal hypoarctic; and (iii) southern hypoarctic tundras (Yurtsev *et al.*, 1978). In floristic respect, this is one of the most thoroughly studied areas of the Arctic. It is part of the Yamal-Gydan subprovince of the East Siberian floristic province. "The subprovince is characterized by a general paucity and a sharply negative peculiarity of flora, based on disjunction of areas of species many mountain (mainly, East Siberian species and the absence of many eastern ("trans-Yenisei") and western (European, amphi-Atlantic, and other) species which reached the Urals"(Yurtsev *et al.*, 1978).

In paleobotanical respect, late Pleistocene and Holocene deposits of the area are poorly studied. Investigations, undertaken in the middle Yuribei River Basin in connection with the find of a mammoth (Gorlova, 1982; Stanishcheva, 1982; Ukraintseva, 1982a) compliment data, obtained by Zubkov as early as 1931; taken together these data give an idea of the character of flora, vegetation and climate and their variations during the last 10,000 years.

At the end of late Pleistocene and the onset of Holocene, the northern areas of the region (70-71°N) were under arctic tundras, resembling recent tundras of this type (Ukraintseva, 1982b). About 10,000 yr B.P. they gave way to hypoarctic tundras (Stanishcheva, 1982; Ukraintseva, 1982b; see Chapter 4.6). During the climatic optimum when temperature was 2.0-2.5° higher as compared to that of the present and no ice cover was present in the Arctic (Borisov, 1975), birch forest tundra was spread at the latitude of the mammoth's find and to the north. This is suggested by data, obtained from the study of a relict peat bog;, situated a few kilometers north-east of the site where the mammoth was found. The peat bog occurs on river terrace III (15 m high), built up of sand alluvium. Peat, 0.5-3.5 m thick, rests on grey clay. Flat-lying trunks, radical parts, boughs and bark of Betula alba s.l., the woody birch, were found in abundance at the contact of peat with underlying clay (Zubkov, 1931). The sphagnum (upper) layer of the peat bog is made up of stems and leaflets of *Sphagnum riparium, S. angustifolium, S. subsecundum*. Remains of *Equisetum limosum* are rare. Peat is quite mineralized. The horse tail - true moss (middle) layer includes up to 5% of humified particles; it is composed of stems of *Equisetum limosum*, leaflet of the mosses (*Drepanocladus exanmelatus, D. intermedius,* and *D. revolvens*) with some admixture of remains of *Carex rotundata* (sacculuses, rootlets), and bark and sprigs of *Betula alba* s.l. A well preserved stem of *Lycopodium annotinum* and stems of horse tail with very long whorls of leaves, resembling those of *Equisetum sylvaticum* were also found in the layer. According to Zubkov (1931), the peat bog formed because of swamping of a moist valley birch forest. The horse tail - true moss bog, formed at first, later turned into sphagnum bog; then the process of peat formation ceased and the moss cover began to degrade.

At present, peat formation is known to be very slow in the Arctic region, outside the range of woody plants, and sphagnum peat bogs with deposits of sphagnum peat occur southward only. Tikhomirov (1941) also related the formation of relatively thick (1.2-2.0 m) peat bogs, studied by him at Yamal, to the Holocene climatic optimum. This was caused by more favorable global climatic conditions which gave impetus to the advance of tree species northward and promoted the processes of swamping and peat formation in the Arctic.

Data of Zubkov (1931), Tikhomirov (1941), and Levkovskaya (1977) indicate that during the Holocene climatic optimum (Atlantic-subboreal time) woody plants advanced northward for 500-400 km, to the middle part of the present typical tundra subzone. Data on the northern Ob-Yenisei part of West Siberia (Levkovskaya, 1977) suggest that during the Holocene climatic optimum the limit of distribution of woody plants in West Siberia lay in the middle part of the present typical tundra subzone. The bush tundra subzone and the southern part of typical tundra were the area of distribution of boreal taiga forests. In the Holocene, the boreal/middle taiga boundary lay near the present southern boundary of the light spruce-larch forest subzone. Hence, the range of the past tundra afforestation was beyond the boundaries of the subzone and reached 450 km in width.

Thus, the above data indicate that during the Holocene climatic optimum practically the whole northern portion of Western Siberia, including Gydan Peninsula and Yamal, had already been afforested and therefore this vast region had been unsuitable for dwelling of large herbivorous mammals.

6.1.2. Taimyr area

An enormous territory of Taimyr Peninsula (about 820,000 km^2) is dominated by tundra landscapes, namely, typical and arctic tundras. Only small areas at Cape Chelyuskin are occupied by polar deserts (Geobotanical map of the USSR, 1954; Aleksandrova, 1977; Matveeva, 1979a).

On the central North Siberian lowland, a complex of modern physical-geographical conditions (continentality, relatively warm summer and others) are beneficial for the northernmost distribution of forest vegetation on the earth. The northern limit of forest and isotherm +12° practically coincide in outline as if bound by the Taimyr Peninsula on the south. It was found out that the area had developed in the continental regime for an extended time. In both the Zyrianka (early Wisconsin) and Sartan (full Wisconsin) Ice Ages, the area remained more or less free of ice sheets which had come down the Byrranga and

Putorana mountains but had not merged in the center of North Siberian lowland. During the Karginsky transgression, even at its maximum, the territory, including the Bolshaya Balakhnya River Basin (73°30'N), was not flooded by sea water (Andreeva, 1980; Belorusova and Ukraintseva, 1980; Isaeva, *et al.* 1980; Kind and Leonov, eds, 1982). Thus, for a relatively long time period, at least during the last 55,000 years, the region was a kind of refugium for the preservation of genofond of plants and animals in the extremal stages of environmental development of Taimyr. It is precisely this fact that explains the presence of late Pleistocene exposures in the Novaya River Basin in which the environmental evolution of the region is recorded. Studies of some natural exposures and peat bogs (Kultina (Ukraintseva) *et al.*, 1974; Lilironenko and Savina, 1975; Belorusova and Ukraintseva, 1980; Ukraintseva *et al.*, 1981) and studies of Pleistocene and Holocene deposits in other areas of Taimyr (Zaklinskaya, 1954; Berdovskaya *et al.*, 1970; Danilov *et al.*, 1971; Andreeva, 1980; Nikolskaya, 1980; Nikolskaya *et al.*, 1980; Kind and Leonov, eds, 1982) allowed tracing the emplacement and the history of biogocenoses of Taimyr for the last 55,000-60,000 years against the background of climatic variations, thus forming a notion of living conditions of representatives of the "mammoth" faunal complex such as mammoth, two species of horses, Pleistocene bison, musk ox and other animals in the region. In the Novaya River Basin, mammoths (earlier type) dwelt almost 53,000 yr B.P., during the last phases of Murukta (early Zyrianka) Ice Age which lasted, according to data available, from $110,000 \pm 117,000$ to 50,000(55,000) yr B.P. (Arkhipov, 1983). Like other areas of Siberia, Eurasia and America, Taimyr was inhabited by mammoths of later type as early as the end of late Pleistocene. A complete skeleton of the animal which died $11,450 \pm 450$ yr B.P. was found on the Marnonta River, Northwestern Taimyr. Teeth, tusks and ribs of the animal are common in river and lake scarps. Inclusions of bone remains of specimens of the "mammoth" faunal complex are very characteristic of lacustrine deposits of Lake Taimyr. "They locally compose whole intercalations" in outcrops of its later terraces (Kind and Leonov, eds, 1982, p. 134). ^{14}C ages of the finds of specimens of the "mammoth" faunal complex, including dated recently obtained by L.D. Sulerzhitsky[20] on bone remains of a mammoth, two horses, some musk oxen, and an elk suggest that representatives of the "mammoth" faunal complex inhabited Taimyr for the past 50,000-55,000 years at least. As late as 9670 ± 60 yr B.P. mammoths lived in the Liizhnyaya Tunguska River Basin (L.D. Sulerzhitsky, oral communication); musk oxen lived at Cape Chelyuskin 2920 ± 50 yr B.P. (2945-GIN, Sulerzhitsky). Horses were the first to leave the area; they inhabited the Bolshaya Balakhnya River Basin and the vicinity of Lake Taimyr $36,300 \pm 900$ yr B.P. (3119-GIN, Sulerzhitsky) and 2150 ± 200 yr B.P. (2744-GIN, Sulerzhitsky), (Sulerzhitsky 1976) respectively.

Let's consider, at length, living conditions of the above-mentioned animals and, initially, the character of vegetation and climate at that time interval.

In Zyrianka time (early Wisconsin), paleobiogeocenoses of the polar desert type existed even in the Novaya River Basin; this is reliably established by palynological data, geologo-geomorphological analysis of the territory, and radiometric dates (Ukraintseva *et al.*, 1981). In character, they are similar or close to the present-day arctic deserts of Cape Chely-uskin (Matveeva, 1979a,b). At that time, the territory under study represented barren areas or a vegetational cover, if any, was so thinned out that the influx of pollen and spores, produced by separate plants,into the sediments on which they grew was extremely low. Soils which were poorly sodded or completely barren were readily subject to denudation in summer seasons due to active snow and permafrost thawing. This is suggested by a high percentage (59-92%) of pollen and spores, redeposited from older (Liesozoic-Neogene) sediments, particularly, in outcrop No. 1 in the Bolshaya Lesnaya Rassokha River mouth (beds 1-3, Fig. 10). According to Matveeva (1979a,b),the ground cover ranges from O to 15% in the present-day polar deserts of Cape Chelyuskin. The vegetational cover is made up of no more than 10 species of flowering plants and 5 to 12 species of mosses. A paucity of flowering plants and mosses is no doubt related to the extremely severe climate. Two months only, July and August[21], which have positive mean monthly temperatures, namely, +1.5° and +0.8°, respectively, fall on the "warm" part of the year; the positive temperature sum for July and August totals +45.6°C only. The highest precipitation accounting for 28% of total annual precipitation falls on the two months (Fig. 55). Analysis of climatographs infers that a very low temperature provision and a relative precipitation abundance during the vegetation period are factors which limit the development of plants, their distribution and the formation of a stable plant cover in the polar deserts.

About 55,000 yr B.P. a climate like that of the present at Cape Chelyuskin was typical of the central areas of the present day North Siberian lowland; this was related to extremely severe conditions caused by glaciation of the Byrranga and Putorana mountains (Makeev, 1970, cited in Berdovskaya *et al.*, 1970; Makeev, *et al.* 1979; Kind and Leonov, eds, 1982). These new data are in agreement with data of R.M. Votakh (Arkhipov *et al.*, 1977) which indicate that Arctic tundras were developed in the Ob River area, farther south-west of the area under study at the latitude of Salekhard (64°30'N), in the areas not covered by glacial ice.

At the end of Zyrianha (Murukta) Ice Age, grass-forb and sedge-grass phytocenoses which served as pastures for mammoths and other herbivorous animals (late Pleistocene horse, musk oxen, bisons and others) were spread in the center of North Siberian lowland and in adjacent areas free of ice. In the Movaya River Basin, similar phytocenoses containing *Poa* sp., *Poaceae, Cyperaceae, Polygonum bistorta, Rumex arcticus, Caryophyllaceae, Papaver* sp., *Ranunculaceae, Hedysarum hedysaroides, Valeriana capitata, Artemisia* sp. and others in grass stands were developed about 53,170 yr B.P.[22] (Ukraintseva *et al.*, 1981). At that time the alder (*Alnus fruticosa*) and the dwarf birch (*Betula nana*) had not been predominant yet in the plant cover of the area as suggested by rare pollen grains of the plants in the mammoth-enclosing deposits.

According to Nikolskaya (1980), grass-sedge communities, including *Kobresia* sp., *Cyperaceae, Poaceae, Artemisia* sp. and others existed in the Zakharova Rassokha River Basin more than 48,000 yr B.P.

The period succeeding Karginsky time (50,000-25,000 yr B.P.) was marked by warming and sea transgression (Andreeva, 1980; Kind and Leonov, eds, 1982). In the continental area, warming was manifested in higher summer temperatures, intense thermokarsting, widened lake areas and it was accompanied by heavy afforestation of previously treeless areas and creation of swamps. The study of Karginsky terrestrial deposits showed that during that time interval the climate was not even over the vast territory of Taimyr Peninsula, but consists of alternations of three more or less warm stages (early, Malaya Kheta, Lipovsko-Novosyelovskoe), separated by two cold stages (early, Konoshel); this advocated Kind's notion (1973) of multiple climatic fluctuations in the north of Siberia in Karginsky time.

Radiocarbon dating of organic remains of that time (peat, wood) allowed us to tie up these waves of warmth and cold and to gain an objective, i.e. historically true notion of the dynamics of vegetation of the vast regions in time and space (Table 21).

During warmer stages of that time, continental areas of Taimyr were dominated by forest formations, namely, larch, spruce-larch forests with admixture of large woody birch; spruce forests, composed of *Picea obovata*, locally occur in suitable edaphic conditions. Grass-sedge and grass-sedge-forb communities, shrub formations (*Alnus fruticosa, Salix* sp. sp.) were developed along the rivers and around the lakes. At climatic optimum (Lipovsko-Novosyelovskoe warming), the range of *Picea obovata*, the spruce, reached the right bank of Bolshaya Balakhnya River, 73°30'N (Nikolskaya, 1980; Kind and Leonov, eds, 1982). The radiocarbon ages of Karginsky wood in the Ary-Mas urochishche are in the range of 46,000±2200 to 29,820±470 yr B.P. An abundance of wood remains in Karginsky deposits in the Ary-Mas area and in the Bolshaya Bolshnya River Basin cannot be explained by transportation of wood from the south by the river, as is the case with wood inclusions in similar deposits on the longitudinal Siberian rivers. According to Belorusova (1978), the river system had already acquired the present contours by Karginsky time; the lower left-hand streams of Khatanga River, including the Novaya River, had already stretched sublatitudinally along the strike of the North Siberian lowland and crosses forest landscapes.

The final stages of climatic optimum were characterized by processes of swamp formation and the disappearance of spruce, large woody birch and associated boreal elements from the forest composition. Even the central areas of North Siberian lowland were dominated by larch forests and open woodlands. This is reliably evidenced by new paleobotanical data, dated by that time, on the lower Bolshaya Lesnaya Rassokha River, the Novaya River Basin. As early as 35,000 yr B.P., polygonal bogs were probably distributed in the area. This is suggested by a ^{14}C value of 34,730 years B.P., obtained on a peat sample, taken

Table 21

Basic diagram of vegetational variations of Taimyr in Late Pleistocene
and Holocene time against the background of climatic variations.

		Ary-Mas, 72°30'N	Bolshaya Lesnaya Rassokha River, 72°35'N	Zakharova Rassokha, 73°50'N	Bolshaya Balakhnya River, 73°30'N	Mamonta River, 75°15'N
Warmth	Cold					
			Low bush and bush tundras, 2175±200 yr B.P. (IM-672)			
4 Late Holocene cooling				Larch forests with birch and spruce, 5180±150 yr B.P.		
7 Optimum		Larch forests with birch and spruce (*Picea obovata, Betula alba* s.), 5860±yr B.P. 6670±90 yr B.P. (IM)		Light boreal taiga forests, 5990±50 yr B.P. (GIN-1460)	Light boreal taiga larch forests, 8310±yr B.P. (GIN-774)	Bush tundra with Betula exilis; dwarf willow formation (*Salix lanata* and others); grass-forb communities; larch along the valley(?)
8 L. Sanchugovka cooling warming						

To be continued

Table 21 (cont)

	Warmth	Cold	Ary-Mas, 72°30'N	Bolshaya Lesnaya Rassokha River, 72°35'N	Zakharova Rassokha, 73°50'N	Bolshaya Balakhnya River, 73°30'N	Mamonta River, 75°15'N
9							
10	Pit-Mgarkin cooling Warming					Low bush and bush tundras with a moss layer, 10,480±250 yr B.P. (GIN-792)	*Cotton grass-sedge tundras; moss tundras with* Betula exilis, Salix polaris, *11,450±450 yr B.P.*
11	Norilsk stage			Larch forests 10,500±500 yr B.P. (IM-671)			
12	Taimyr warming						
13	Kokorevka warming		Various tundras; sedge-grass and grass-sedge-forb communities				
14	Nyapan stage?						
15							
20	Gydan stage		Light larch forests, 23,250±300 yr B.P. (IM)				

To be continued

Table 21 (cont)

Warmth	Cold	Ary-Mas, 72°30'N	Bolshaya Lesnaya Rassokha River, 72°35'N	Zakharova Rassokha, 73°50'N	Bolshaya Balakhnya River, 73°30'N	Mamonta River, 75°15'N
25						
Lipovsko-Novosyelovskoe warming		Spruce-larch forests with *Betula alba s.*			Larch forests with spruce, 26,600±1000 yr B.P. (GIN-999)	
30						
Konoshel cooling		Larch (fringing) forests, 29,820±470 yr B.P. (IM); Low bush and moss tundras				

To be continued

Table 21 (cont)

	Warmth	Cold	Ary-Mas, 72°30'N	Bolshaya Lesnaya Rassokha River, 72°35'N	Zakharova Rassokha, 73°50'N	Bolshaya Balakhnya River, 73°30'N	Mamonta River, 75°15'N
35				Sedge-larch forests, 34,730 yr B.P. (LU-1189). Low bush tundras with *Betula nana* ssp. exilis; polygonal bogs with *Carex aquatilis,* ssp. stans, *Eriophorum* sp., *Drepanocladus vernicosus* sp., *Bryales* sp.			
40	Malaya Kheta warming (optimum)				Larch forests with spruce, 39,000±1000 yr B.P. (GIN-1441)	Larch forests with spruce, 39,900±1500 yr B.P. (GIN-784)	
45	Early cooling Early warming		Larch forests with spruce, 43,800±1000 yr B.P. (GIN-1154)				
50				Grass-forb and sedge-grass communities, 53,170 yr B.P. (LU-1057)	Grass-sedge (*Kobresia* sp., Cyperaceae, Poaceae) communities with Artemisia sp. and forbs, 48,000 yr B.P. (GIN-1217)		

Sources: 1) Tikhomirov, 1950, 1958; Zaklinskaya, 1954; 2) Kultina (Ukraintseva) et al., 1974; 3) Belorusova and Ukraintseva, 1980; 4) Nikolskaya, 1980, 1982; 5) Ukraintseva et al., 1981; 6) Anthropogene of Taimyr, 1982.

Note: (X) marks 14C age, obtained on one of seven stumps, found by the author on the surface of (present) terrace II above the flood plain of the Bolshaya Lesnaya Rassokha River.

from the top of peat deposit, 45 cm thick. It took at least 300-400 yr to form the deposit, if the rate of peat accumulation is taken not to exceed 1.5 mm per year[23]. Macroremains of plants which took part in the formation of phytocenoses of that time, such as *Carex aquatilis* ssp. *stans, Eriophorum* sp., *Drepanocladus vernicosus, Drepanocladus* sp., were recognized by M.S. Boch from botanical analysis of the peat deposit (Ukraintseva *et al.*, 1981). At the initial stages of peat formation, the composition of both flowering plants and mosses was poorer than that of plants which form the present-day polygonal bogs in the area. Even mosses which make up the moss layer were then represented by no more than three species; this is supported by both gross composition and spore composition. It should be noted that at the initial stages of peat accumulation, shrubs and undershrubs did not dominate in the composition of plant cover. they were more common during the formation of the middle part of peat deposit (Fig. 10). At the final stage of peat accumulation (about 34,730 yr B.P.), sedge-larch forests, composed of *Larix gmelinii*, were distributed in the Novaya River Basin. however, neither birch nor spruce participated in forest composition in the area. The northern limits of the above-mentioned trees may have been close to their present limits as suggested by their rare, doubtless adventitious, pollen grains in spectrum of sample 13 (Fig. 10). Nevertheless, the climate during the formation of the upper part of peat deposit in the Bolshaya Lesnaya Kassokha Basin and adjacent areas was somewhat warmer than that of today, since forests, although monodominant, reached the area which is treeless at present and, possibly, extended even farther north.

There is good reason to believe that the climate of that time, i.e. about 35,000 yr B.P., in the Bolshaya Lesnaya Rassokhca Basin, was similar or close to that of the present in the Ary-Mas area. The area displays all the main types of plant cover typical of the Khatanga area of southern Taimyr, namely, light larch forests, nano-polygonal "spotty" low bush and sedge tundras, low willow and bushy tundras, polygonal bogs (Norin, 1978b).

Climatic conditions of the Ary-Mas, the most northerly forest tract on the earth, can be characterized by observational data, recorded by the nearby Khatanga meteorological station. in the station area, mean annual temperature is -13.4°; July mean temperature is as high as +12.3°C; the highest precipitation, accounting for 17% of the annual sum total also falls on July. According to multi-year data, positive temperature sum totals +83.7°C and ranges from +13.2°C in June to +45°C in September with a peak of +38.1 °C falling on July (Fig. 55) An almost "explosive" character of the growth of larch in height, determined by Lovelius (1979) at the most northerly limit of its distribution, falls just on the second half of July when the sum of positive temperatures reaches maximum and then drastically reduces in August and early September (Fig. 55). The climatograph given in Figure 55 vividly characterizes conditions which are necessary and sufficient for the growth of monodominant larch forests at the most northerly limit of their distribution on the earth. The composition of paleofloras and the composition of paleovegetation suggest that similar

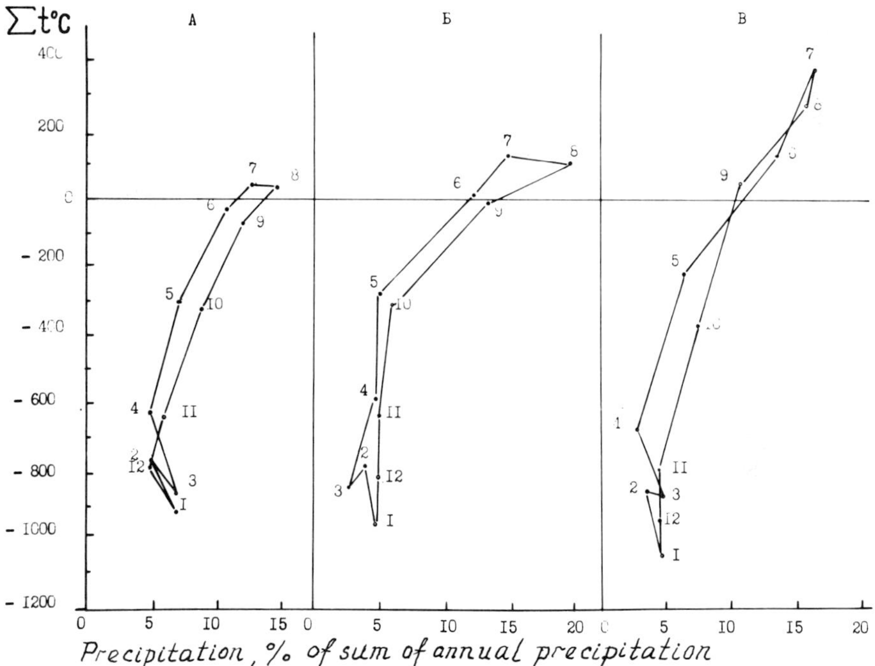

Figure 55. Climagraph characterizing the annual course of temperature and moisture conditions of some areas of Taimyr. (A) vicinity of Cape Chelyuskin, (B) vicinity of Maria Pronchishcheva Bay, (C) vicinity of the settlement of Khatanga.

climagraphs can characterize paleobiogeocenoses of monodominant larch forests and open woodlands which existed in various areas of Taimyr at certain stages of the Pleistocene and in the Holocene.

As suggested by palynological data, cooling which followed the climatic optimum (35,000-30,000 yr B.P.) was relatively inclement. Its signs are ubiquitous. At that time, tundras were widely distributed in the north of the region. Given the central areas of North Siberian lowland, in particular, the Ary-Mas area, were also dominated by tundras, however, the larch may have survived under favorable conditions. To the south, the basins of the Kotui, and Kheta, Coganida rivers were dominated by forest-tundra which contained birch and spruce.

Palynological analysis of the boundary "wood" layer (branches and trunks, up to 10 cm in diameter which yielded a ^{14}C date of 29,820±470 yr B.P.) and overlying deposits showed that during the following, final phase of Karginsky Interglacial, Lipovsko-Novosyelovskoe, which is believed to be not the warmest phase, the Novaya River Basin was occupied by birch-larch forests and bogs. Spruce forests grew in the areas, characterized by the most favorable moisture conditions. Pollen of tree species of the upper section is dominated by that of *Picea obovata*, the Siberian spruce; its amount in the tree-shrub group accounts for (9.2) 16.8-24% (Fig. 56).

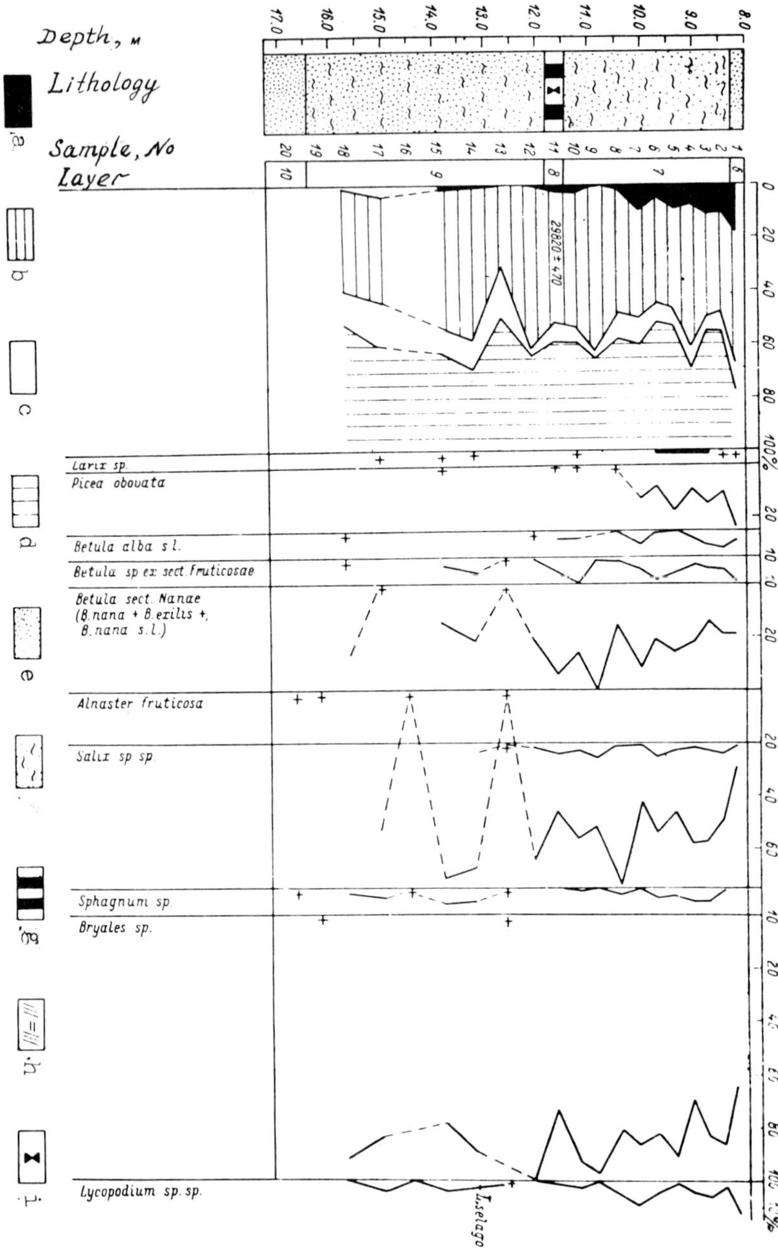

Figure 56. Spore-pollen diagram of outcrop No. 12, Ary-Mas, right bank of the Novaya River, East Taimyr. After Belorusova and Ukraintseva, 1980. Pollen of: a) trees, b) shrubs and low shrubs, c) grasses; d) spores of Bryophyta and Pteridophyta, e) sand, f) sandy loam, g) peat, h) surficial layer of sod, and i) the site where samples were taken for ^{14}C analysis.

In the west of North Siberian lowland, woody plants were not distributed farther north than in the East Taimyr region. According to Troitsky (1967), woody remains in the Karginsky sediments are extremely rare there (in the area of Ust-Yenisei port); they cannot evidence for forest existence as they occur in the river valley which flowed water from the taiga zone. in the east of North Siberian lowland, in the Popigai and the Anabar river valleys, Karginsky deposits contain inclusions of peat, accumulations of boughs and large fragments of shrubs (Saks, 1953). However, the absence of type section and absolute age dates permits no firm conclusion regarding the forest distribution in the area at that time, but does not exclude it.

Thus, unquestionable traces of the farthest forest advance to the north during both the Karginsky warm maximum (Malaya Kheta phase, 42,000-35,000 yr B.P.) and its final stage (30,000-25,000 yr B.P.) are fixed in the central North Siberian lowland only (on eastern Taimyr *in sensu lato*). Of interest are reconstructions of some paleoclimatic characteristics, made for the first Karginsky warming (50,000-45,000 yr B.P.) in the Zakharova Rassokha River Basin. According to Nikolskaya (1980), a boreal light taiga existed there at that time. Lean annual temperatures are determined by V.A. Klimanov for that time at -12° (-12°, -16°), mean temperature of January -34° (-34°, -35°), mean temperature of July elevated to 14° (Kind and Leonov, eds, 1382).

A turning-point in the evolution of natural environment of Taimyr occurred about 25,000 yr B.P. the last phase of Karginsky warming came to an end a transition to maximal cooling about 20,000 yr B.P. and to the development of glaciation in the Byrranga and Putorana mountains took place. According to the modern notions (Kind and Leonov, eds, 1382), glaciation of Taimyr was incomplete. A small ice shield only formed on the Putorana Plateau; the shield gave birth to piedmont placiers at the northern plateau foot on the north Siberian lowland. Glaciers of the Anabar center did not reach the North Siberian lowland at all. Field studies, carried out by Zh.M Belorusova revealed no evidence of Sartan Glaciation either on the eastern North Siberian lowland or in the lower Khatanga River area (Belorusova *et al.*, 1977). No continental glaciation of the region is shown on the landscape map of the epoch of maximal cooling of the late Pleistocene (Gerasimov and Velichko, 1984). A glaciation of the valley-reticulated type developed on northeastern Taimyr, in its highest mountainous part (Makeev *et al*, 1975). Maximal advance of the Sartan glaciation southwards took place on western Taimyr.

During the Sartan cooling, the climate was rigorous. Open, treeless landscapes, i.e. Various tundras, dominated by mosses and lichens, were widely developed. This is suggested by a very low content of pollen and spores in Sartan deposits and a very poor species composition of plants which grew at that time (Belorusova and Ukraintseva, 1980; Nikolskaya, 1980; Kind and Leonov, eds, 1982). Palynological spectra of that time are dominated by spores of cryptogamic plants (mainly, true mosses) and pollen of herbs (sedges, grasses, forbs the latter being poor in species); pollen of shrubs and low shrubs is rare (essentially, low birch, willow, alder); pollen of trees, if any, is represented by scanty grains of larch, large woody birch, Siberian pine, spruce which are undoubtedly adventitious, except for pollen of larch which is known not to be transported long-distance.

An arid and severe climate is also suggested by multiple eolian landforms, formed at that time (Kind and Leonov, eds, 1982), as well as by remains of specimens of the "mammoth" faunal complex, dated by Sartan cooling (Glaciation), in particular, musk ox, horse, mammoth, bison and others. Of particular paleogeographic interest are a series of ^{14}C dated, recently obtained by L.D. Sulerzhitsky on Ovibos remains from Taimyr, as they allow us to trace the paths and time of migration of the animal as the climate and the character of vegetation varied in this vast region. About 42,000 yr B.P., *Ovibos* populations dwelt in the Putorana Mountains where 2 skull of the animal, dated at 41,900±1000 yr B.P. (GIN-2327)[24], was found. During the Konoshel cooling (35,000-30,000 yr B.P.) the *Ovibos* penetrated into the Khatanga River Basin, as evidenced by a skull, dated at 31,000 yr B.P. (GIN-3012). There are three dates on skulls of *Ovibos* from: the Nizhnyaya Taimyra River, 17,800±300 yr B.P. (GIN-1815); the Bolshaya Balakhnya River, 12,500±40 yr B.P. (GlN); Cape Chelyuskin, above-mentioned date 2,920±50 yr B.P. (GIN-2945).

During the 1975-1976 field seasons in the Ary-Mas area, on the Novaya River sloping beach, the author discovered: remains of the horse (*Equus lenensis* Grom.[25]), namely, a skull of very good preservation and limb bones; a horn of musk ox; teeth and limb bones of mammoth of the later type.

According to palynological data, the onset of Sartan cooling in the Ary-Mas area was characterized by progressive floristic impoverishment and change in vegetation. Light forests thinned out and turned into tundra with galeria forests located along the rivers. Tundra plant associations, dominated by mosses and lichens, became ubiquitous in time. The Sartan periglacial complex undoubtedly included herbaceous (sedge-grass) communities, developed in the site of drained lakes, which were gradually replaced by moss vegetation (Belorusova and Ukraintseva, 1980). The following, Holocene, stage was marked by warming of climate. Starting from the early Holocene, warming was fixed even in the most northerly areas of the northeastern mountainous part of Taimyr where a rapid glacier retreat and a raising of sea level took place (Berdovskaya *et al.*, 1970).

The Holocene warming is, first of all, evidenced by rather large accumulations of fossil wood in the recent tundra zone and the most northerly finds thereof in Eurasia.

Miroshnikov (1958) generalized information, obtained by geologists, on the finds of large wood remains on Taimyr, including stumps and trunks of larch, birch, and even spruce. All wood remains were found in deposits of terraces or at their surface. Spruce trunks were discovered on the western shore of Lake Taimyr, in lacustrine deposits of a 25-30 m terrace, at 74° 30′ N. The most northerly finds of wood (ten vertically standing stumps and seven trunks of larch, up to 15 cm in diameter) were found in the extreme north of the peninsula, 80-100 km south of Cape Chelyuskin (76° 50'N). Dating of these wood remains was not done.

Two absolute age values only are known for the most northerly areas of Taimyr (Garutt, 1965). One of them characterizes a semidecomposed mammoth corpse (11,450°250 yr B.P.), found on the Mamonta River, in the Shrenk River Basin (75°27'N); the other gives an age (11,700±300 yr B.P.) of the wood from accumulation deposits of the northern shore of Lake Taimyr.

The North Siberian lowland (basins of the Novaya, Boganida, Bolshaya Balakhnya rivers) is one of the areas of Taimyr where fossil wood is strikingly abundant; it is determined by a considerable number of radiocarbon dates. As early as 1867, when A.F. Middendorf was back from his travel through the North of Siberia, he stated that "antediluvian wood" is extremely common in the Novaya River Basin. The amount of fossil wood has substantially reduced since as it has been used as fire-wood, nevertheless, large wood remains can be found at different geomorphological levels to the present day.

Important paleobioclimatic information is provided, in addition to the stumps and trunks of trees, by flood-plain and terrace sandy loam and loam deposits, interbedded with peat and detritus, as well as by peat bogs, found in the river valleys and in watersheds of Taimyr. Practically all the chronostratigraphic units of the Holocene are rich in plant remains, frequently dated by ^{14}C method in the process of palynological studies. Wood and peat from Holocene deposits belong to all the Holocene units namely, early (10,500±500 yr B.P.), middle (5970±70 yr B.P., 5180±150 yr B.P.), and late (2175±200 yr B.P.).

One of the earliest Holocene finds on Taimyr was made by the author in 1978 in the lower part of a right tributary of the Novaya River when surveying; the area adjacent to the Khatanga mammoth burial site (Fig. 7). At the surface of Bolshaya Lesnaya Rassokha River terrace II above the flood plain, the author found remains of trees once growing in the area which is treeless at present. They were seven stumps of larch, 45 to 55 cm high. Six stumps were closely spaced in a low bush, small hummocky cotton grass-sedge tundra (Fig. 57), and the seventh stump was located at a distance of about 2 km from the rest, in polygonal bog (Fig. 57). The nearest forest, Ary-Mas, which is the most northerly "forest island" on earth, is situated at a distance of 25-30 km.

This summer an attempt was made to dig out two of the seven stumps. As soil thawed out, more than 50 cm of, ground was taken off around each stump, however, the root systems were never reached. The field season came to an end and the work had to be abandoned. The total height of one of the stumps was 0.98 cm, and the diameter 12.5 cm; the height and the diameter of the other stump were 1 m and 18 cm, respectively. The latter was sawn off. Its lower part, well preserved in permafrost, was used for radiocarbon analysis which showed that one of the larches growing on terrace II above the flood plain died out 10,500±500 yr B.P., i.e. in a transition time from the late Pleistocene to Holocene or at the very beginning of Eoholocene (Early Holocene). Hence, the

Figure 57. Bush small-hummocky tundras, Bolshaya Lesnaya Rassokha terrace II above the flood plain. Part of larch trunk died off, according to ^{14}C analysis, at 10,500$\pm$500 yr. B.P. Photo by V.V. Ukraintseva.

Figure 58.   Polygonal bogs on Bolshaya Lesnaya Rassokha River terrace II above the flood plain. Fossil larch trunk in the foreground. Photo by V.V. Ukraintseva.

stumps found in the area witnessed the earliest forest phase of Holocene on Taimyr. This find, unique for the high latitudes, suggests that as early as the Eoholocene the range of larch in the region advanced north of the present-day limit of forest and woody plants at all. As shown by Belorusova and coworkers (1987), the time of larch growth almost coincided with the first Holocene warming, but slightly forestalled it. If the wood is assumed to be related to the late Dryas, then it must be admitted that the larch easily survived the Norilsk stage of Sartan cooling (Fig. 58).

Second warming is recognized at the end of Eoholocene (Kind, 1373). It is fixed by a peat bog on the Malaya Kheta River and dated by 14C method at 8500 ± 200 yr B.P. (GIN-26). According to Khotinsky (1981), the Malaya Kheta peat bog reflects the forest phase and the main maximum of spruce in Siberia. In his opinion, the Boreal period was the warmest and the most humid in Siberia. The eastern North Siberian lowland (Kolshaya Balakhnya River area) is provided by earlier radiocarbon values, obtained on samples of peat and wood, namely, 9410 ± 70 yr B.P. (GIN-1322a), 3300 ± 100 yr B.P. (GIN-1322), 9280 ± 60 yr B.P. (GIN-1451), 3830 ± 100 yr B.P. (GIN-776). Peat bogs are situated there in sinks and lake depressions of the watershed area, 60-80 m high (Kind and Leonov, eds, 1982). The data led the author to the conclusion that intensive swamping and peat accumulation lasted almost 1000 years on East Taimyr. Thermokarsting and self-development of forest biocenoses may have conduced to the spreading of moss cover and an increase in area of bogs which existed along with forest communities.

A Mesoholocene (mid-Holocene) warming is beyond any doubt. Wood and peat dated by this time (climatic optimum) occur in the basins of Pyasina, Novaya, Boganida, Bolshaya Balahnya, Malakhai-Tari and other rivers. An unique accumulation of wood remains, formed by Middendorf (1860) and described by Tyulina (1937), occurs on a sand beach of the left bank of Novaya River, in its lower streams. Zh.M. Belorusova and N.V. Lovelius made a complete inventory of fossil wood in the "Tyulina fossil forest" which appears to occupy an area of 340 X 15 m. it includes 70 stumps, up to 50 cm in diameter; 22 trunks, up to 30 cm in diameter; 30 root systems, up to 50 cm in diameter, 102 long roots (up to 270 cm). A radiocarbon date of the death of one of the trees, 5970 ± 70 yr B.P., was obtained on wood of the root. At the present, larch grows as small beds in the vicinity of the "Tyulina fossil forest", and the diameter of trees does not exceed 7-10 cm at the level of root collar. In the Ary-Mas forest the diameter of some trees only reaches 20-30 cm.

North of the Ary-Mas area, fossil wood is exposed on the flood plain of the Zakharova Rassokha River, a left tributary of the Novaya River (Kultina *et al.*, 1974). A bed of wood remains, 0.15 cm thick, is made up of stumps and trunks of trees, bark, branches and overlain by a bed of peat, 0.55 cm thick. The stumps have well preserved main roots and fine rootlets. An age of the wood bed, 5180 ± 150 yr B.P., was obtained on one of larch roots. Mesoholocene wood remains were found both on the flood plain and at higher geomorphological levels of East Taimyr. Wood, dated at 6580 ± 100 yr B.P., was discovered on a lake terrace, 60 m high, in the Malaya Balakhnya River Basin (Kind and Leonov, eds, 1982).

Interesting information on the Holocene climatic optimum is provided by fossil peat bogs. They occur as small (0.2-0.7 m) beds of peat on flood-plain deposits or as peat deposits, up to 4-5 m thick, on even surfaces of high terraces or local divides. One of the peat bogs was studied by the author in the Ary-Mas area. The peat bog occupies a small flat depression in a poorly drained interfluve. Its surface is broken into hummocks by frost clefts, widened by thermokarst, nival and erosional processes. Peat in hummocks, up to 4 m high, is best exposed on southern slope aspects. Apparent thickness is about 2.5 m. The peat is characterized by a low degree of decomposition: separate mosses, sedge stems, twigs are distinctly seen in it. It was stated (Belorusova and Ukraintseva, 1980) that the formation of peat deposit continued for 1200 years, from 6695 ± 80 yr B.P. to 5495 ± 80 yr B.P.; the mean rate of peat accumulation was 1.5-1.6 mm per year.

A notion of flora, and the main stages of vegetational and climatic variations is provided by results of palynological analysis of samples, taken from the peat deposit (Fig. 59, Table 22). Spectra of the middle part of peat deposit marked an improvement of climatic conditions and the distribution of woody-plants. The phase of optimal climatic conditions is dated at 5860 ± 60 yr B.P. In spectra, dated by that time, the amount of pollen of tree species drastically increases, reaching a maximum of 12 % in spectrum of sample No. 6. Pollen of trees is dominated by pollen of Larix cf. gmelinii (6-12 %); pollen of the spruce (*Picea obovata*) is permanent but minor. The middle part of diagram (Fig. 59) displays a total decrease in pollen of *Betula* sp. ex sect. *Nanae* in the group of shrubs and low shrubs (no more than 25 %) and an increase in pollen of *Betula* sp. ex sect. *Betula* (up to 7 %) in the group of tree species. These data suggest the distribution of larch forests, containing spruce and birch and associate boreal elements of the undershrub-herb group of plants, over the area under study during that time interval (Table 22). During the climatic optimum of the Holocene (Atlantic time), the role of spruce in pnytocenosis markedly reduced, as compared to the final phases of Karginsky Interglacial; this is suggested by a decrease of its pollen in spectra, dated by that time. In the author's opinion, this was related to an increase in continentality of climate at that time interval because of an increase in area of land, as compared to Karginsky time.

Nikolskaya (1980) who studied a great quantity of sections of Holocene (alluvial, lacustrine-bog and bog) deposits on the central North Siberian lowland states that pollen and spores are absent in sediments of the first half of the Atlantic, whereas remains of trees, stumps and trunks of larch, dated by that time (Sulerzhitsky, 1976), are relatively common throughout the region. In the author's opinion, this can be explained by the circumstance that at the onset of the Atlantic the deposits were underflowed at rivers on low terraces and even in local watershed areas, and, as the result, pollen and spores were washed out of them.

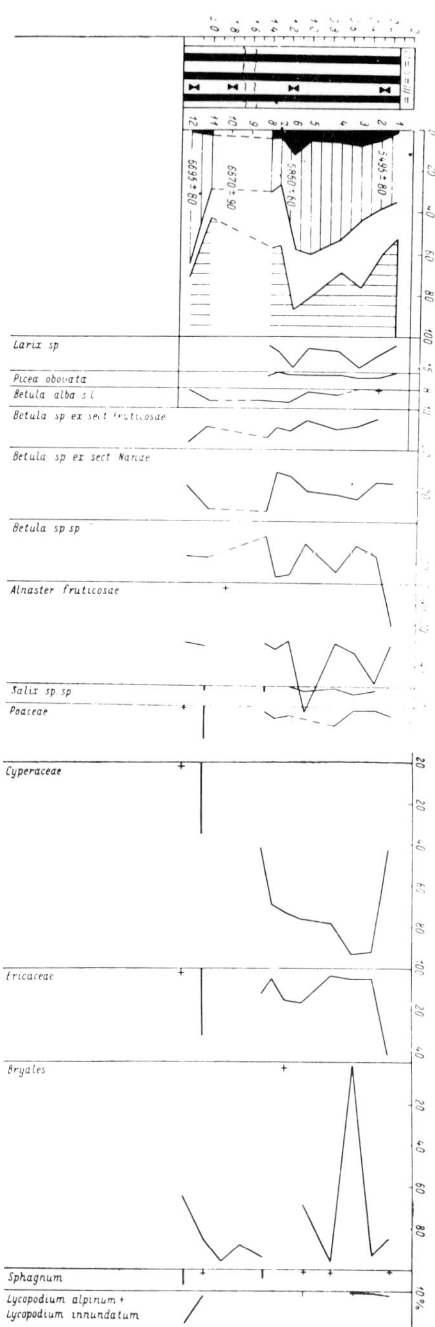

Figure 59. Spore-pollen diagram of outcrop No. 25, Ary-Mas, right bank of the Novaya River, East Taimyr. After Belorusova and Ukraintseva, 1980. (For legend see Figure 56).

Table 22

Results of palynological analysis of samples from Outcrop 25, Ary-Mas area

Sample #, bed #, depth (m), lithology Plant	Site, animal, time of death											
	Sample 1, Bed 1, 0.01-0.2, sod		Sample 2, Bed 2 , 0.4-0.6, peat 5495±80 yr B.P.		Sample 3, Bed 2, 0.4-0.6, peat		Sample 4, Bed 3, 0.6-0.8, peat		Sample 5, Bed 4, 1.1-1.2, peat		Sample 6, Bed 4, 1.1-1.2, peat	
1	2		3		4		5		6		7	
	abs.	%	abs.	%	abs.	%	abs.	%	abs.	%	abs.	%
Larix sp.	4	4.0	10	9.2	9	14.6	12	7.8	20	6.3	13	15.4
Picea obovata + *Picea* sp.		-	1	0.9	1	1.7	1	0.6	2	0.6	1	1.2
Pinus sibirica		-	2	1.8	1	1.7	1	0.6		-		-
Pinus sp.		-		-		-	1	0.6		-		-
Betula sp. (sect. *Betula*)		-	1	0.9		-	4	2.6	3	1.9	5	6.0
Betula sp. (sect. *Fruticosae*)		-	6	5.5	5	8.5	15	9.8	18	5.7	9	10.5
Betula exilis	3	3.0	3	2.7		-		-	26	8.0	11	12.9
Betula tundrarum		-		-	1	1.17		-	1	0.3		-

To be continued

Table 22 (cont)

1	2		3		4		5		6		7	
	abs.	%	abs.	%	abs.	%	abs.	%	abs.	%	abs.	%
Betula sp. (sect. *Nanae*)	13	12.7	14	12.8	13	22.0	33	12.7	36	11.3	22	25.6
Betula sp.	51	51.0	18	16.5	7	12.0	39	25.4	-		-	
Alnus glutinosa	-		-		-		2	1.3	-		3	3.6
A. fruticosa	31	30.3	52	47.9	20	34.0	44	29.0	209	63.3	21	24.7
Salix sp. sp.	-		2	1.8	2	3.4	1	0.6	4	2.5	-	
Poa sp.	-		1	1.5	-		-		-		-	
Poaceae	2	3.7	-		1	2.3	4	8.7	-		2	4.5
Cyperaceae	22	42.5	59	90.8	40	93.1	37	78.2	74	76.7	33	73.0
Urtica sp.	-		-		-		-		3	3.3	2	4.5
Chenopodium sp.	2	3.7	-		-		-		1	1.1	-	
Caryophyllaceae	1	1.9	-		-		-		-		-	
Parnaseia sp.	-		-		-		-		-		-	
Saxifraga sp.	-		1	1.5	-		3	6.6	1	1.1	1	2.3
Potentilla sp.	-		-		-		-		-		-	

To be continued

Table 22 (cont)

1	2		3		4		5		6		7	
	abs.	%	abs.	%	abs.	%	abs.	%	abs.	%	abs.	%
Rosaceae	1	1.9	-		-		-		-		-	
Chamaerion latifolium	-		-		-		-		-		-	
Cassiope tetragona	-		-		-		-		5	5.2	3	6.7
Ericaceae	22	42.5	3	4.7	2	4.6	2	4.3	11	11.5	4	9.0
Menyanthes trifoliata	1	1.9	-		-		-		-		-	
Pedicularis sp.	-		-		-		-		-		-	
Artemisia sp.	-		1	1.5	-		1	2.2	-		-	
Dicotyledoneae indet.	1	1.9	-		-		-		-		-	
Monocotyledoneae indet.	-		-		-		-		-		-	
Hepaticae	-		1	0.9	-		-		-		-	
Sphagnum sp.	2	1.5	-		-		1	1.1	1	1.2	-	
Dicranum sp.	-		-		-		-		3	5.0	-	
Bryales sp.	109	85.2	106	93.6	28		88	95.6	56	68.2	15	
Botrychium lunaria	-		-		-		-		-		-	
Polypodiaceae	-		-		-		-		-		-	
Equisetum sp.	15	11.7	4	4.6	2		3	3.3	20	24.4	5	

To be continued

Table 22 (cont)

1	2		3		4		5		6		7	
	abs.	%	abs.	%	abs.	%	abs.	%	abs.	%	abs.	%
Lycopodium clavatum	-		1	0.9	-		-		-		-	
L. innundatum	1	0.8	-		-		-		-		-	
Lycopodium sp.	-		-		-		-		-		-	
Total amount of pollen and spores	282		286		134		292		487		150	
Including those of:												
trees	4	1.4	14	4.9	11	8.2	19	6.5	25	5.1	19	12.5
shrubs and low shrubs	98	34.8	95	33.0	48	35.8	134	46.5	294	56.5	66	44.0
grasses	52	18.4	65	22.7	43	32.0	47	16.0	96	19.6	45	30.0
cryptogamous plants	128	45.4	112	39.4	32	24.0	92	31.0	82	18.8	20	13.5

To be continued

Table 22 (cont)

Sample #, bed #, depth (m), lithology Plant	Site, animal, time of death									
	Sample 7, Bed 4, 1.2-1.3, 5860±60 yr B.P.		Sample 8, Bed 4 , 1.5-1.7, peat		Sample10, Bed 6, 1.9-2.1, peat		Sample 11, Bed 6, 2.1-2.2, peat		Sample12, Bed 7, 2.2-2.35	
1	8		9		10		11		12	
	abs.	%	abs.	%	abs.	%	abs.	%	abs.	%
Larix sp.	8	7.6	6	4.5	-		2	1.4	1	0.8
Picea obovata + *Picea* sp.	-		2	1.5	-		2	1.4	-	
Pinus sibirica	1	0.9	-		-		1	0.7	1	0.8
Pinus sp.	-		1	0.7	-		1	0.7	-	
Betula sp. (sect. Betula)	7	6.0	8	6.0	-		9	6.3	1	0.8
Betula sp. (sect. *Fruticosae*)	11	9.5	19	14.2	-		13	9.2	-	
Betula exilis	12	10.4	6	4.5	-		16	1.2	-	
Betula tundrarum	1	0.9	1	0.7	-		-		1	0.8

To be continued

Table 22 (cont)

1	8		9		10		11		12	
	abs.	%	abs.	%	abs.	%	abs.	%	abs.	%
Betula sp. (sect. *Nanae*)	-		36	27.0	4		26	18.2	65	52.4
Betula sp.	31	27.0	10	7.5	8	26	26	18.2	22	17.4
Alnus glutinosa	1	0.9	-		-		-		-	
A. fruticosa	36	30.8	39	29.0	7		33	30.6	36	28.6
Salix sp.	7	6.0	4	3.0	-		3	2.1	-	
Poa sp.	-		-		-		-		-	
Poaceae	8	6.2	3	2.5	-		11	16.0	1	
Cyperaceae	89	68.9	49	41.1	-		24	34.0	1	
Urtica sp.	-		-		-		-		-	
Chenopodium sp.	-		-		-		-		-	
Caryophyllaceae	-		-		-		2	2.9	-	
Parnaseia sp.	1	0.8	-		-		-		-	
Saxifraga sp.	14	10.8	30	25.0	-		-		1	
Potentilla sp.	4	3.1	9	8.0	-		1	1.4	-	

To be continued

Table 22 (cont)

1	8		9		10		11		12	
	abs.	%	abs.	%	abs.	%	abs.	%	abs.	%
Rosaceae	-		-		-		-		-	
Chamaerion latifolium	1	0.8	-		-		-		-	
Cassiope tetragona	-		-		-		-		-	
Ericaceae	7	5.4	14	11.7		22	22	32.0	5	
Menyanthes trifoliata	-		-		-		3	2.2	3	2.2
Pedicularis sp.	-		-		-		1	1.4	-	
Artemisia sp.	-		-		-		-		1	
Dicotyledon indeter.	2	1.5	13	10.9	1		4	5.7	4	
Monocotyledon indeter.	3	2.5	-		-		1	1.4	1	
Hepaticae	-		-		-		-		1	1.8
Sphagnum sp.	-		7	3.7	-		2	1.4	4	7.2
Dicranum sp.	-		-		-		-		-	
Bryales sp.	179	94.2	168	87.9	100		129	84.9	35	64.0
Botrychium lunaria	-		-		-		1	0.7	-	
Polypodiaceae	-		-		-		-		2	3.6
Equisetum sp.	11	5.8	16	8.4	4	17	17	11.0	5	9.0

To be continued

Table 22 (cont)

1	8		9		10		11		12	
	abs.	%	abs.	%	abs.	%	abs.	%	abs.	%
Lycopodium clavatum	-		-		-		3	2.0	-	
L. innundatum	-		-		-		-		6	10.8
Lycopodium sp.	-		-		-		-		2	3.6
Total amount of pollen and spores	434		444		124		490		194	
Including those of:										
trees	17	3.9	17	3.9	-		15	3.0	2	1.3
shrubs and low shrubs	98	22.6	117	26.4	19		127	26.0	124	63.6
grasses	129	29.6	119	26.8	1		69	14.0	13	6.7
cryptogamous plants	130	43.9	191	42.9	104		152	57.0	55	28.4

A similar situation took place about 6600 yr B.P. in the peat deposit (Bogatyr River Basin, outcrop No. 25), studied by the author. The fact that the peat deposit was underflowed at that time is evidenced by a very low content of pollen and spores in peat and an obviously washed sand band in at a depth interval 2.0-1.4 m. Water, saturated with sand particles in suspension, may have washed through the peat deposit; as time result, the sand band formed at a depth of 1.6-1.7 m (Fig. 59) and pollen and spores were almost completely washed away. At the beginning; of Atlantic, similar situations were probably rather common over the vast territories of North Siberian lowland. In this context, the above-described peat bog which provided paleobotanical characteristics of much of Atlantic time is of particular paleobiogeographical importance; it can be treated as a type section for the central North Siberian lowland.

Palynological analysis of peats from outcrop on the Zakharov a Rassokha River, underlain by wood remains with an age of 5180±150 yr B.P., points to the existence of thinned-out birch-larch forests, containing spruce, and their replacement by open tundra or bog landscapes (Kultina *et al.*, 1974). A plan spectrum, obtained from a sample of peat, containing abundant wood remains, pollen of: woody plants (*Betula* sp. ex sect. *Betula, Picea obovata, Pinus sibirica, Larix* cf. *gmelenii*) accounts for 18%; shrubs and low shrubs 17%; herbs 50%; spores of cryptogamous plants (Bryophyta and Pteridophyta) total 15%; Pollen of herbs is dominated by that of sedges amounting to 57%; pollen of grasses and *Ericaceae* accounts for 17.5% and 9.5%, respectively. Scanty pollen grains of *Polygonum viviparum, Polemonium boreale, Potentilla* sp., *Chenopodium* sp.1, sp.2, and Asteraceae were noted. The group of spores is dominated by spores of true mosses, coming to 62%. Spores of bog moss and horse-tail occur in approximately equal proportions, accounting for 12% and 13.8%, respectively. Such a high percentage of sedge pollen and moss spores suggests not only a progressive inundation and swamping of the area, but ecological compatibility of forest communities and back bogs. The total composition of spore and pollen spectra which characterize the overlying peats are dominated by pollen of herbs (15-16%) and spores (27-71%). Pollen of woody and shrubby plants decreases to 15-21%. As before, pollen of herbaceous plants is dominated by sedges (34-37%), and the group of spores displays the greatest proportion of true mosses (52-91%). Spores of *Lycopodium annotinum, L. appressum,* and *Equisetum* sp., associate plants of boreal forests, are present in spectra throughout the section.

In the Pyasina River Basin, western Taimyr, peat deposits, dated by the Holocene climatic optimum, are as thick as 3-5 m. The study of one of the peat deposits, formed at the surface of a 20-25 m high terrace above the flood plain in the middle Pyasina River area (73°N) showed that the onset of its formation which is synchronous to the Holocene climatic optimum coincided with the distribution of forest-tundra landscapes, with forests containing spruce, cedar pine, large woody birch and larch, in the region. Light forests alternated with tundra-mire vegetation and alder brakes, existed along valleys of rivers and streams (Danilov *et al.*, 1971).

According to Zaklinskaya (1954), no forests were present on the northwestern coast of Taimyr, in the Mamonta River Basin (75°27′N) during the Holocene climatic optimum. Judging from analysis of recent and fossil flora of the area, Tikhomirov (1950a) assumed the presence of larch beds on well drained warmed slopes among herbaceous formations during the Holocene climatic optimum . On northeastern Taimyr, warming began to manifest as early as the Holocene (Berdovskaya *et al.*, 1970). However, in this area the warming maximum also fell on the epoch of climatic optimum. Last glaciers, survived in the mountainous area since Zyrianka time, disappeared at that time. Active swamping and peat accumulation, locally sufficiently intense, occurred in the river valleys.

At present, it is reliably established that during the Holocene climatic optimum the range of larch almost reached the present-day northern coast of Taimyr peninsula, as evidenced by its stumps and trunks, found in tundra and in deposits of river terraces (Miroshnikov, 1958), and by palynological data (Berdovskaya *et al.*, 1970). As far as the author knows, no specific assignment of these fossil remains has been determined. However, taking into account ecological features of larch, it was probably *Larix* cf. *gmelinii*.

The growth of woody plants and the formation of high moors in high latitudes are known to determine a stable (no less than 60 days) period when the ambient air temperature exceeds 10°C and the Sum of active temperatures is in excess of 30.0°. Judging from wood finds and palynological data, a deviation of the July isotherm from the present-day values for Taimyr was no less than 8-10°, and the mean temperature of the hottest month reached 10-12° at Cape Chelyuskin. Anomalously high summer temperatures, accompanied by excessive humidification during the vegetational period, were favorable for the development of forest and bog landscapes on Taimyr. During the maximal warm phase, the northern limit of forest reached the latitude of Lake Taimyr, and. the northern limit of woody plants was bounded by the continental coast. Larch forests are common in the continental sector; birch arid spruce forests are local and rare, respectively. As a rule, spruce grew in birch-larch forests, in river valleys; it did not form pure spruce forests as spruce is sensitive to rigorous continental climate and demands much of atmospheric humidification.

A maximum anomaly of summer temperatures fell on eastern Taimyr, including the Novaya River Basin and the lower Khatanga River (Belorusova *et al.*, 1987). Deviations of summer temperatures from the present-day values were the highest in the Arctic and subarctic regions; they were caused by the geographical position of the region in the intracontinental sector, low relief, protection against summer north winds by the Byrranga Mountains.

Climatic conditions deteriorated in the Neoholocene (late Holocene). Though the upper Holocene deposits contain abundant organic remains, they are essentially represented by vegetable detritus, less commonly, peat and rare wood. Two finds of wood were made by: L.D. Sulerzhitsky in the Boganida River Basin; they yielded two similar ^{14}C values, 1865±40 yr B.P.(GIN-817,

GIN-820). A third ^{14}C date, 2175±200 yr B.P. (IM-672), was obtained by V.V. Kostyukhevich in rootlets and twigs, taken by the author from deposits of a 6 m high flood plain of the Bolshaya Lesnaya Rassokha River, Novaya River Basin. Palynological analysis revealed poor spore and pollen spectra for the whole strata of deposits: rare pollen grains of trees[26] (*Larix* sp., *Picea obovata*, *Pinus sibirica*, *Betula* sp. (sect. *Betula*), *Tilia* sp.); shrubs and low shrubs (*Betula. exilis*, *Betulus* sp. ex sect *Nanae* (underdeveloped, fine pollen grains), *Alnus fruticosa* (primarily underdeveloped as well), *Salix* sp.; grasses and herbs (*Cyperaceae*, *Poaceae*, *Ranunculus* sp., *Polygonum bistorta*, *Stellaria* sp., *Caryophyllaceae*, *Ericaceae*, *Artemisia* sp., *Asteraceae*); sparse spores of mosses (*Bryales*, *Sphagnum*), common club-moss (*Lycopodium clavatum* and *Selaginella huperzia selago*), as well as horse-tail (*Equisetum* sp.). All the samples studied show abundant redeposited forms of Neogene-Mesozoic age. Data obtained give a notion of rewashing of sediments, accumulated at that time, rather than the character of vegetational cover.

Available data suggest that the second half of Holocene (about 4000 yr B.P.) witnessed forest degradation, caused by both cooling and spreading of moss cover owing to the self-development of forest formations.

At the Norilsk stage of Sartan cooling (10,200-11,300 yr B.P.), the vegetation zones shifted 100-300 km southward (Zubakov and Kind, 1974). A maximum shift of forest limit northward on Taimyr, as compared to its present position, is estimated at 500-600 km. Thus, the amount of shift measured almost 1000 km. Abundant wood remains, dated by ^{14}C, suggest that woody plants did not disappear completely during short cold phases of the late Pleistocene and Holocene. They survived as small forests which served as peculiar advanced posts for forest shift northward in warm phases. Most favorable conditions for forest refuges during the cryochrons of the late Pleistocene existed on eastern North Siberian lowland, in the site of Ary-Mas, the most northerly forest tract on the earth (72°30').

6.1.3. Indigirka-Krolyma area

The area embraces the basins of two largest rivers of the North East of Siberia, the Indigirka and the Kolyma with their numerous tributaries. Most of the area is mountainous. Two vast lowlands only, Indigirka and Kolyma, are located on the north. The area lies in the subpolar and arctic climatic zones. A major, mountainous part of the area belongs to the Yana-Kolyma botanicogeographical macroprovince, characterized by a prominent zoning of the northeastern continental type, manifested in an ascending succession of light monoedificator larch forests, (sub-alpine tundra) cenoses of mountain pine, and typical (alpine tundra) vegetation which occupy vast areas and give way to mountain-stony cold deserts in upper topographic parts (Shumilova, 1962). Its northern, plain part, recognized by L.V. Shumilova as a separate, North Yakutian, province of the Arctic macroprovince, lies in the tundra zone.

Main features of mountain relief of this vast area have undergone no fundamental changes during the last 50,000-40,000 yr B.P. Considerable changes took place in the Promorsky lowland area clue to fluctuations of sea level (Shilo *et al.*, 1983).

This area yielded most finds of specimens of the mammoth faunal complex, first of all, sufficiently well preserved frozen corpses of animals, their full skeletons or fragments (Tikhomirov, 1958; Sher, 1971; Shilo *et al.*, 1983). Of nine finds known at present, six were discovered in the basins of the two rivers (Fig. 1). Even this fact only suggests conditions which promoted, on the one hand, sudden deaths of the animals, and, on the other hand, rapid "conservation" of their corpses. Such conditions include, first of all, thermal ground subsidence and solifluction (Vereshchagin, 1979), as well as heavy swamping and an increase in area of lakes, very boggy ground, permafrost (Ukraintseva, 1979). A continuous existence of perennially frozen grounds on the Prirnorsky lowlands has taken place since the second part of the middle Pleistocene (Kaplina, 1979; Popov, 1982a, b). Fluctuations of geocryolithological conditions, related to climatic variations, have occurred there since. During cold (glacial) epochs, strata of sediments, ranging from 30 to 70-80 m in thickness, accumulated there and in adjacent areas; the whole thicknesses of strata were pierced by reticulated polygonal fissure ice; during warm epochs, ices thawed resulting in active thermokarsting. While radiocarbon method allowed recognition of cold epochs of Zyrianka (older than 50,000 yr B.P.) and Sartan (from 24,000 (23,000) to 12,000 yr B.P.) time, separated by a relatively warm Karginsky interval. Deposits of this, so-called "glacial complex" (Kaplina *et al.*, 1980; Eopov, 1982a, b) are adequately studied in palynological respect; this is especially true of deposits of warm rhythms, hence, they received comprehensive paleobotanical characteristics. Results of examinations of: (i) gastrointestinal contents of fossil herbivores which perished there in various stages of the late Pleistocene, and (ii) deposits which enclosed fossil remains, dated by ^{14}C method, are not only in good agreement with available data, but substantially supplement them thus allowing tracing of variations in the character of vegetation in the area.

In the lower Indigirka River Basin and in the area of Yana Indigirka interfluve, the onset of late Pleistocene was characterized by a wide distribution of forest vegetation of light forest vegetation, including larch, woody birch, dwarf birch and alder. The presence of spruce in phytocenoses is not excluded (Barkova 1969; 1970a, b). On the Kolyma lowland, the first warm phase, associated, with the beginning of late Pleistocene and coincided in time with the Kazantsevo Interglacial[27] of more westerly areas of Siberia, was represented, according to Giterman (1985), by larch-birch forest of the boreal taiga type; spruce as a small admixture may have been present in the forest. Taiga, somewhat resembling the present-day taiga of the Lena-Kirensk Region, where Dahurian larch was admixed with spruce, pine, Siberian larch and cedar pine, existed in the mountainous area, in intermountain basins and wide valleys. Mean annual temperature of about 1-4°C much exceeded that of the present. Mountain slopes were covered by forest-tundra, thickets of mountain pine and alpine tundra (Karavaev and Skryabin, 1971).

A fundamental change in vegetation occurred in the succeeding, Zyrianka, cold (glacial) interval which is believed to have lasted from 100,000 to 55,000 yr B.P. (Alekseev *et al.*, 1984). The change affected the then vegetational cover owing to cooling and alpine glaciation. Many species, first of all, tree species, which are more warm- and moisture-requiring, such as spruce, cedar pine, disappeared from the communities; species of plants which are less warm- and moisture-requiring, such as herbs, low shrubs and shrubs, became common. During the periods of maximal cooling, the predominant type of vegetation were various tundras, for example, different mountain tundras in mountain areas, and tundras, containing xerophytes. The part of woody plants was insignificant at that time; birch and larch only grew in the most suitable habitats. *Selaginella rupestris* was common in the vegetational cover (Giterman, 1963; (Giterman and Golubeva, 1965).

Available paleobotanical data, dated by ^{14}C, fix the development of forest vegetation and climatic warming in a time interval of 50,000-(25,000) 24,000 yr B.P. in the area. However, this warm interval, synchronous to the Karginsky Interglacial of more westerly regions, was not climatically uniform there, it consisted of three warm phases, separated by two phases of relative cooling.

Communities of forest formations were developed during the warm phases (early, Malaya Kheta, Lipovsko-Novosyelovskoe) both on the Primorsky lowland and in the mountainous part of the area (Table 23). During the optimal phases (early, Malaya Kheta warming), light coniferous forests, made up of larch with subordinate spruce and woody birch and with *Alnus fruticosa* and the mountain pine (*Pinus pumila*) in underwood, advanced to the area of present-day tundra. (Rybakova and Piruranova, 1983; Giterman, 1985). In the mountainous part of the area, forests included, in addition to larch, two species of spruce, (*Picea ajanensis, P. obovata*), the Siberian pine (*Pinus sibirica*), woody birches [*Betula* sp. (sect. *Costatae*), B. *pendula*]. *Nuphar pumila*, the spatterdock, continued to grow in dead lakes and bights (Ukraintseva (Kultina), 1977). During the Kirgilyakh (early) and Konoshel cold. phases, thinned-out larch forests with minor *Pinus pumila, Alnus fruticosa, Betula exilis* survived on the river flood plains and low hillsides only. Upslope they give way to forest-tundra and still higher to communities of the subgoletz (sub-alpine tundra) and goletz (alpine tundra) zones (Ukraintseva, 1982; Shilo *et al.*, 1983). Open and semi-open landscapes, various tundras, willow stands, dwarf shrubs became predominant. In the northern plain part of the area, cold phases of the Karginsky Interglacial were at first characterized by the predominance of mesophytic grass tundras, replaced by crypxerophilous associations which were dominated by grasses, wormwoods, and *Selaginella rupestris*.

A notion of the character of temperature and moisture conditions of the region 38,590-42,000 yr B.P. is provided by the main climatic characteristics, reconstructed for: (i) the end of Kirgilyakh (early) cooling in the Berelekh River

Table 23

Basic diagram of vegetational variations of the Kolyma River Basin in Late Pleistocene and Holocene time against the background of climatic variations

14C yr B.P.	Warmth — Cold	Kolyma River		
		Upper	Middle	Lower
		1	2	3
0				
1				
2				Forest-tundra, 3500±180 yr B.P. (IM-303)
3	Late Holocene cooling			
4				
5			Larch forests with birch, 6830±200 (MAG-374)		Larch forests with Birch
6	Optimum			
7				
8	L. Sanchugovka cooling Warm	5360±50 yr B.P. (MAG-441)		Birch forest-tundra, 8400±250 yr B.P.
9	Pit-Mgarkin cooling Warming			Low bush-moss grass tundra
10				
11				
12	Norilsk stage			
	Taimyr warming			
	Kokorevka warming			

To be continued

Table 23 (cont)

	1	2	3
13 14 15 Nyapan stage?			
Gydan stage	Wood, 19,650±200 yr B.P. (MAG-510)		
Lipovskp-Novosyelovskoe warming		Birch forests with larch and mountain pine, 24,550±260 yr B.P. (GIN-160)	
Konoshel cooling			Larch forests with alder, 29,600 ±1400 yr B.P.
Malaya Kheta warming (optimum)	Larch forests with alder and mountain pine		Light larch forests with mountain pine and alder, 36,900±500 (MGU-469), 37,600±1100 (MGU-468)
Early Cooling	Forest-tundra; bush and low bush tundras; communities of mountian steppe type; mountain tundras, 41000±900 yr B.P. (MAG)	Larch forests with birch; meadows along valleys, 44,000±3500 yr B.P.	Larch-birch forests, 42,800±400 yr B.P. (GIN-149)
Early warming			

Basin, upper Kolyma River; (ii) the onset of Malaya Kheta warming in the Shandrin River Basin, lower Indigirka River; and (iii) the optimum of Malaya Kheta warming in the Elga River Basin, Upper Indigirka River (Table 24). The sum of positive temperatures suggests that the warmth amount in the mountainous part during the Karginsky Interglacial optimum was twice that of the present; precipitation was also somewhat greater than at present with summer being the wettest season. All these factors taken together were responsible for a richer species composition of both belt-zonal vegetational communities and azonal communities, namely, meadows, meadowlike associations, meadow-steppes and steppes, riverside aquatic vegetation and the like .

The end of late Pleistocene (12,000-24,000 yr B.P.) was marked by climatic cooling, responsible for the emergence of glaciers in the mountainous parts of the area and in adjacent terrorories of the North East. Although glaciation embraced relatively small areas and showed spotted distribution and small average thickness (Glushkova, 1982), but, nevertheless, vegetation underwent substantial changes. Spore and pollen spectra of deposits of that time point to a primary distribution of various tundras, such as bush-moss, bush-grass, and stony tundras in the mountains; locally tundras gave way to forest-tundras and even larch forests. This is, in particular, evidenced by the date $19,650\pm200$ yr B.P., obtained on the wood of a larch, buried in slope wash at the level of Kirgilyakh Creek terrace I above the flood plain, upper Kolyma. River (Shilo *et al.*, 1983).

Severe cryogenic conditions which caused the formation of thick ground-ice wedges developed at that time on the Primorsky lowland (Velichko, 1973; Kaplina, 1979; Tomirdiaro, 1972, During the formation of lower strata of the glacial complex, the vegetational cover was dominated by tundras similar in character to recent typical tundras (southern type). Spore and pollen spectra, obtained by P.E. Giterman (Sher and Kaplina, eds, 1979) from deposits in the lower Kolyma River and dated at 14,000-15,000 yr B.P., suggest that during this time interval the eastern Primorsky lowland, including the areas of the recent open boreal taiga zone, was under Arctic tundra. Tundra-marking spectra are dominated by pollen of herbs and spores of true mosses; pollen of both trees and shrubs is absent. Similar spectra, characterized by a high content of pollen of herbaceous plants (grasses and forbs) and a practically complete absence of pollen of trees and shrubs, were recognized in deposits of the lower bed of a 10-12 m high Berelekh River terrace (lower Indigirka River) in the area of Berelekh mammoth cemetery. They formed 13,700-12,200 yr B.P. (Lozhkin, 1975, 1977). Both the data presented above. and data, obtained by the author and coworkers (Vereshchagin and Ukraintseva, 1985; Chapter 4.7) suggest that treeless landscapes also existed for a long time at the end of late Pleistocene in the northern Indigirka-Kolyma area; this was associated with rigorous climatic conditions. Nevertheless, a rather large population of mammoths and their associates, as well as primitive men who hunted them, lived there at that time.

Data available infer that first notable climate warming over the territory

Table 24

Degree of climatic variability in the Indigirka-Kolyma area in the late Pleistocene

Characteristic of main climatic elements	Area	Time, Thousand years B.P.			the present	Difference as compared to the present
		41000±900	48350±880	38590±1120		
1	2	3	4	5	6	7
	1	9.0	-	-	13.0	-4.0
July Temperature	2	-	12.0	-	8.0	+4.0
	3	-	-	17.0	11.0	+6.0
	1	-29.0	-	-	-39.0	-10.0
January temperature	2	-	-34.0	-	-31.0	+3.0
	3	-	-	-31.0	-48.0	-17.0
	1	-10.0	-	-	-13.0	-3.0
Mean annual temperature	2	-	-13.0	-	-15.0	-2.0
	3	-	-	-7.0	-21.0	-14.0
	1	680.0	-	-	1096.0	-416.0
Sum of temperatures above 0°C	2	-	936.0	-	513.0	+423.0
	3	-	-	1680.0	833.0	+847.0
	1	281	-	-	266	+15
Sum of annual precipitation, mm	2	-	277	-	185	+92
	3	-	-	377	288	+89
	1	132(47%)	-	-	186(69%)	-54
	2	-	146(53%)	-	96(52%)	+50

1 - middle Berelekh River, upper Kolyma River; 2 - middle Shandrin River, lower Indigirka river; 3 - middle Elga River, upper Indigirka River.

of Yakutia occurred at the close of late Pleistocene, about 11,800-12,000 yr B.P. At that time woody plants and shrubs settled in the Primorsky lowland area. However, data presented by the author suggest that the first late Pleistocene wave of warming, first established by Lozhkin (1977), was virtually the second wave for the lower Indigirka River. The first wave was equivalent in time to the earlier, Kokorevka, warming of Siberia (Kind., 1973) or the Bölling Interstadial of Western and Central Europe (12,200-13,000 yr B.P.). As early as that time, forests and light forests, composed of larch and minor woody birch, developed in the lower Indigirka River Basin. As shown above, this was recorded in spore and pollen spectra of the upper deposit sequence of the 10-12 m high Berelekh River terrace in the area of Berelekh mammoth cemetery (Table 18, Fig. 54; Vereshchagin and Ukraintseva, 1985), as well as in spectra of deposits on the right bank of Chroma River and in Mys-Khaya outcrop on The Yana River (Chapter 4.7).

Spore and pollen spectra which characterize the early Holocene deposits in the northern part of Bolshoi Lyakhovsky Island, Novosibirsky Archipelago (Ukraintseva, 1989), indicate that in a time interval $10,540 \pm 170$ to $10,080 \pm 210$ yr B.P. a continuum of vegetation from subarctic bushy tundra and polygonal bogs, with higher sites grown over with the low alder (*Alnus fruticosa*) and birch (*Betula exilis*), (Levaya Kutta River Basin), to forb-Sedge-grass tundras (lower Malaya Kutta River Basin) existed in the areas, presently, covered by arctic tundras (Aleksandrova, 1963). Some taxa which are absent in the present-day flora of Bolshoi Lyakhovsky Island were recognized by the author in early Holocene flora, reliably dated by geological-geomorphological evidence and radiocarbon analysis of organic remains (Ukraintseva et al., 1989). They are primarily represented by shrubs and undershrubs of *Alnus fruticosa, Betula exilis, Cassiope tetragona,* other specimens of *Ericaceae,* some herbaceous (*Thalictrum* sp., *Chamerion* sp., *Epilobium dahuricum, Polemonium boreale, Polemonium* sp. and *P.* cryptogamous (*Dryopteris* sp., *Polypodium* sp., *Lycopodium annotinum, Selaginella rupestris*) plants which are now common in the subarctic and alpine tundra zone, in forest-tundra, and in the boreal taiga zone. The amount of pollen of the alder (*Alnus fruticosa*) and the birch (*Betula exilis*) in spore and pollen spectra of outcrops Nos 533 and 509A (Fig.s 60, 61) suggests that the above species were common in the most northerly part of the island (74°N, 141°30′E) about 10,540-10,080 yr B.P. spectra of deposits of outcrop No. 509, situated almost 4 km north-east of outcrop No. 533, point that $10,540 \pm 170$ yr B.P. the limit of northern distribution of the plants on the island lay in the Malaya Kutta River Basin. As differentiated from coeval spectra of outcrops Nos 533 and 509A, a spectra of outcrop No. 509 pollen of alder and low birch is represented by scanty grains which were air-transported from the nearest sites in the Levaya Kutta River Basin where they grew. Thus, the composition of paleofloras and the character of vegetation, reconstructed for the Levaya Kutta River Basin, relatively high (more than 0.2 cm/yr) rates of peat bog accumulation 10,540-10,080 yr B.P., and its thickness (about 1m), sufficiently

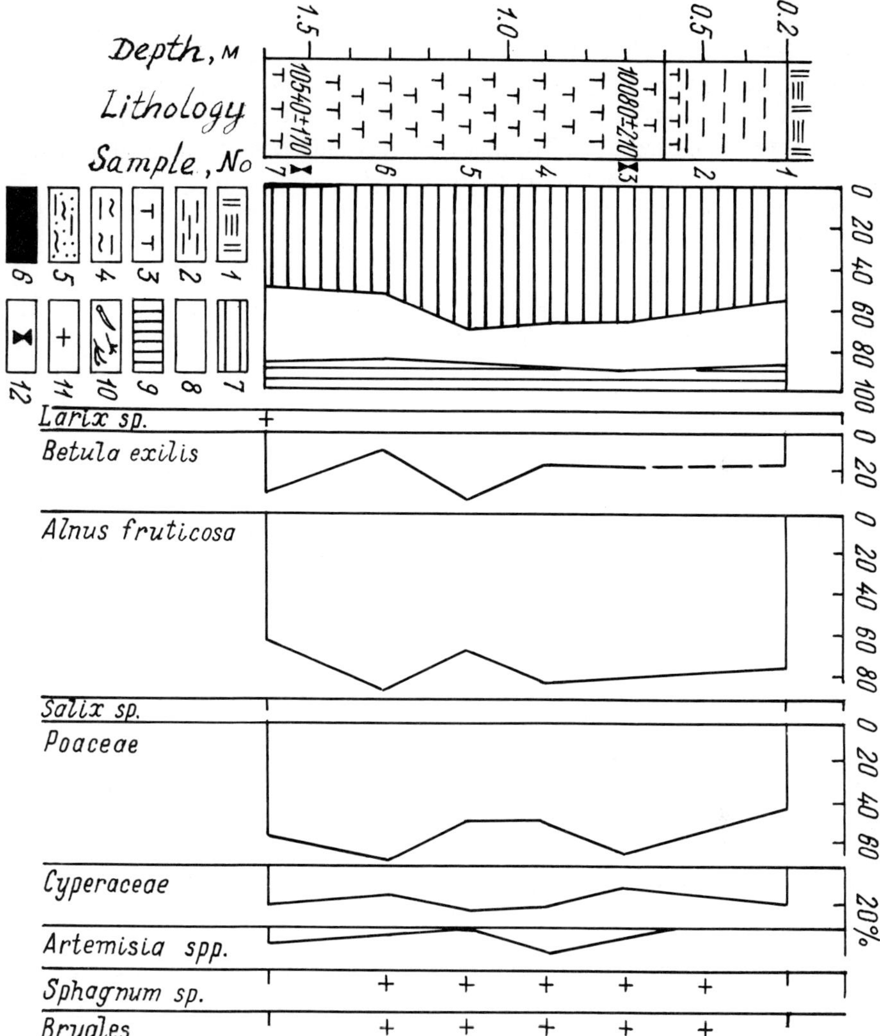

Figure 60. Spore and pollen diagram of early Holocene deposits of Levaya Kutta River terrace I above the flood plain Bolshoi Lyakhovsky Island, outcrop No. 533: 1) soil-vegetational layer, 2) silt, 3) peat, 4) loam, 5) sandy loam, 6) pollen of trees, 7) pollen of shrubs and low shrubs, 8) pollen of grasses and forbs, 9) spores of spore-bearing plants, 10) macroremains of plants, 11) taxon amounts less than 1%, and 12) the site of sampling for ^{14}C analysis.

great for the latitude (74°N), indicate a warmer climate than that of today. According to Aleksandrova (1963), the thickness of present-day peat bogs on Bolshoi Lyakhovsky Island does not exceed 15-22 cm.

On Kotelny Island, situated to the north (75-76°N, 137-145°E), the first climate warming, the beginning of peat bog formation, the appearance of low

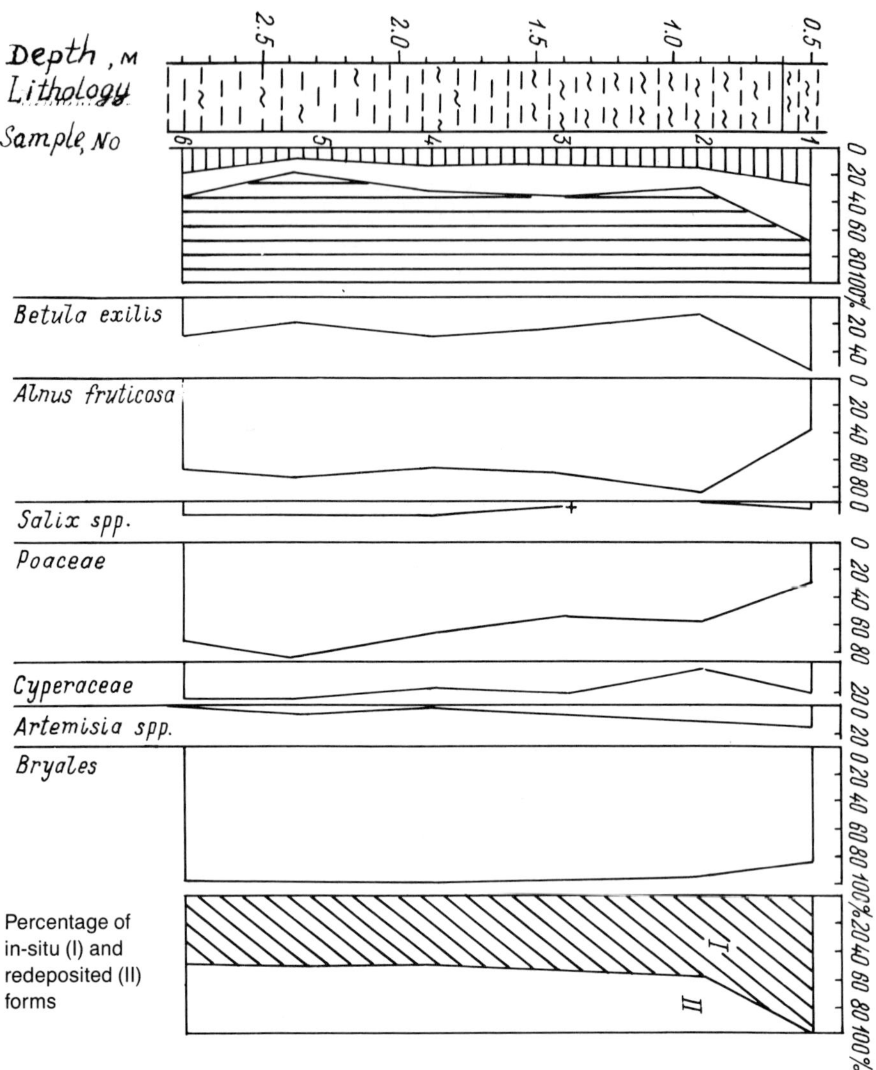

Figure 61.   Spore and pollen diagram of deposits on the left bank of Malaya Kutta River, Bolshoi Lyakhovsky Island, outcrop No. 509A. (For legend see Figure 60).

shrubs took place 12,200-10,000 yr B.P. (Makeev and Ponomareva, 1988). Warming reached a maximum 10,000-9,000 yr B.P. It was accompanied by a relatively fast growth of peat bogs (up to 0.5 cm/yr) and a wide distribution of shrubs and low shrubs, containing some elements of southern tundras.

The above new palynological data on the Novosibirsk Archipelago islands suggest that temperature conditions of this high latitude Arctic region about 10,000 yr B.P. were similar to those of areas of the recent distribution of subarctic bushy tundras in the inland part of the continent (Ukraintseva, 1988; Ukraintseva *et al.*, 1989).

The recurrence of cold at the very beginning of early Holocene (9,500-10,300 yr B.P.) led to thinning of forests, reduction of their areas, and increase of the role of various tundras in the vegetational cover of Primorsky lowland. According to data presented here, about 9480 ± 300 yr B.P. the vegetational cover in the mountainous part of the Indigirka River Basin was dominated by shrub and undershrub associations, including *Pinus pumila, Alnus fruticosa, Betula exilis* and others. However, the presence of larch and larch light forests at that time cannot be ruled out (Belorusova *et al.*, 1977).

A relatively appreciable climate warming set in Boreal time with its optimum occurring about 8500 yr B.P. when the range of woody birch and alder advanced to Kotelny Island (75-76°N, 137-145°E), i.e. the northern limit of woody birch was situated more than 600 km north of its present position, and Siberian spruce appeared on the southern margin of Indigirka-Kolyma lowland. Such environment survived to about 6500 yr B.P. The second half of the Atlantic was accompanied by cooling when landscape zoning acquired its present appearance. No further essential changes of landscapes occurred, although subboreal time may have witnessed warming. Landscapes of territory evolved following the "continental type," recognized for Siberia by Khotinsky (1977).

6.2. Main stages of the history of vegetation and Environment of Siberia and their significance in understanding the history of "mammoth" fauna

The above data on the history of vegetation and the character of climatic variations in three major key regions of Siberia which yielded most specimens of the "mammoth" fauna, dated by the method of ^{14}C analysis, and information, reported in literature on Kamchatka, Chukotka, the Penzhina Bay area (Boyarskaya and Malaeva, 1967; Veinbergs *et al.*, 1976a, b; Bespalyi *et al.*, 1982), the North East of siberia and Yakutia (Giterman, 1963, 1985; Giterman *et al.*, 1968; Tomskaya, 1981), Western Central, Southern Siberia, and the Transbaikalina area (Grichuk, 1961, 1973; Volkova, 1977; Volkova *et al.*, 1982,1984; Arkhipov *et al.*, 1977; Belova, 1983; Belov and Belova, 1984) outline two main stages in the post-Anthropogene history of vegetation and environment of Siberia, namely, late Pleistocene and Holocene stages.

Late Pleistocene stage (112,000-115,000(130,000) yr B.P.) is subdivided into four substages, characterized by alternations of interglacial (Kazantsevo, Karginsky) and glacial (Zyrianka, Sartan) epochs. Deposits of lower horizons of this interval yielded the following age values: $110,000\pm27,000$ yr B.P. for Western Siberia and $112,000\pm28,000$ yr B.P. for Eastern Siberia (Alekseev *et al.*, 1984).

Kazantsevo Interglacial (100,000-115,000 yr B.P.) is equivalent to the Sangomon Interglacial of North America, Mikulino Interglacial of Eastern Europe, Eemian and Ipswich of Northern Europe and England, respectively.

The notion of the character of geographical zones in the optimal phases of this interglacial can be gained from maps of vegetation and main components of landscapes, complied on the basis of numerous paleopalynological and paleobotanical data both for various area of Siberia (Giterman *et al.*, 1968; Giterman, 1985), and the whole USSR territory (Tikhomirov, 1962; Boyarskaya, 1965; Gerasimov and Velichko, 1984). They vividly show that during the Kazantsevo Interglacial optimum the vegetational cover of Northern Eurasia had a well developed, climatically related zonal structure with forest formations playing a leading role in the vegetation. As a whole, the then zonal structure resembled the present one. A main difference amounts to their present position. The vast territory of Siberia was dominated by boreal forests of different types. However, they substantially differed from recent forests of this type in composition (Boyarskaya, 1965). The most prominent were zones of: (i) broad-leaved forests in the European part of the USSR, in the Caucasus, and Far East; (ii) coniferous-broad-leaved forests in the north of the European part of the USSR; and (iii) coniferous forests with some broad-leaved elements in some areas of the south of Siberia and the Transbaikalian area. Zones of tundras, steppes, and deserts were subordinate. According to Tikhomirov (1962), hillocky tundras and tundra-marsh communities, analogous or close to the present southern tundras, were widely developed in the north of Eastern Siberia. According to Grichuk (1982), no typical tundra zone and steppe zone were present in the European territory. The forest vegetation was represented by two types: nemorse and boreal (primarily, birch forests of Scandinavia). The role of communities of azonal character (meadows, meadow-like communities, dwarf willow and birch formations of flood plains and creeks, bogs, riverside-aquatic vegetation) considerably reduced at that time due to a wider distribution of zonal communities of the forest type.

At that time, the most favorable conditions for the lie of big herbivorous mammals existed in the south of Central Siberia, in the north of Kazakhstan, in the south of East European Plain and the Danube Lowland, i.e. in the areas under steppes and forest-steppes.

Zyrianka (early Würm, early Valdai, early Wisconsin) Ice Age (55,000-100,000 yr B.P.) is considered as the time of maximal development of glacial processes not only in the northern but also southern areas of the region in question. On Taimyr, the Murukta (Zyrianka) glaciation was continental, and at its early, North Siberian stage the ice cover embraced the North Siberian lowland from the Yenisei River to the Popigai River mouth. It advanced onto the lowland from three centers, namely, the Northern (presumably, the Kara Sea shelf), the Putorana, and the Anabar centers. Coalescence of the ice covers occurred in the valley of Kheta and Khatanga Rivers. During the second, North Kokorevka stage, the ice covers of the northern and southern alimentation centers did not coalesce in the eastern and central part of North Siberian lowland (Kind and Leonov, eds, 1982). This is also suggested by data on the basins of Bolshaya Lesnaya Rassokha River, right tributary of the Khatanga River (Ukraintseva *et al.*, 1981). Just at that time, at the final stages of this ice age, mammoth penetrated to this part of Taimyr. One specimen died in the valley of Bolshaya Lesnaya Rassokha River mouth more than 53,000 yr B.P. (see Chapter 4.1). The glaciation was alpine in the mountains of the North East of Siberia (Glushkova, 1982; Shilo *et al.*, 1983, and others).

As early as the onset of Zyrianka time, a considerable degradation of forests took place and the then zonation of vegetational cover was disturbed. The forest zone was broken into separate "island" forests and open woodlands, alternated with the area of paludal tundras. The boundaries of tundra and forest-tundra moved far south. the steppe zone considerably reduced also. Much of the territory of Siberia was occupied by treeless landscapes, covered by periglacial vegetation (Boyarskaya, 1965; Giterman *et al.*, 1968). According to Giterman (1963, 1985) and Tomskaya (1981), tundra-steppes were widely distributed in northeastern Siberia and central Yakutia. Abundant woody plants are fixed at the early and final phases of glaciation in the territory of Western siberia. In the middle Ob River Basin, the growth of birch and coniferous forests and meadow steppes are assigned to the initial and the final phases of glaciation, respectively (Grichuk, 1961). At the cold maximum when the climate became drier, the presence of mesophytes and tundra elements (*Betula exilis, Alnus fruticosa, Ericales,* and others) reduced and the role of xerophytes increased. At that time, grass-wormwood and wormwood-forb associations containing oraches and ephedra were abundant in various areas, and the above associations including *Selaginella rupestris* were common in East Siberia and Chukotka (Giterman *et al.*, 1968 and others).

From the climatic point of view, the Zyrianka Ice Age was moister and warmer as compared to the succeeding Sartan Ice Age. According to Belova (1983), in the periglacial zone the sum of positive temperatures totalled 1000°C, mean annual precipitation 250-300 mm, the duration of frostless period 45-50 days, mean temperature of January -35°C, July 10-15°C. In large depressions of the Baikalian type, mean temperatures of July were as high as 10-13°C, mean annual precipitation 600-800 mm. the above characteristics are believed not to rule out the growth of larch forests and light forests in the intermontane basins in the south of Siberia.

Karginsky Interglacial (25,000-50,000 yr B.P.). Unlike the earlier, Kazantsevo Interglacial, the Karginsky Interglacial was characterized by essential climatic variations; warm phases (early, Malaya Kheta, i.e. optimum, Lipovsko-Novosyelovskoe warm phases) gave way to cold phases (early, Kirgilyakh, Konoshchel), traced over the whole of Siberia and correlated with appropriate phases of the middle Würm, Valdai in Europe and the middle Wisconsin in North America (Kind, 1974). In the optimal phases (35,000-42,000 yr B.P.) most of Siberia was forested. Tundra and forest-tundra may have formed a narrow zone, primarily, on the Primorsky lowlands. Open Larch forests with undergrowth, composed of mountain pine, were distributed on Chukotka; bushy tundra occupied some localities only (Boyarskaya, 1980). Richer vegetation, as compared to that of the present, covered the areas adjacent to the Penzhina Bay. Even as late as the close of optimal phase (34,720±560 yr B.P.) the landscapes were dominated by birch forests (*Betula ermanii*), similar to the recent birch forests of Kamchatka Peninsula. A zone of mountain pine was also better developed than that of today. The presence of larch forests and sphagnum bogs cannot be ruled out (Bespalyi et al., 1982). As early as the very beginning of optimal phases (40,350±880 yr B.P.), highly paludal forests and light forests, composed of larix Gmelinn, alternated with sites, occupied by tundra of various types in the lower Indigirka river Basin (70° N) (Solonevich *et al.*, 1977; Gorlova, 1982; Chapter 4.3, present work). Larch-birch forests, similar to boreal light taiga with dwarf birch, alder and mountain pine growing in underwood, existed in optimum on the Kolyma lowland. Sphagnum and moss bogs were widely distributed (Giterman, 1985). Farther south, in the mountain part of the Kolyma River Basin, in the area where the Kirgilyakh mammoth was found, similar forests took place 38,000±880 yr B.P. (Shilo *et al.*, 1983). West of the area, in the upper Indigirka River Basin (64°30′N), larch, the main forest-forming species, admixed with birch (*Betula pendula, Betula* sp. (sect. *Costatae*), spruce (*Picea ajanensis, P. obovata*), Siberian pine (*Pinus sibirica*), alder (*Alnus hirsuta*); juniper, mountain pine, dwarf alder and birch, willow of different types and other bushes grew in under growth. Communities of azonal character such as meadows, steppe meadows, aggregations of steppe-like character and riverside-aquatic aggregations which served as pastures for herbivores developed under proper edaphic conditions (Ukraintseva (Kultina), 1977; Chapter 4.4, present work). Along with the larch forests, steppe associations, namely grass-forb and wormwood-forb associations were common in central Yakutia, particularly in the Lena-Amga Interfluve (Tomskaya, 1981). Coniferous-broad-leaved and birch forests existed in the territory of Far East (Giterman *et al.*, 1968). In the valleys of large tributaries of the Amur River, ranges of broad-leaved species, such as elm and lime, existed considerably north than at present and reached the south boundaries of Yakutia and, possibly, more northerly areas. This is suggested by pollen of elm and lime found in food remains of the Selerikhan fossil horse, and by pollen of elm in food remains of the bison. These new data are in good agreement with those on the south of Western and Central Siberia, as well as on the Transbaikalina area. In the south of Central Siberia, a notable warming is fixed 50,000-33,000 yr B.P. On the left bank of lower Tungusska River, the vegetation of early Karginsky time was represented by pine-larch forests including dwarf alder and birch. The valley complex included spruce. Mid-taiga forests of similar composition were spread as far as the middle Lena river Basin. Somewhat further south (in the basin of the upper Surinda and Verkhnii Chunk rivers), the vegetation was made up of pine forests which contained spruce, silver fir larch and broad-leaved species (elm, lime, hazel-nut). Pine-birch grass forests were common in the region stretching from the sublatitudinal stream of Angara River southward to the Irkut-Cheremekha Valley. The tree layer was composed of elm, oak, hazel-nut. Pine-larch and birch-pine forests with some subordinate broad-leaved species and with participation of steppe elements occurred in the east of the region, in the western Transbaikalian area and, particularly, in the depressions of the baikalian type. On the south, these forests containing some steppe elements, completely gave way to steppes and even semi-desert aggregations (Belova, 1983). In the south of Western Siberia, in the Tunkin Depression, 26,250±300 yr B.P. the hillsides were covered by spruce-cedar taiga, admixed with silver fir; lime, oak, hazel-nut appeared downslope. Hemlock was minor. the depression bottom was occupied by pine forests with subordinate broad-leaved species, combined with steppes and elm groves (Belov and Belova, 1984). On Taimyr, according to Nikolskaya (1980), during the optimal phase the larch forests, including woody birch and spruce, reached the right bank of Bolshaya Balakhnya River (73° 30′N). About 35,000 yr B.P. monodominate larch forests existed even farther south, in the Novaya river basin (Ukrainsteva *et al.*, 1981). In the north of Western Siberian lowland, the dark coniferous forests in places gave way to birch and larch forests; the part of Novosibirsk and Omsk, the forests changed to forest-steppe and, farther south, steppe. In the Karginsky Interglacial optimum, the boundaries of forest-steppe and steppe were close to their present limits (Giterman *et al.*, 1968).

At that period, the European part of USSR was also dominated by forests. Tundra and forest-tundra existed only in the areas adjacent to the North Sea; coniferous forests with subordinate birch were ubiquitous in the Pechora River Basin, i.e. in the areas now covered by forest-tundra; coniferous forests embraced the Northwestern and Central areas.

Thus, paleobotanical data on various areas of Siberia and the European part of USSR suggests a much warmer climate in the period of Karginsky warm phase, especially in its optimum in Siberia, than that over the European part of the continent; this may have been related to the then position of geomagnetic pole mainly in the northwestern Pacific and in the north of Central Siberia. Although the Karginsky warm phase embraced the circumpolar regions - the arctic, subarctic and middle latitudes of the Northern Hemisphere, its scale and results were different in various regions.

In North America, a warm phase equivalent to the Karginsky warm phase in Eurasia, has not been so marked. Because of a smaller size of the continent, as compared to Eurasia, the North American glacier proved to have been more resistant and survived in the Hudson Bay area during the whole Pleistocene and the commencement of Holocene. Therefore, no warm phase similar to Plum-Point has been recognized in Canada.

Sartan Ice Age (12,000(13,000) - 23,000(25,000) yr B.P.). the final stage of Pleistocene was parked by the most essential cooling, related to the Last (Sartan, late valdai, late Wisconsin) Glaciation in all glaciated areas. Numerous radiocarbon dates for both various regions of Siberia and the whole of Northern Hemisphere indicate that the cooling initiated 22,000-23,000 yr B.P.; in the middle and northern latitudes of Northern Eurasia and North America, it reached a maximum 18,000-20,000 yr B.P. (Velichko, 1973; Avenarius *et al.*, 1978; Gerasimov and Velichko, 1984). According to Van Campo (1984), in Western

Europe the cooling maximum occurred 15,000 yr B.P. The cooling was so severe that the Arctic, the Pacific, and the Atlantic were ice-covered. This caused a profound reorganization of the zonal structure of vegetation zones, formed earlier. At that time, most of extratropical Eurasia was covered by open, treeless landscapes, which were named the hyper-zone by Velichko (1973). Small localities of light coniferous and birch forests were situated farther south only (Gerasimov and Velichko, 1984). Arctic deserts existed in the north of Eurasia, on the shelf free of sea water (Gorodkov, 1944); they gave way to tundras of different type to the south and to periglacial steppes farther south (at the latitude of recent forest-steppe and steppe) with tundra-steppes occurring in the transitional area only (Lavrenko, 1981). At that time, ranges of tundra and steppe animals were contiguous or even overlapped each other. Steppe animals considerably widened their ranges northwards. All the above-given coenoses are marked by palynological spectra which reflect the treeless and almost shrubless nature of vegetational cover. The existence of coenoses which produced the above spectra raises no doubt, but their taxonomic composition at present is a matter of speculation. The author holds the opinion, stated by Ager (1982), that such pollen spectra can represent types of vegetation which both have recent equivalents (for example, polar deserts, alpine grass tundra) and have no equivalents (arctic steppes or tundra-steppe). The author believes that specific definitions of pollen and spores only will enable us to solve this complicated problem which is of prime scientific and applied importance.

Avenarius and co-workers (1978) hold another opinion on the character of zonation in the late Pleistocene cooling. They believe that the vast USSR territory contained three geographical belts, namely, arctic, subarctic, and temperate belts with the boundaries highly shifted to the south as compared to their present positions. It is natural that the boundaries of geographical zones were also shifted southwards. Over the vast territories of the north of Eurasia, the zone of arctic deserts was developed and the tundra zone essentially extended; a common cooling led to the suppression of forest vegetation; however, the forest zone did not disappear completely in the territory of Russian Plain and Western Siberia.

As shown above, a substantial climatic change occurred 12,000 yr B.P. which was marked by a reduction of the area of periglacial landscapes. The time when these peculiar landscapes disappeared completely, giving way to forests, is estimated at 10,200-10,300 yr B.P. and regarded as the onset of Holocene (Fig. 63). According to Neishtadt and Steklov (1982), it is precisely this time that was transitional between the late Glaciation and the Holocene.

Holocene stage. Though the Holocene spanned not a long geological time period (past 10,000-12,000 yr B.P.), it witnessed notable climatic variations, as suggested by the history of vegetation in Northern Eurasia at that time interval (Neishtadt, 1957; Khotinsky, 1977, 1981), as well as new data on Taimyr and the Kolyma-Indigirka area (Chapter 6.1.2 and 6.1.3). Three warmest stages, coinciding with three thermal maxima of global scale, are recognized in the north

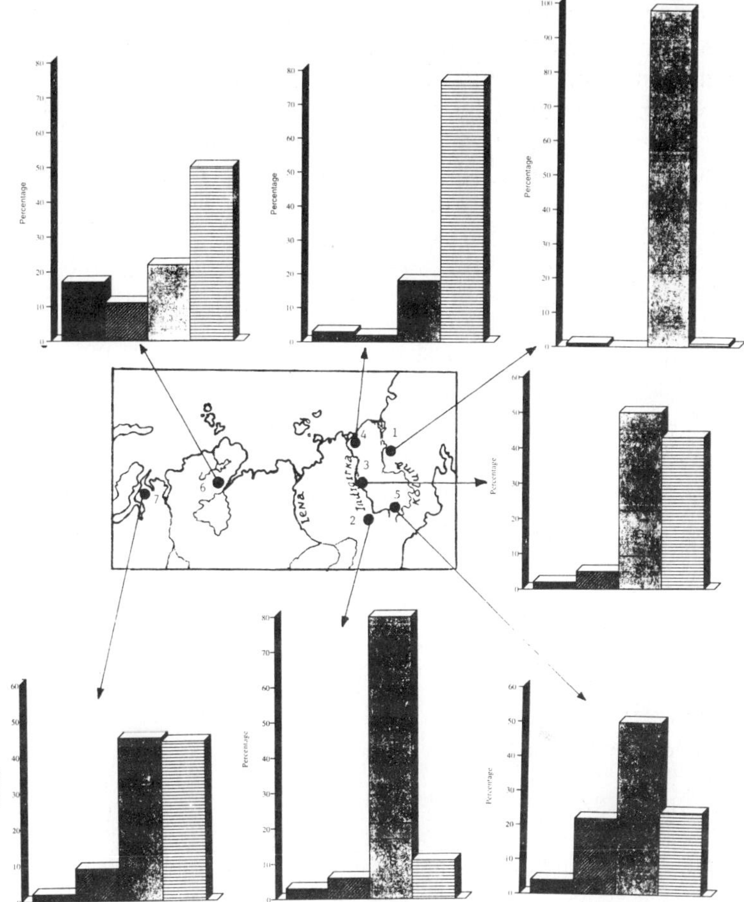

Figure 62. Percentage of main plant groups in the contents of gastrointestinal tracts of fossil animals of Siberia from palynological data: 1) Berezovka mammoth (Tikhomirov and Kupriyanova, 1954), 2) Selerikhan horse, 3) Mylakhchin bison, 4) Shandrin mammoth, 5) Kirgilyakh (Magadan) mammoth, 6) Vereshchagin mammoth, 7) Yuribei mammoth. Designations: (S) sedges, (G) grasses, (F) forbs. (For other designations see Figure 30).

of Eurasia (Khotinsky, 1981). Ample paleobotanical data indicate that the first warming in post-glacial time is fixed 8500 yr B.P. Two other maxima took place in the middle Holocene, namely, in Atlantic (5000 yr B.P.) and subboreal time (3500 yr B.P.). The warming of Atlantic time was maximal: the northern limit of forests lay then near the coast of the northern seas. However, results of palynological studies of lacustrine deposits and some peat bogs, formed in some regions of the North East of Siberia (Vereshchagin and Ukraintseva, 1985), some islands of the Novosibirsk Archipelago (Makeev and Ponomareva, 1988; Ukraintseva, 1988, Ukraintseva *et al.*, 1989), Alaska and Canada (Ritchie *et al.*,

1983; Ritchie, 1984, *et al.*, 1989; Matthews,1989) suggest that the first warm wave occurred in early Holocene, about 10,000 yr B.P. Calculations made by Milankovitch for the period indicate that the influx of solar radiation in high latitudes of the Northern Hemisphere could be 9-10% higher than that of today (cited from Ritchie *et al.*, 1983). This early Holocene warming is evidenced by a small paleoforest, represented by seven larch stumps buried alive; the forest was found by seven larch stumps buried alive; the forest was found by the author on Taimyr (Ukraintseva, 1985). As shown above, one of the larches died about 10,500±500 yr B.P., i.e. in the time interval intermediate between the late Pleistocene and the Holocene or at the very beginning of early Holocene. Thus, these fossil trees represent the first evidence in the world of the existence of larch forests north of the present forest limit not only on Taimyr but also at high latitudes of the entire Arctic in early Holocene time.

The principal conclusions drawn from the above paleobotanical and paleogeographical data can be summarized as follows:

1. Two main stages - late Pleistocene and Holocene - are traced in the history of vegetation and environment in various areas of Siberia in the late Anthropogene (Quaternary). The main tendency of the late Pleistocene stage is towards cyclic climatic variations with alternating glacial (cold) and interglacial (warm) ages. Representatives of the "mammoth" faunal complex which had evolved morpho-functional adaptations to the life in dry and cold climate had to exist under conditions of these cyclic variations and related reorganizations in the character of climate, flora and vegetation, and, hence, diet.

2. The vegetational cover of interglacial (Kazantsevo, Karginsky) ages had a zonal, climatically related, structure, dominated by forest formations. An increase in afforestation and swamping during the interglacial ages led to a dramatic reduction of treeless associations of both zonal (tundra, steppes) and intrazonal (various meadows, steppe meadows, riverside shrub formations) character. The reduction of suitable areas for herbivores dwelling resulted in breaking up populations into smaller ones which had doubtless vital consequences.

3. Important natural consistent patterns of the warm ages became evident. The Kazantsevo (Mikulino) Interglacial was warmer over the whole of the European part of the continent. Unlike the Kazantsevo Interglacial, the later, Karginsky Interglacial was: (i) noted for essential climatic variations: warm phases (early, Malaya Kheta (optimum), Lipovsko-Novosyelovskoe) gave way to relatively cold phases (early, Kirgilyakh, Konoshchel), traced over the whole of Siberia and correlated with proper phases of middle Würm, Valdai in Europe and middle Wisconsin in North America; (ii) warmer in Siberia, as compared to the European part of the continent. This phenomenon of metachronism in the evolution of environment of Eurasia, marked by a number of scientists, was first supported by the results of examination of food remains of the fossil herbivorous mammals and deposits containing faunal remains.

4. The Zyrianka and Sartan Ice Ages were characterized by open treeless

landscapes (Arctic meadows, various tundras, periglacial steppes); this led to the expansion of ranges of animals, first of all, large herbivores. During these periods, vast territories of Eurasia and North America were occupied by their large continuous intrabreeding populations. However, the woody vegetation did not disappear completely in the cold epochs. In the valleys of large rivers and in other favorable refugiums, the vegetation remained as small forests and individual trees which served as peculiar outposts of forest advance northward in the warm ages.

5. A substantial climatic change took place about 12,000 yr B.P., manifested in a reduction of periglacial landscapes. The time when these peculiar landscapes entirely vanished, again giving way to forests, is estimated at 10,200-10,300 yr B.P. and taken as the onset of Holocene by most of investigators. The Holocene stage is noted for a gradual emplacement of the present-day vegetational cover and present-day latitudinal zonation.

6. The above given paleobiological and paleogeographical data are the most convincing evidence of possible emergence of new global, climatically related landscape reorganizations in the future which will no doubt produce a pronounced effect on human environment over vast territories.

Chapter 7

What, in fact, Did Cause the Extinction of Mammoths?
"The problem of extinction of alive creatures is as old as the hills, but recently it has assumed a special character."

(Vereshchagin, 1979, p. 168)

The Tertiary is generally called the "era of mammals" since they dominated on land throughout the period. However, in the late Pleistocene this era (epoch) came to an end, or the beginning of an end: mammals began to die out on a mass scale. The extinction embraced groups of different taxonomic rank, starting from separate species giant caribou (*Megaloceros gigantheus*) to superspecies groups (saber-toothed cats, giant armadillo, mastodons, mammoths, woolly rhinoceros, North American wild horses, and others). The question of the causes of a considerable change in fauna of middle latitudes in the Pleistocene and the extinction of both some indicator species and entire groups of animals on a mass scale has been widely discussed since the end of the past century. A thorough review of the notions concerning the causes of extinction was first made in this country by Pavlova (1924) and later by Davitashvili (1969) and Vereshchagin (1979), Vereshchagin and Baryshnikov (1985), hence they will not be reviewed here. It should be noted that most investigators have recognized the influence of climatic variations on the extinction of certain groups of organisms. In so doing, some authors (I.D. Chersky, V.I. Gromov, among others) have related the extinction of some species of large herbivorous mammals to climate cooling in the Anthropogene, (Quaternary) whereas others (Vangengeim, 1961; Garrutt, 1965; Tomirdiaro, 1972, 1980; Velichko, 1973; Vereshchagin, 1977, 1979; Kuzmina, 1977; Ukraintseva, 1979, among others) relate it to warming. Some of writers believe that the main cause of extinction of large herbivorous mammals was associated with the destructive activity of man (Budyko, 1967; Martin, 1973, cited from Guthrie, 1976; Martin, 1982; Agenbroad, 1982, 1984a; Liu Dongxin and Li Xingguo, 1982, 1984 among others). A contrary opinion was advocated by Trofimov (1955), Flerov (1955, 1965, 1979), Butzer (1964), Velichko (1973), Guthrie (1976, 1982), Shilo and co-workers (1983) among others. There has been a claim for Holocene survival on Wrangel Island (Vartanyan *et al.*, 1993)

222

In order to understand objective reasons of the extinction of certain organisms, we should be aware of their life, their ecology. Scientists have made sure long ago that "it is impossible to study animals without understanding conditions under which they lived and biological environment which some way exerts influence on changes or organs and can cause their further evolution or death" (Pavlova, 1924, p. 14).

As shown in Chapter 4, the study of gastrointestinal contents of herbivores is of particular interest for reconstructing habitats of fossil animals. Of no less value is the study of paleocoprolites and paleoexcrements of some animals (Martin, 1961; Spaulding and Martin, 1979; Hansen, 1980; Agenbroad, 1984; Agenbroad *et al.*, 1984, among others). Although results of the studies provide a static pattern of landscape environment, fixed at the moment of the death of a given specimen (Shilo *et al.*, 1983), nevertheless, such finds and studies thereof can contribute towards the understanding of the habitat of certain fossil organisms and their ecology. In order to gain a more penetrating insight into environment of not separate organisms, but that of populations and species[28], we must understand the character of global environmental variations in time and space.

Current ideas of environment of a fauna of megamammals in the Anthropogene were generalized in this country by Velichko (1973), Vereshchagin (1979), and Shilo and coworkers (1983). It should be noted that discussion of the problem of the causes of an essential change in animal kingdom at the end of Pleistocene is confined, as a rule, to qualitative considerations which are difficult to prove or disprove without attempts of quantitative treatment of the problem. The present level of our awareness of environments of some specimens of the "mammoth" faunal complex, understanding of natural process pattern in the Pleistocene and its structure (Velichko, 1973, 1981) allow us to tackle the problem of an essential change in animal world at the close of Pleistocene and the problem of the causes of extinction of some cold-resistant animal species, including such an indicator species as mammoth, on quantitative ground. Being aware that quantitative methods are not the aim, but only the tool of study which provides, according to L.D. Armand, reliability, validity and demonstrativeness thereof, the author proposes to assume the following criteria as quantitative to tackle the problem at the present stage:

I. The dates of death of animals, obtained by the method of radiocarbon analysis of forage mass, muscle and bone tissues of the animals which perished under these or those circumstances. Percentages of main groups of plants, namely, trees, shrubs and undershrubs, herbs, cryptogamous plants in the composition of forage massof gastrointestinal tracts of the fossil animals and in the composition of palynological spectra synchronous to the animal life.

II. Duration of cold (glacial) and warm (interglacial) rhythms (epochs) and their relation in the Pleistocene. It should be noted that in tackling the problem of the causes of extinction, the size of animal populations is one of the most important criteria, along with the above-enumerated ones, however, at the present stage the author has restricted herself in considering this criteria to its qualitative aspect. Let's dwell at length on all the above criteria.

III. [14]C values of the death of some specimens of the "mammoth" faunal complex and their analysis[29]. Analysis of the dates of death of some specimens of the "mammoth" faunal complex following ^{14}C method (Heinz and Garutt, 1964; Arslanov, 1982; Arslanov and Chernov,1977; Arslanov *et al.*, 1980, 1981, 1982; Evseev *et al.*, 1982; Lozhkin, 1977; Makeev *et al.*, 1979; Shilo and Lozhkin, 1981; Shilo *et al.*, 1983) indicates that:

1. Time interval 55,000-(8000) 9000 yr B.P. is the period when animals of the "mammoth" faunal complex were widely developed in Siberia and their populations were abundant as suggested by numerous remains, complete or almost complete skeletons or their fragments (skulls, tusks, teeth), complete or almost complete corpses of the animals (Chersky, 1891; Pavlova, 1906, 1910; Popov, 1950; Flerov, 1955, 1965; Gromov, 1950; Sher, 1971; Vereshchagin, 1975, 1977, 1979; Kuzmina, 1977, among others).

2. Time interval (47,000) 45,000-30,000 yr B.P. is the period of death of specimens of the "mammoth" faunal complex in Siberia on a mass scale, i.e. the most difficult period for the mammoths and some of their associates (Ukraintseva, 1973, 1981); this is evidenced by complete or almost complete corpses of the animals and their skeletons, recently discovered in Siberia. Of the nine animals whose frozen carcasses and skeletons were found in the basins of Kolyma and Indigirka rivers, on Taimyr and Gydan Peninsulas, six perished in different stages of this time interval (Fig. 1, Table 19) equivalent, according to Kind (1973), to the Karginsky Interglacial of Siberia. These finds suggest that because of global warming in this period the masses of animals such as mammoths, bisons, musk oxen and other mammals, adjusted to live under cold climatic conditions, rushed for the northern and northeastern regions of Siberia. At that time they were highly increasing in number in Siberia, but their mortality was greatly increasing too owing to abiotic (in increase in swamping and areal extents of lakes, theormokarst topography, excessive bogginess of grounds near waterbodies) and biotic (less nutritious and less rich forage) factors (Ukraintseva, 1979). Owing to the above circumstances, by the end of the period, i.e. about 25,000 yr B.P. the quantities of populations of mammoths and woolly rhinoceros, i.e. the most narrowly specialized animals of the "glacial" faunal complex had become so low that, in fact, it was in this period that they proved to be on the verge of extinction. A great number of mammoths which once lived and died in various areas of Siberia is vividly demonstrated by little known data, presented by Middendorf (1860), on the amount of ivory of mammoth's tusks thrown on the market every year. According to Middendorf (1860), about 40,000 pounds of fossil ivory, i.e. at least around a hundred individual mammoth's tusks were annually delivered from North Siberia. Hence, during two centuries of our contact with North Siberia, no less than 20,000 mammoths took part in provision of the market with ivory. However, Middendorf (1860) had doubts as to this estimation and proposed to double it proceeding from the assumption that stories about finds of mammoth's tusks weighing 120 pounds

(about 200 kg) had been grossly exaggerated and this could be the weight of both tusks as suggested by specimens demonstrated at the Moscow Museum.

3. Time interval (23,000) 24,000-13,000 yr B.P., the period of pre-Sartan cooling and Sartan Glaciation, is indicated by a few ^{14}C values, obtained on separate skeletal fragments of specimens of the "mammoth" faunal complex which died during the period; this probably points to a reduction of the complex, in number, in that time interval. Over 21,000 yr B.P., small populations of mammoths may have dwelled in the Lena River Basin and, in particular, on the Molodo River, a left tributary of Lena, as evidenced by a jaw and two teeth of a mammoth of later type, found in the area (the fragments were identified by E.G. Vangengeim, (1977), the animal died about 21,260±310 yr B.P. About 20,000 yr B.P., mammoths reached the islands of Severnaya Zemlya Archipelago (Makeev et al., 1979). This is indicated by fragments of their skeletons found on Oktyabrskoi Revolyutsii Island (80°N). The fragments yielded the following absolute values: 19,600±330 yr (on a tusk); 19,970±110 yr (on a tooth); 11,500±60 yr (on a tusk). According to the authors, "all the four dates undoubtedly indicate that the mammoths inhabited the island at the end of late Pleistocene, i.e. at the time when the north of Europe, including not only land but also the Barents Sea shelf, was covered by thick ice shields" (Makeev et al., 1979, p. 422). This was caused by a global cooling with maximum occurring 18,000 yr B.P. (Avenarius et al., 1978). Cooling and glaciation, in turn, gave an impetus to the retreat of the World Ocean which reached a value of about 100 m in the time interval 15,000-20,000 yr B.P., thus increasing the area of land in the north of Siberia to 1000-2000 km as compared to the present-day area. At that period the islands of northern water basins were joined to the continent, and the shelf free of sea water was inhabited by pioneer but probably highly productive herbaceous (sedge-grass or grass-sedge-forb) communities which served as pastures for representatives of the "mammoth" faunal complex.

Only the most northerly peripheral parts of these vast dry areas of the shelf represented "arctic desert plains" (Avenarius et al., 1978, p. 33). Ice shields which covered the Severnaya Zemlya islands may have embraced lesser areas than those of today, otherwise, for lack of forage, the mammoths did not rush there at that time (Makeev et al., 1979). In the light of these new findings, the scheme of paleogeographic landscapes of the epoch 18,000 yr B.P. in the territory of USSR is, in principle, correct, but it requires more accurate definitions on the northern margins of the region. About 14,000 yr B.P., a rather large population of mammoths and their associates may have existed on the lower Indigirka River as suggested by the Berelekh mammoth cemetery, one of the largest in Siberia (Vereshchagin, 1971, 1975, 1977; Vereshchagin and Ukraintseva, 1985; Baryshnikov et al., 1977; Zherekhova, 1977). [14]C dates, (13,700±400 yr Lozhkin, 1977 and 12,240±160 yr LU-149, Arslanov et al., 1980) were obtained on the remains of 2 of the 140 mammoths dwelling there.

4. Time interval 13,000-(10,000) 9,000 yr B.P. is the last fatal period of existence of specimens of the genus *Mammuthus* in the territory of Siberia. A

new global warming about 13,000 yr B.P. turned out to be disastrous for the last, few surviving specimens. It was the time of extinction of the last representatives of the genus in Siberia, such as the mammoth on the Berelekh River, lower Indigirka River, which died about $12,240\pm160$ yr B.P.; the mammoth on the Mamonta River, Taimyr, which perished about $11,450\pm450$ yr B.P.; one of the "last of the Mohicans" of the genus may have been the Yuribei mammoth which perished on Gydan Peninsula on the middle Yuribei River about $10,000\pm70$ yr B.P.

As yet no discoveries dated at the Holocene climatic optimum have been reported from Siberia, thus the available information suggests that the mammoths had not survived to the Holocene climatic optimum in North Siberia. Hence, dated finds available at present indicate that the extinction of mammoths in Siberia falls in the period of about 10,000 yr B.P. At the same period, representatives of the genus (*Mammuthus columbi* and *M. primigenius*) died out in the New World (Agenbroad, 1984). Analysis of the dates of death of the specimens of the "mammoth" faunal complex points on the following consistent pattern which is crucial for paleogeographical reconstructions and stratigraphic correlations, namely, ^{14}C values, obtained on the entire corpses and complete skeletons generally suggest that the animals died during warm intervals of the Pleistocene and at the onset of Holocene (see Fig. 63); however, ^{14}C dates, obtained on some fragments (tusks, teeth, separate bones) of the skeleton point that the animals died at cold stages (intervals) of the Pleistocene. From this it is inferred that conditions propitious for burial of even such large animals as mammoth, rhinoceros, bison existed in Siberia during warm epochs (stages) of the Pleistocene (Ukraintseva, 1979, 1981). Conditions include increased swamps and an increase in lake areas, very boggy grounds, permafrost which occurred at a greater depth as compared to that of the present, particularly near banks of rivers, lakes, on bog margins which increased bogginess and, hence, the risk of fatal accidents for animals. Therefore during warm periods, summer mortality of the animals was extremely high which resulted in a sharp decrease of their populations. Permafrost has prevented decomposition and thus preserved the frozen carcasses of animals, which perished in summer seasons of Pleistocene warm intervals up to the present. On the contrary, cold epochs (intervals) of the Pleistocene were unsuitable for the burial of the animals which perished under some circumstances, as their corpses remained on the surface and they were eaten by predators or subjected to decomposition. In this context, living conditions of the animals which existed in cold epochs of the Pleistocene can be characterized only on the basis of paleobotanical study of enclosing deposits, the redeposition of which is ruled out and the position *in situ* is established. In this respect, of great paleogeographical significance are remains of the mammoth buried in Zyrianka deposits of Bolshaya Lesnaya Rassokha River terrace III above the flood plain, in southeastern Taimyr; the animal died more than 53,170 years B.P. (see Chapter 4.1).

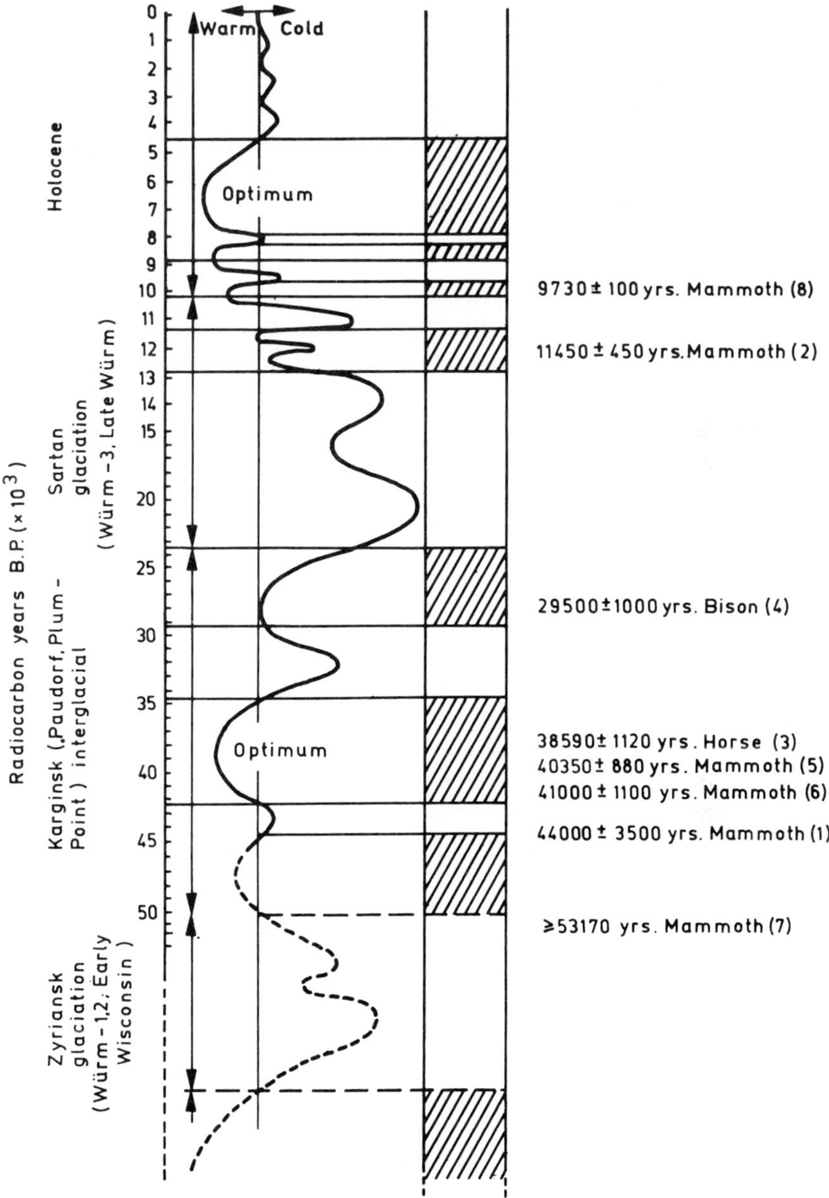

Figure 63.   Scheme displaying the causes of relatively fast extinction of mammoth and some of its "associates"[X]. Warm intervals are shown by hachures; ^{14}C dates of death of some representatives of the "mammoth" faunal complex are designated by numbers. The localities with dates of discovery are: 1) Berezovka mammoth, 1900; 2) Taimyr mammoth, 1948; 3) Selerikhan horse, 1968; 4) Mylakhchin bison, 1971; 5) Shandrin mammoth, 1971; 7) Kirgilyakh mammoth ("Dima"), 1977; 8) Vereshchagin's mammoth, 1978; 9) Yuribei mammoth, 1979.

II. Feed composition

Table 20 gives the plant composition revealed from integrated studies of gastrointestinal contents of some large herbivores which perished in various areas of Siberia during the time interval 53,000-10,000 yr B.P. A total list contains 221 taxa of flowering and cryptogamic plants with identifiable species, genus and family. The data presented in Table 20 indicate that the composition of plants, determined in each case, varies, thus reflecting, on the one hand, feed ration patterns of the animals, caused by their selectivity for these or those fodders, and, on the other hand, floristic and vegetational patterns of the areas where the animals died. This is convincingly illustrated by the following examples. 264 species of flowering plants were determined in a flora occurring near the burial site of the fossil horse (Sokolova, 1977). The area is one of the most vital for horse breeding and outdoor wintering in Yakutia. Forage plants of the Yakut horse number near 80-100 species with 40 of them only representing valuable feed. The 40 species form four groups, namely, grasses, sedges, forbs, shrubs and low shrubs (*Chosenia*, willow). Of 230 species of plants, represented in the flora of the Bikada station on Taimyr where the musk ox has been reacclimatized since the 1970's, 107 species (47 % of floristic composition) serve as feed for the animals (Rapota,1981) with plants of four families, such as grasses, sedges, willows and legumes showing the highest forage values. Willow, cotton grass, grass, sedge, dryad, forb form forage of the musk ox in summer. In winter, the animals mainly eat *Carex bigelowii* ssp. *arctisibirica*, *Eriophorum polystachion*, *Salix pulchra*, and *Arctagrostis* and *latifolia* (Rapota, 1981).

Comparison of the frequency of occurrences of macroremains of plants, their pollen and spores in forage mass of each of fossil animals, with no allowance for their abundance (Table 20, Chapter 5), shows that the palynological method provided the most complete taxonomic composition of the plants; as shown above, this is associated with resistance of pollen and spore exine to the action of ferments of gastric auice. Calculations of the frequency of occurrence of macroremains, pollen and spores in per cent indicate that the percentages of herbs, shrubs and undershrubs, determined in forage mass from macroremains (Solonevich and Vikhireva-Vasilkova, 1977) and pollen (Ukraintseva, 1977) are very similar (Fig. 30); this reflects a true role of these plant groups in feeding of the animals. it should be noted just here that percentages of main fodder plant groups were determined by the investigators independently and they became known only when published.

Despite some diversity of the plants, determined in the gastrointestinal contents of each of fossil animals, they basically lived on herbs and mosses in the areas where the latter were abundant in the vegetational cover. Palynological spectra of their forage mass are dominated by pollen and spores of these plants (Fig. 62). More or less thin twigs and sprigs of shrubs and undershrubs, lower branches of trees were also used by the animals, especially in autumn-winter-spring seasons, therefore pollen of shrubs and undershrubs accounts for 6-14 % in spectra of forage mass[30], whereas pollen of trees does not exceed 2 %.

A notion of the role and percentages of main fodder plants such as grasses, sedges, forbs, shrubs and undershrubs in feed of the animals can be provided by percentages of pollen and spores of the above plants in forage mass (Fig. 62).[31] As histograms presented indicate, the percentages vary, what is quite natural, as they reflect not only the role of above-mentioned plant groups in feed ration of the animals, but also patterns of vegetational communities which served as pastures to them. Abundant macroremains of mosses and their spores in the gastrointestinal tract of the Shandrin mammoth (Solonevich *et al.*, 1977) and bison (Ukraintseva *et al.*, 1978) suggest that these plants which are of low nutritious value and, in fact, ballast feed (Koshkina, 1961) were, nevertheless, eaten by the animals as the plants were abundant in the vegetational cover of the areas where the animals lived and died. This inference is confirmed by results of studies of macro-and microremains of forage mass of the Yuribei mammoth, obtained as the result of differentiated study of the contents of different parts of the gastrointestinal tract. According to Stanishcheva (1982), the bulk of the sample of the upper part of its intestine was composed of sedges and grasses. In addition, analysis yielded 20 pieces of terminal sprigs, leaves and caluptras of true mosses. Stanishcheva (1982) identified remains of leaflets and sprig apexes of two or three species of true mosses31 in the sample, taken from the colon. The author studied the samples, taken from four sections of the gastrointestinal tract of the Yuribei mammoth. The study showed that the percentages of pollen and spores, and, hence, producing plants vary in different parts of the animal's intestine (Table 14). The spectrum of the sample from the stomach contained equal proportions of herb pollen and moss spores (45-3% and 44.0%, respectively); pollen of shrubs and low shrubs accounts for 8.7%. The group of herbaceous plants is dominated by pollen of grasses (67.9%); pollen of sedges comes to 27.0%; pollen of forbs is represented by rare grains of Valeriana capitata, *Artemisia* sp., *Caryophyllaceae*. Spectra of the samples, taken from the stomach and the rectum are similar in composition; they are dominated by spores of true mosses (56.0% and 63.8% respectively). Pollen of herbs accounts for 20.1% and 16.0%, respectively. The samples taken from the stomach and rectum are dominated by pollen of grasses and sedges, respectively. Only the spectrum of the sample, taken from the colon is dominated by pollen of herbs, accounting for 62.9%. Spores of true mosses and pollen of shrubs and undershrubs amount to 24.0% and 9.5%, respectively. The group of herbs is dominated by pollen of grasses (87.3%), pollen of sedges comes to only 5.8%. Scanty pollen grains of *Dryas punctata* and some specimens of herbs such as *Ranunculus* sp., *Valeriana capitata, Artemisia* sp. were encountered. The abundance of spores of true mosses in three of the four samples in question (Table 14) and their macroremains suggest a great proportion of mosses in the vegetational cover in the area where the animal perished; the mammoth ate them together with sedges, grasses and forbs, the latter being rather poor in species composition.

A summary of the plants, recognized as the result of integrated study of forage masses of the fossil animals, and percentages of main plant groups, revealed in forage masses from palynological data (Fig. 62). They suggest that the animals which died in various areas of Siberia in warm intervals of the upper Pleistocene (horse, bison, Shandrin mammoth) and at the beginning of Holocene (Yuribei mammoth) mainly ate plants of wet and swampy habitats, i.e. sedges, cotton grasses, true and peat mosses. They significantly differ from plants of dry habitats and meadow forbs in the content of main nutrient components (protein, albumens, fats) and mineral composition (calcium, potassium, phosphorus and others) (Larin, 1958). True mosses are known to be mass forage but they are of low nutritious value (Koshkina, 1961). The protein content is one of main nutrient index of feed (Andreev, ed, 1974). Sedges and grasses whose remains dominated in the gastrointestinal tracts of the Shandrin mammoth, the Yuribei mammoth, and the bison are almost equal in nutritious value and content of mineral elements. Sedges of wet habitats are considerably inferior in nutrient content and mineral composition, in particular, calcium content, to sedges of dry (desert) and mountain habitats (Larin, 1958). The cotton grass (*Eriophorum vaginatum*), one of the most abundant and significant spring fodders of reindeer, contains 0.33% calcium and 2.70% potash black ash only, whereas *Cladonia alpestris*, the main winter feed of the animal, contains 0.13% and 0.54%, respectively (Egorov, 1960). *Arctophilla fulva* is one of the most nutritious food plants in the subarctic zone (Aleksandrova *et al.*, 1964; Shvedchikov, 1975). it is very nutritious when young: at the beginning of growth, the content of protein and albumen is its leaves ranges according to ambient conditions within 18.7-23.9% and 17.7-18.9%, respectively. At that time, protein digestibility is also high: the amount of digestible protein varies from 131 to 196 g/kg and from 68 to 106 g/kg in leaves and stems, respectively. Arctophilla shows a relatively high ash content as well. Phosphorus and potassium are rather high in its composition, particularly in summer. On the contrary, calcium is low at the beginning of growing period and in autumn-winter season. In most cases, silica is high, especially in samples taken in winter and autumn.

The significance of calcium, potassium, phosphorus and other elements for large herbivorous mammals is unquestionable. A low supply of the elements in the organisms leads to a decrease in birth rate and frequent loss of animals in winter. The seasonal dynamics of nutrients and mineral elements is believed to be of great importance for the extinct animals such as mammoth, rhinoceros and others, and recent large herbivores. The content of protein and albumen is known to drastically decrease in pasture forage in autumn-winter-spring seasons (Andreev, ed., 1974). In the upper Yana area, the content of protein decreases in main types of forage herbs in winter as compared to summer: grasses and sedges show a 3.5-6.9 and 2.6-6.4 times decrease, respectively. In the middle Kolyma area, the content of protein and albumen in plants decreases by winter to a lesser degree: grasses and sedges have the content of protein and albumen less than 2.4-3.5 and 2.4-5.2 times their contents in summer. Variations in protein content is accompanied by seasonal variations in its qualitative composition. for instance, the total content of nine essential amino acids (with the exception of triptophane) in forage herbs, studied in the upper Yana and middle Kolyma areas ranges from 19 to 37 g/kg air-dried matter and from 7 to 11 g/kg dry matter in summer and winter, respectively. A sharp drop in protein-albumen nutritiousness of winter forage affects albumen exchange and productivity of the animals, in particular, horses.

In estimating the role of the feed composition to large herbivores, much though should be given to calcium and phosphorus as they are vital for normal bone formation and development of animals. Legumes, forbs, willows and other woody and shrubby plants are known to be relatively rich in calcium and phosphorus. The highest contents of calcium and magnesium in *Arctophilla* fall in winter; the elements are contained in woody forage, composed of branches with well preserved leaves, mainly eaten by the animals in autumn-winter-early spring seasons. Winter samples show a 3 - 5 times increase in calcium as compared to summer samples. In most cases, silica is higher, especially in winter and autumn samples. Sedges are characterized by normal ash but higher and medium silica contents.

Cotton grasses are almost similar to sedges in content of calcium, magnesium and phosphorus with a few exceptions. The fact that sedges are richer in silica as compared to cotton grasses should be stressed here as it is essential for inferences regarding the understanding of the ecology of habitats of specimens of the "mammoth" faunal complex.

At the initial period of vegetation, the forb group shows a low silica absorption; later it increases and reaches a maximum in autumn-winter seasons. Similar variations are observed in calcium, magnesium, and total ash contents (Andreev, ed., 1974). Swamp horsetail displays an increase in separate ash elements and ash content in winter period; ash is 1.5-2, magnesium 2.2, calcium 7-8, silica 5 times their contents in early summer. In general, swamp horsetail shows an extremely high ash content and high magnesium accumulation even in winter.

It should be noted that silica and its compounds have been regarded as inert matter for a long time. Until recently, the role of silica in vital activity has been ignored. It was known only that its content in living matter amounts to about 109 tons and its concentration varies for some reasons in different parts of the organisms from tens to hundreds times. For instance, physicians noted approximately 50 times increase in silica content at fractures in men and animals. Moreover, it has been recorded long ago that monkeys which grew bald in winter greedily eat clay and their wool becomes thick and bright. Clay is rich in silica. Besides, there is evidence that in remote regions of Brazil, people are not ill with cancer, and the evil of the century -cardiovascular disease is almost unknown there. This is explained by an extremely high content of silica in soil and water of local springs (Anonymous, 1979).

The scientists of the Institute of Organic Chemistry, Siberian Branch of the USSR Academy of Sciences, Irkutsk, in cooperation with researchers studying organic synthesis have recently determined biological activity of organic silicon compounds: it has been proven that silica and its compounds stimulate growth, strengthen cell membranes to such a degree that they form a barrier to some harmful substances and even to proboscis of insects (Anonymous, 1979). Significant in this respect are two experiments carried out at the Institute: silica was excluded from the feed ration of newly hatched chickens and they developed to almost bare weaklings. Upon addition of only 0.003 % of silicium to feed the chickens got stronger and their growth rate increased considerably. In another experiment: scientists kept guinea-pigs under conditions conductive to baldness, they began to add one of preparations containing organic silicon compounds to feed. As a result, wool started growing at such a rate that the guinea-pigs merely turned into hairy balls and their offspring preserved the quality, i.e. they were hairy as well.

The above data on the role of organic silicon compounds in vital activity and on the high content of silica shown by some plants, particularly specimens of forbs, sedges and some other plants in autumn-winter seasons elucidate the role of autumn-winter plants in feeding ration of representatives of the "glacial" fauna which have thick and long hair coats with undercoats, composed of fine but relatively long hair, namely, mammoth, bison, musk ox, woolly rhinoceros, yak and others. Plants rich in calcium, potassium, phosphorus, and, especially, silica which stimulates the growth of hair were undoubtedly more important in feed rations of the animals for maintenance of their normal vital activity and the formation of "warm coats". If low contents of protein, carbohydrates and fats in autumn-winter feed could be compensated by their higher consumption, summer forages could not meet the requirements of the organisms of these large mammals in vital mineral elements such as calcium, phosphorus, potassium, and, especially, silica. This suggests that at cold stages of the Pleistocene when representatives of the "mammoth" fauna were common throughout extratropical Eurasia, the seasonal dynamics of nutrients and mineral substances in feed composition somewhat differed from that at warm (interglacial) stages, but it was suitable for large animals such as mammoth, bison, yak, and others which required not only great amounts of feed[32], but feed should be of high nutritious value. Naturally, during cold stages (rhythms) of the Pleistocene with their long and cold winters, the feed rations of the animals were dominated by winter plants which are known to be of sufficiently high nutritious value (Davydova, 1937, 1951, cited from Sheludyakova, 1961 ; Sheludyakova, 1961; Andreev, ed., 1974). L. N. Davydova who thoroughly studied habitats of horses wintering in steppe showed that feeding quality of winter plants is not inferior to hay. According to Sheludyakova (1961), meadow forbs left unmown are poorer in albuminous substances, nevertheless, they contain, on average, 7.82 % protein, 44.5 % nitrogen-free extractive substances, 29.2 % cellulose, 6.5 % ash; variegated horsetail contains up to 17.0 % protein. Multiyear observations led L.N. Sheludyakova to the conclusion that dry autumn and early frosts are beneficial to horse wintering outdoors, as in this case forbs, caught by early frost when they are green, do not lose nutritious value throughout the winter. In the case of a long and wet autumn, forbs turn yellow and lose their feeding quality. It was found that horses wintering in steppe do not lose in weight, but remain in a good state. L.N. Sheludyakova believes that outdoor wintering makes the horse strong and capable of great endurance.

III. Duration of cold (glacial) and warm (interglacial) epochs (rhythms), their relation and role in the life of mammals.

A climatic fluctuation towards cooling, followed by glaciations in the Pleistocene, induced the emergence of cold-enduring types of organisms, both plants and animals, including such a cold-enduring stenoecic species as a mammoth. Analysis of the character of global climatic and environmental variations in the Pleistocene (Gerasimov, 1961; Markov *et al.*, 1968; Arkhipov, 1971, 1983; Velichko, 1973, 1981; Kind, 1973, 1974; Zubakov and Borzenkova, 1983, among others) leads to an inference of great-significance that the epoch of mammoth, i.e. the time of existence of the mammoth as a genus. Other specimens of the "mammoth" faunal complex were characterized by a cyclic, oscillating character of environmental variations, namely, alternations of cold (glacial) and warm (interglacial) epochs (rhythms). Data recently obtained on the duration and relation between cold and warm epochs (rhythms) in the Pleistocene (Velichko,1973, 1981) are of fundamental importance for understanding of the causes of a relatively fast extinction of some cold-enduring species of the animals, including such a representing species of the "glacial" fauna as the woolly mammoth, in the Pleistocene. On the basis of study of rates of loess accumulation which represents peculiar "watch" of cold epochs, A.A. Velichko arrived at the conclusion that the duration of separate cold epochs did not exceed 50,000 yr. The total duration of cold epochs (glaciations) ranging from Oka to Valdai time is estimated at 500,000 yr. "The Pleistocene whose salient features are believed to nave been glacial epochs, turned out to have been, in fact, an essentially warm "interglacial" period. Over two-thirds of its entire time duration falls on interglacial conditions. Cold epochs are thought to represent relatively short pulses against the background of general warm conditions which dominated in the Pleistocene" (Velichko, 1981, p. 229).

The conclusion made by A.A. Velichko that cold epochs (rhythms) of the Pleistocene were much shorter than warm epochs and close to each other in duration is in line with data, obtained by Emiliani (1966, cited from Velichko, 1981) from the study of deep-sea sediments. Calculations carried out by Emiliani indicate that the Riss Epoch covered all in all 25,000 yr (between 100,000 and 125,000 yr B.P.); the Mindel and Gcold epochs continued about 30,000 yr each (between 180,000 and 210,000 yr B.P., and 265,000 and 290,000 yr B.P., respectively). The data further "narrow" the cold epochs and further emphasize the conclusion made by A.A. Velichko from the study of loess. Noteworthy that even at the coldest stage of Pleistocene (W Valdai epochs), warm rhythms lasted longer than cold rhythms, for instance, three warm rhythms (early warming, Malaya Kheta warming, Lipovsko-Novosyelovskoe warming), 5000-8000 yr long each, and two cold rhythms (early cooling and Konoshel cooling),

2000-4500 yr long each, were recognized within the Karginsky Interglacial, time interval 50,000-25,000 yr B.P. (Kind, 1973, 1374). The duration of warm rhythms in the time interval 50,000-10,000 yr B.P. totals about 22,000 year, i.e. slightly more than half the interval (Fig. 63). The above data indicate: *first*, that specimens of the "glacial" fauna which, in the process of evolution, adjusted to life under cold conditions (Gromov, 1948; Garutt, 1951, 1965; Farrand, 1961; Vereshchagin and Baryshnikov, 1980; Kubiak, 1982) had to exist under conditions of cyclic changes of cold and warm rhythms; and *second*, that for much of the time (totalling two-thirds) they had to exist in environment unsuitable to their morpho-functional adjustments, i.e. under conditions of warm epochs which were unfavorable, harmful to them (Ukraintseva, 1979, 1981, 1986).

IV. Population size[33]

The important role of a population as a factor determining the rate and the character of evolution was discussed by Wright (1931), cited from Simpson (1944), Simpson (1944), Shmalgausen (1876). However, it is very difficult to determine the quantitative sizes of populations, even current ones. Although such attempts were made, Simpson (1944) justly regarded them as useless. The notions "large" and "small" populations, generally used by geneticists, are quite acceptable for the purposes of this study. The author uses the notions to consider the evolutionary potential of populations and the interrelations between the populations of animals of the above types and their habitats. As shown above, cyclic variations in natural environments, i.e. alternations of cold (glacial) and warm (interglacial) rhythms (epochs) are distinctive features of environment of Pleistocene mammals.

According to Shmalgausen (1876), under varying conditions, competition comes to that of separate lines, populations, subspecies and species not only in quality but in rate of adaptations. In cold (glacial) intervals of the Pleistocene when the vast territory of North Eurasia was dominated by open treeless landscapes (see Chapter 6), large continuously intrabreeding populations of herbivorous mammals were developed as suggested by multiple bone remains of their specimens in Eurasia and North America. Warm (interglacial) intervals were dominated by medium and small populations. This was associated with high afforestation and swamping which resulted in a sharp reduction in areas suitable to the dwelling of the animals and, first of all, areas under natural meadowlands and, correspondingly, in food reserve; as previously mentioned, the qualitative composition of forage also deteriorated. Taken together, these factors led to a drop in birth rate. At the same time, mortality of animals built up in both winter and summer (Ukraintseva, 1979). The above factors conduced to a reduction and division of the population into medium and small populations and groups.

Large continuously intrabreeding populations. According to Grant (1977), in such a population, selection poorly acts on recessive alleles which occur at low frequency and on most of new mutations while a compromise between selection and drift is quite ineffective. As a result of wide intrabreeding, new gene combinations do not consolidate, therefore in environment which varies at fast rate, this kind of population does not keep up with environmental variations and, hence, cannot react adequately on intensive selection. In this case, it falls into small relict populations (sequoia) or dies out (mammoth).

Small population in an environment which varies at fast rate proves to be uncapable of adequate evolutionary reaction as this is prevented by lack of variability. In the population, most of polymorphic genes consolidate under the action of drift or drift and selection. As a result, the store of genetic variability which could allow adaptation to new situations in the future proves to be inadequate. The population can migrate to some refuge or it is subject to extinction.

Hence, in both large and small populations living in variable environment, conditions were unfavorable to maintain high rates of evolution. For Pleistocene animals, high rates of evolution were dictated by those of environment. In this case, the rate of alternations of cold and warm rhythms grew and their duration dropped (Velichko, 1973) and, as a result, the quality and the rate of adaptations of the animals were upset. Hence, they could not keep up with environmental evolution. Because of the cyclic, oscillating and, to be more precise, unstable character of natural environmen of the Pleistocene, the animals, initially the narrowly specialized animals, were unable to adapt to certain environmental conditions. They did not possess the strategy of adjustment of the species to natural environment, since its variations were relatively sharp. Cold and warm rhythms replaced each other at such fast rates (on geological time scale) that "readjustment of correlation relations in the organisms of animals which can be realized little by little in evolution" (Zavadsky and Kolchinsky, 1977) was practically excluded. A "backward" evolution was ruled out for the Pleistocene animals because they merely had no time to readjust to new, warm conditions, and, in this context, for most of them, especially narrowly specialized mammals such as mammoth, woolly rhinoceros, cave bear, the extinction was only a question of time, as the law of irreversibility of evolution entered in force. Eisley (1942) had good reasons to believe that large Pleistocene animals could die out because of a degree of adaptation to glacial conditions which they had gained, whereas *Bison bison* could survive because it had not attained such a degree of adaptation.

Thus, the extinction of mammoth, woolly rhinoceros, cave bear and some other specimens of the "mammoth" faunal complex is "the result of unsolvable contradictions between the necessity for a sharp change in correlation structure of the organism due to varying environment and the necessity of life for integrity of the organism (Schmidt, 1979, p. 82). The following generalizations have emerged from all the above:

1. Mammoth, woolly rhinoceros, cave bear and some other representatives of the "mammoth" faunal complex are stenoecic species adjusted to life in cold and dry climate.

2. Their emergence was related to global cooling and fundamental change

in natural environment of the Pleistocene. Therefore cold and glacial intervals (epochs) when the vast territory of North Eurasia were under treeless landscapes such as cold steppes, grass-, bush- and low bush tundras, arctic meadows were more auspicious to their life and settling.

3. Herbaceous plants, leaves and sprigs of shrubs and undershrubs played a leading part in feed ration of representatives of the "mammoth" faunal complex. At cold periods of the Pleistocene, their food was dominated by winter plants which are not inferior to hay in nutritious value.

4. Cyclic variations of environment, i.e. alternations of cold (glacial) and warm (interglacial) rhythms (epochs) were salient features of natural environment of Pleistocene animals. Approaching the present, the rate of rhythm alternations built up and their duration reduced.

5. Neither very large, nor small and medium-sized populations of animals could withstand such high rates of evolution of natural environment, since the quality and the rate of their adaptations could not keep up with the high rates of environmental variations.

6. Because of the cyclic, oscillating or (to be more exact) unstable character of natural environment of the Pleistocene, the animals had no common strategy of adaptation to certain conditions, therefore for many Pleistocene species, first of all, some mammals narrowly specialized to cold or warmth, extinction was only a question of time.

7. Thus, the process of extinction of some groups of organisms, both plants and animals, cold-enduring and warm-requiring, is a function of the rate of evolutionary potential of populations of various types and the rate of environmental evolution.

8. Short-term (on geological time scale) existence of some cold-enduring species of animals was related to: (i) the cyclically varying character of the natural environment of the Pleistocene; (ii) the relation between the duration of cold and warm rhythms; (iii) the fact that cold-enduring Pleistocene animals existed for much (totalling two-thirds) of the time in environments unsuitable to their morpho-functional adjustments, i.e. under conditions of warm rhythms (epochs). Warm rhythms which, from man's viewpoint, are more favorable to life, proved to be unpropitious, adverse to the representatives of the "glacial" fauna. At periods, the territory of Northern Eurasia was primarily covered by forests, various bogs and other communities, similar or analogous to recent ones in composition. At that time the composition of forage basically varied: feed was dominated by plants of wet and swampy habitats such as sedges, cotton grasses,grasses, mosses which are poorer in the content of main nutrients (protein, albumen, fats) and minerals (calcium, potassium, phosphorus and others), as compared to plants of dry habitats and meadow forbs. A lower qualitative composition of forage, and its shortage during warm stages were responsible for a lower birth rate, and more common losses of animals in winter. This resulted in a sharp reduction in number of animals, division of populations into smaller ones which, because of their low evolutionary potential, were subject to extinc-

tion in a varying environment.

9. The present level of knowledge in the paleogeography of the Pleistocene and recent data on environment and the ecology of mammoth and some of its associates (woolly rhinoceros, horse, bison, and others) do not suggest a sharp extinction of mammoth and some other species at the Pleistocene/Holocene boundary. As shown above, the extinction was permanent-cyclic in character but proceeded at relatively fast rates on a geological time scale.

10. There are no reasons to accuse the primeval man in the disappearance of mammoth and some other species of animals from the face of the earth, although primeval hunters contributed to the reduction in their number, especially at the last stages of the Pleistocene and in the Holocene.

11. Some specimens of the "mammoth" faunal complex such as bison, reindeer, musk ox, arctic fox, lemming and some others which were specialized to a lesser degree, have survived to the present. Their ranges are situated in the areas of intensive industrial development. Lest the animals should share the fate of mammoth and some of its contemporaries, it is necessary to make the most efficient use of, and to conserve, where it is required, the areas where the animals dwell. In particular, it is necessary to reasonably use reindeer pastures, i.e. to reveal new and to make the efficient use of old pastures.

CHAPTER 8:
Summary and Conclusions

Finds of frozen carcasses and complete skeletons of fossil herbivorous mammals with sufficiently well preserved contents of their gastrointestinal tracts provide valuable information on ecosystems of the past. Information can be obtained from study of: (i) remains of plants once eaten by the animals; (ii) the deposits which enclosed fauna remains, and/or coeval deposits; (iii) the fossil animal or animals as is the case with the Berelekh cemetery of mammoths and their associates.

Food remains of the fossil herbivores represent unique objects of paleobotanical study. They are collectors of pollen and spores of plants once growing in habitats of the fossil animals. All the finds known up to the present show that pollen and spores are practically not subjected to digestion and decomposition in the gastrointestinal tracts. Seeds are slightly digested; in places they show signs of mechanical damage such as cracks, breaks and the like. This makes all of them specifically and generically determinable; as for plant remains proper, practically none of them are preserved well enough for it to be identified either specifically or generically. The foregoing makes the palynological method the key tool of investigations. Nevertheless, the composition of plants once growing in habitats of the fossil animals can be thoroughly determined only when the investigator uses a whole store of paleobotanical methods, namely, the method of paleopalynofloras, spore-pollen analysis, carpological and anatomo-morphological techniques. The floristic composition, determined as the result of integrated studies can be treated as local flora reconstructed more or less completely. The lifetime of such floras is determined by ^{14}C dating of food remains, fragments of skeletons and/or soft tissues of the animals. As for their abundance, it is determined, first of all, by abundance of the local floras which existed in the past in habitats of the fossil animals. Their completeness is a function, first of all, of the season when the animals died, the habitats which served as pastures to them, selectivity of the animals to these or those plants, and, at last, the degree of digestion of the plants.

Floras, reconstructed from findings of integrated investigations of gastrointestinal contents of some fossil animals, found in various areas of Siberia, are late Anthropogene in age. Five floras, recognized in the basins of Kolyma and Indigirka rivers, are dated as different stages of the Karginsky Interglacial (25,000-50,000 yr B.P.), equivalent to the middle Wisconsin of North America; only one flora, determined in the middle Yuribei River (Gydan Peninsula) is dated as early Holocene (Table 20). The flora of the optimal phases of Karginsky Interglacial (Elga River in the upper Indigirka River Basin) is the richest in taxonomic composition. The core of flowering plants which grew there as early as 38,590±1120 yr B.P. was composed of species survived there up to the present; the flora was dominated by boreal and hypoarctic species. Part of the species, first of all, tree species (*Picea ajanensis, P. obovata, Pinus sylvestris, Betula sp. ex sect. costatae*), aquatic and riverside aquatic (*Nuphar pumila, Typha latifolia*) and herbaceous (*Kobresia capilliformis, K. filifolia* and others) plants are absent from the recent flora. The above-mentioned and some other species suggest that the flora is related to recent floras of Ayan, Lena-Aldan type, as well as floras of the Stanovoi Upland and Northern Mongolia. The fact that the flora contained some species of steppe communities, steppe meadows and dry meadows, first recognized from palynological data, suggests their occurrence in the upper Indigirka River Basin since the optima of Karginsky Interglacial at least. The flora of the Middle Shandrin River (lower Indigirka River) is the poorest of all the floras recognized; it also characterizes floras of the optima of the Karginsky Interglacial. This is caused by both its northernmost zonal position (70°N) and the season when the animal died (early spring). The taxonomic composition of the paleofloras of the optima of Karginsky Interglacial suggests a distinct flora differentiation and subzonal vegetational cover differentiation in the Indigirka River Basin 38,590-40,350 yr B.P.

Reconstructed floras are dated by ^{14}C analysis and geologogeomorphological data. They are the best studied paleofloras of Siberia for the established time "sections". With good reason they can be regarded as standard for the areas where they were recognized.

Analysis of all the reconstructed floras indicates that they underwent no catastrophic variations throughout the territory of Siberia for the last 55,000 yr, at least. A steady climate cooling led to a gradual impoverishment and transformation of late Anthropogene (Quaternary) floras and the emergence of recent floras. The cyclic nature of climatic variations imparted an undulating, pulsating character to the unidirectional course of evolution.

Joint spore-pollen analysis (performed for the first time) of gastrointestinal contents of the fossil animals, the deposits which enclosed the fauna remains and/or coeval deposits, and the surficial samples, taken in communities of recent key vegetation, indicates that spectra of intestinal contents are averaged, similar to spectra of all the above-mentioned sample types. A single system of estimation, i.e. the division into four groups corresponding to four groups of life forms of plants (pollen of trees, shrubs and low shrubs, herbs and forbs, spores of cryptogamic plants) allowed the author to recognize their zonal and local components and thus reliably reconstruct communities of zonal paleovegetation 21 types synchronous to the life of fossil animals. The local component provided a quite correct notion of the percentages of main plant groups in the feed ration of the animals such as herbs, including grasses, sedges, forbs, shrubs and low shrubs. This was supported by results of anatomo-morphological study of food remains and spore-pollen analysis of the deposits which enclosed fauna remains.

Integrated investigations threw light on environments of the concrete specimens of fossil animals discussed here and the "mammoth" fauna as a whole. Most of the animals died in warm periods during the last 53,000-10,000 years when the vast territory of Siberia was covered by various forests, paludal communities and bogs, whereas communities of azonal character (meadows, steppe meadows and others) which served as pastures to the mammals sharply reduced in area. Only two of the nine fossil animals lived and died at cold periods (glaciations) of the late Pleistocene, namely, during the Zyrianka and Kirgilyakh (early Karginsky) Ice Ages. As for environments of the "mammoth" fauna as a whole, they were characterized by cyclic changes of glacial (Zyrianka, Sartan) and interglacial (Kazantgevo, Karginsky) epochs. On the basis of data of the author and other investigators, the epochs are distinctly traced not only in three key areas, namely, the Indigirka-Kolyma Basin, Taimyr, and Gydan Peninsula where the specimens of the "mammoth" faunal complex were found, but throughout Siberia. Latitudinal zoning of natural landscapes was the most typical feature of warm epochs; in the mountainous areas, the structure of altitudinal zoning became more complicated at that time. Cold (glacial) epochs were dominated by open, treeless landscapes such as arctic meadows, various tundras, periglacial steppes. Remains of fossil trees, dated by 14C analysis, suggest that woody plants did not disappear completely during short cold phases of the late Pleistocene and Holocene. They remained as small forests and separate trees in the river valleys and serves as peculiar advanced posts for forest advance northwards at warm epochs.

In both cold (glacial) and warm (interglacial) epochs of the late Pleistocene, herbaceous plants, leaves and sprigs of shrubs and low shrubs played a leading part in the feed ration of specimens of the "mammoth" faunal complex. The cold (glacial) periods were dominated by plants of drier habitats and winter fodder, whereas warm (interglacial) periods were marked by the predominance of plants of wet and swampy habitats such sedges, cotton grasses, grasses, mosses which are much inferior to the plants of dry habitats and meadow forbs in nutritious value.

Paleoclimatic reconstructions made by the author showed that sums of temperatures above 0°C and precipitation for a period of positive temperatures, i.e. the characteristics which determine the distribution of communities of zonal-type vegetation, along with mean monthly characteristics of air temperature and precipitation, must be used to estimate the character of paleoclimates in high latitudes and, especially, in mountainous areas. This approach gives the lowest skewness in estimates of paleoclimates.

Another important conclusion is inferred from the data presented in Chapter 7: the initial cause of relatively fast extinction of many species of both cold-enduring and warm-requiring mammals in the Pleistocene were cyclic alternations of cold and warm rhythms (epochs) and related fundamental changes in the character of vegetation and climate, these main components of natural environment which had both direct and indirect effect on the life of animals.

Because of the unstable character of natural environment of the Pleistocene, the populations of animals had no common strategy of adaptation to certain conditions thus dooming separate species, genera and entire groups to a fast extinction. Undoubtedly, man also made his contribution to a reduction of the number of some species, especially large herbivorous mammals and thus accelerated the process of natural extinction. The aggregate of new paleobotanical data, obtained as the result of integrated studies provides a reliable support to mineral exploration and a necessary historical backward glance at long-term evaluation in making plans for the growth of industry in certain regions of Siberia.

The problem of study of food remains of the fossil herbivores and the deposits which enclosed the fauna remains is a separate complex problem the realization of which is of theoretical and applied significance. The methods applied in the study are intercontrolable; the integrated approach only ensures a successful realization of the problem as a whole. This allows a solution of some questions of particular interest (reconstruction of the composition of forage, the character of pastures, the season when the animals died), general (reconstruction of the composition of floras, the character of vegetation, landscapes, paleoclimate; substantiation of the age of deposits, climate-stratigraphic modeling, correlation of simultaneous natural phenomena and processes), and general biological significance (causes of extinction of plants and animals).

Footnotes

Chapter 4

1   The finds made in the period 1900-1979 are meant here. The author is unaware of other such finds. (p. 29)

2   The section is described by the author.

3   P.E. Giterman who studied the samples, taken by I.A. Dubrovo, also arrived at the conclusion that spectra of the deposits overlying the mammoth differ from those of the intestinal tract. (p. 52)

4   Identified by A.L. Abramova (Botanical Institute, USSR Academy of Sciences, Leningrad). (p. 54)

5   A.V. Logachev, bulldozer operator, carefully collected the remains of all plants, put them into a polyethyle bag, placed it into a refrigerator and then gave the bag to the author. (p. 54)

6   They were cut off by I.A. Dubrovo (Paleontological Institute, Moscow). (p. 55)

7   Diatoms were identified by E.M. Vishnevskaya (State University, Leningrad). (p. 56)

8   Identified by A.L. Abramova, V.L. Komarov Botanical Institute, Leningrad. (p. 84)

9   Five pollen grains of very good preservation were encountered in the preparations. (p. 87)

10   A value of 5280 ± 140 yr B.P. was obtained on one of the specimens. (p. 93)

11   The date was obtained on a bone fragment of the female bison by L.V. Firsov at the Laboratory of Geochronology, Institute of Geology and Geophysics, Siberian Branch of the USSR Academy of Sciences, Novosibirsk. (p. 99)

12   Pollen of some trees such as Pinus sibirica, Betula pendula (platyphylla), Alnus hirsula, Ulmus sp., represented in the palynological spectrum by sporadic grains should be regarded as long-distance transported pollen. (p. 107)

13   The section was described by the author in the process of taking samples for palynological analysis and supplemented by data of I.A. Dubrovo. (p. 111)

14   Geological conditions of the mammoth's location were studied in detail by Evseev and co-workers (1982). (p. 112)

15   Palynological spectra are characterized in detail in text, therefore the diagram displays the general composition of spectrum only. (p. 118)

16   Cretaceous, Neogene – early Quaternary sediments may have served as a source of redeposition. (p. 118)

17   In depth interval 0-8 m from the water edge in a low water period. (p. 146)

18   Rybakova (1979) explains a high contect of pollen of woody birch, willow, alder by fortuious causes such as long transportation and redeposition, but data presented here are not in line with the conclusion. (p. 146)

Chapter 5

19   Report presented at the 27th International Geological Congress, Moscow, USSR, August 5, 1984. (p. 170)

Chapter 6

[20] Oral Communication at the VI all-union meeting on the study of mammoth and mammoth fauna (May 1983, Leningrad). (p. 175)

[21] No frostless period is actually present, as the temperature frequently drops below 0^0 C in summer. (p. 176)

[22] The date was obtained on a skin fragment of a mammoth whose remains were buried above the flood plain of the Bolshaya Lesnaya Rossokha River, in terrace III. (p. 176)

[23] It was found that the rate of formation of the lower beds of Holocene peat bogs in the Novaya River basin was 1.5-1.6 mm per year (Belorusova and Ukraintseva, 1980). (p. 182)

[28] ^{14}C dates obtained by L.D. Sulerzhitsky (oral communication). (p. 186)

[25] The remains were identified by I.E. Kuzmina. The skull was transferred to the Zoological Institute of the USSR Academy of Sciences. No radio-carbon determinations were made. (p. 186)

[26] Long-distance transported pollen grains. (p. 203)

[27] Sangamon intirglacial of North America. (p. 204)

[28] As stated by S.S. Chetvyarikov (cited from llyin and Smirnov, 1973) a population as a integral system, but not a separate organism, is a unit of evaluation. (p. 223)

[29] Verchagin (1977) and Agenbroad (1982, 1984a) were the first to use this criteria to reveal some of the causes of extinction of the animals. (p. 224)

Chapter 7

[30] Woody forage ranges from 1 to 3 kg in duirnal feed of the Caucasian auroch (Sokolov, ed., 1979). (p. 228)

[31] Unfortunately, the abundance (frequently of occurance and percentage) of all the remains have not been calculated. (p. 229)

[32] Like elephants, mammoths required no less than 300 kg to 700 kg of raw forage daily, according to Guenter (1987). (p. 232)

[33] "Population size" essentially means the effective breeding population. (p. 234)

References Cited

Abramov, I.I. and A. L. Abramova. 1981. Mosses from the Deposits of Magadan Baby Mammoth Site, in Vereshchagin, N.K. and Mikhelson, V.M., eds. Magadan Baby Mammoth. Nauka, Leningrad, pp. 247-253. (Russian)

Agenbroad, L.D. 1982. "The Distribution and Chronology of Mammoth in the New World." A paper presented at the III International Therilogical Congress, Helsinki, Finland, August 15-20.

Agenbroad, L.D. 1984a. New World Mammoth Distribution, in Martin P.S. and Klein, R.G., eds. Quaternary Extinction. A Prehistoric Revolution. Tucson, Arizona, pp. 90-108.

Agenbroad, L.D., P.S. Martin, J.I. Mead, and O.K. Davis. 1984b. Chronology and diet of Mammuthus from Eurasia and Temperate America. Paper presented at the 24th International Geological Congress, Moscow, USSR. August 4-14, 1984. 24 pp.

Agenbroad, L.D. 1985. the Distribution and chronology of Mammoth in the New World. Acta Zoologica Fennica 170: 221-224.

Ager, Th.A. 1982. Vegetational History of Western Alaska during the Wisconsin Glacial Interval and the Holocene, in Hopkins, D.M. et al., eds. Paleoecology of Beringia. Academic Press, Inc., pp. 75-93.

Aleksandrova, V.D. 1963. An Outline of Flora and Vegetation of Bolshoi Lyakhovsky Island, in Rutilevsky, G.L. and Sisko, R.K., eds. Novosibirskie Islands. Morskoi Transport, Leningrad, pp. 6-36. (Russian).

Aleksandrova, V.D. 1977. the Arctic and Antarctic: their Division into Geobotanical Areas. Nauka, Leningrad. 188 pp. (Russian).

Aleksandrova, V.D., V.N. Andreev, T.V. Vakhtin, R.A. Dydina, G.I. Kazov, V.V. Petrovsky, and V.F. Shamurin. 1964. Feed Characteristics of Plants of the Far North. Moscow-Leningrad. 480 pp. (Russian).

Alekseev, M.N., E.V. Devyatkin, S.A. Arkhipov, S.A. Laukhin, O.V. Grinenko, and V.A. Kamaletdinov. 1984. Problems of Quaternary Geology of Siberia, in Nikiforova, K.V. et al., eds. Reports presented at the 27th International Geological Congress, Moscow, USSR. August 4-14, 1984. Nauka, Noscow. Vol. 3:3-12. (Russian).

Andreev, V.N. (ed.). 1974. Winter Pastures of Northeastern Yakutia. Yakut. Knizhn. Izd-vo, Yakutsk. 245 pp. (Russian).

Andreeva, S.M. 1980. North Siberian Lowland in Karginsky Time. Paleogeography, Radiocarbon Chronology of the Quaternary Period. Nauka, Moscow, pp. 183-191. (Russian).

Anonymous. 1979. Very Active Silicium. Znanie-sila 3:4. (Russian).

Arkhipov, S.A. 1971. Quaternary Period in Western Siberia. Nauka, Novosibirsk. 329 pp. (Russian).

Arkhipov, S.A. 1983. Correlation of Quaternary Glaciations of Siberia and the North East, in Arkhipov, S.A. et al., eds. Glaciations and Pleoclimates in the Pleistocene. Nauka, Novosibirsk, pp. 4-18. (Russian).

Arkhipov, S.A., R.M. Votakh, A.V. Golbert, V.I. Gudina, L.A. Dovgel, and A.I. Yidkevich. 1977. Last Glaciation in the Lower Ob River Area. Nauka, Siberian Branch, Novosibirsk. 213 pp. (Russian).

Arslanov, Kh.A. and S.B. Chernov. 1977. Absolute Age of the Selerikhan Horse, in Skarlato, O.A., ed. Anthropogene Fauna and Flora of the Siberian North East. Nauka, Leningrad, pp. 76-78. (Russian).

Arslanov, Kh.A., V.V. Lyadov, and T.V. Tertichnaya. 1981. On the Absolute Age of the Baby Mammoth, in Vereshchagin, N.K., and Mikhelson, V.M,. eds, Magadan Baby Mammoth. Nauka, Leningrad, pp. 50-51. (Russian).

Arslanov, Kh.A., N.K. Vereshchagin, V.V. Lyadov, and V.V. Ukraintseva. 1980. On the Dating of the Karginsky Interglacial Period and Siberian Landscape Reconstruction as Resulting from the Investigation of Mammoth Corpses and their "Satellites", in Ivanova, I.K., and Kind, N.V., eds. Geochronology of the Quaternary Period. Nauka, Moscow, pp. 208-213. (Russian).

Avenarius, I.G., M.V. Muratova, and I.S. Spasskaya. 1978. Paleogeography of North Eurasia in the Late Pleistocene-Holocene and Geographical Prediction. Nauka, Moscow. 76 pp. (Russian).

Azimov, G.I., D. Ya. Krintsyn, and N.F. Popov. 1954. Physiology of Farm Animals. Selkhozizdat, Moscow, pp. 1-15. (Russian).

Barkova, M.V. 1969. Periglacial Vegetation of the Zyrianka Cooling Epoch in the Yana-Indigirka Lowland Area. Trudy NIIGA 25:73-76. Leningrad. (Russian).

Barkova, M.V. 1970a. Spore and Pollen Complexes of Middle Pleistocene Deposits of the Yana-Indigirla Lowland. Uchenye Zapiski NIIGA 30:40-46. Leningrad. (Russian)

Barkova, M.V. 1970b. Spore and Pollen Complexes from Deposits of Preglacial and First Glacial Epochs of the Late Pleistocene of the Yana-Indigirka Interfluve. Uchenye Zapiski NIIGA, Paleontologia i Stratigrafia 31:73-85. Leningrad. (Russian).

Baryshnikov, G.F., I.E. Kuzmina, and V.M. Khrabry. 1977. Results of Measurements of Tubular Bones of Mammoths of the Berelekh Cemetery, in Skarlato, O.A., ed. Mammoth Fauna of the Russian Plain and Eastern Siberia. Trudy Zoologicheskogo Instituta 72:58-67. Leningrad. (Russian).

Belaya, B.V. and Zh.B. Kisterova. 1978. Palynological Data Obtained from the Baby Mammoth Find of the Summer of 1977, in Shilo, N.A., ed. Materials on Geology and Mineral Resources of the North East of the USSR. Magadan, pp. 250-251. (Russian).

Belorusova, Zh.M. 1964. Physico-geographical Characteristics and Quaternary Paleogeography of Tasovsky Peninsula. Thesis of Dissertation, A.I. Gerzen Leningrad State Pedagogical Institute. 22 pp. (Russian).

Belorusova, Zh.M. 1977. Geomorphological Essay on the Selerikhan River (Yakutia), in Skarlato, O.A., ed. Anthropogene Fauna and Flora of the Siberian North East. Nauka, Leningrad, pp. 63-75. (Russian).

Belorusova, Zh.M. 1978. Geology and Geomorphology, in Norin, B.N., ed. Ary-Mas. Natural Environmen, Flora and Vegetation. Nauka, Leningrad, pp. 5-16. (Russian).

Belorusova, Zh.m. and V.V. Ukraintseva. 1980. Paleogeography of the Late Pleistocene and Holocene of the Novaya River Basin (Taimyr). Botanical Journal 65(3):368-379. (Russian).

Belorusova, Zh.M., N.V. Lovelius, and V.V. Ukraintseva. 1977. Paleogeography of the Late Pleistocene and Holocene in the Area of Selerikhan Horse Find, in Skarlato, O.A., ed. Anthropogene Fauna and Flora of the Siberian North East. Nauka, Leningrad, pp. 265-276. (Russian).

Belorusova, Zh.M., N.V. Lovelius, and V.V. Ukraintseva. 1987. Regional Features in the Taimyr Nature Alternations in the Holocene. Botanical Journal 72(5): 610-618. (Russian).

Belov, A.V. and V.A. Belova. 1984. Main Stages of the Development of Vegetatioal Cover in the Late Cenozoic, in Malyshev, L.I., ed. History of Vegetational Cover of North Asia. Nauka, Siberian Branch, Novosibirsk, pp. 42-55. (Russian).

Belova, V.A. 1983. Dynamics of Flora, Vegetation and Climate of the Late Cenozoic in the South of East Siberia, in Arkhipov, S.A. et al., eds. Glaciations and Paleoclimates of Siberia in the Pleistocene. Nauka, Siberian Branch, Novosibirsk, pp. 69-78. (Russian).

Berdovskaya, G.N., N.A. Gei, and V.M. Makeev. 1970. Paleogeography of Northeastern Taimyr in Quaternary Time (from Geological and Palynological Data), in Tolmachev, A.N., ed. The Artic Ocean and its Coast in the Pleistocene. Gidrometeoizdat, Leningrad, pp. 440-446. (Russian).

Bespalyi, V.G., T.D. Davidovich, V.F. Ivanov, and A.V. Lozhkin. 1982. Natural Environment of the Last Glaciation Epoch in the Penega Inlet Area, in Velichko, A.A., ed. Evolution of Nature of the USSR Territory in the Late Pleistocene and Holocene. Nauka, Moscow, pp. 32-40. (Russian).

Boch, M.S. and V.T. Tsareva. 1974. On the Flora of the Lower Indigirka River Area (within the Tundra Zone). Botanical Journal 59 (6): 839-849. (Russian).

Borisov, A.A. 1975. Climate of the USSR in the Past, Present and Future. Izd-vo LGU, Leningrad. 431 pp. (Russian).

Boyarskaya, T.D. 1965. Vegetation of the USSR during the Maximal Glaciation and the Mga Interglacial, in Markov, K.K., ed. Paleo (Russian).

Boyarskaya, T.D. and E.M. Malaeva. 1967. Development of Vegetation in Siberia and the Far East in the Quaternary. Nauka, Moscow. 22 pp. (Russian).

Budyko, M.I. 1967. On the Causes of Extinction of Some Animals at the End of Pleistocene. Izv. Akad. Nauk SSSR, Ser. Georg. 2: 28-36. (Russian).

Byrashnikova, T.A., M.V. Muratova, and I.A. Suetova. 1982. Climatic Model of the USSR Territory During the Holocene Optimum, in Velichko, A.A., ed. Evolution of Nature of the USSR Territory in the Late Pleistocene and Holocene. Nauka, Moscow, pp. 245-251. (Russian).

Chersky, I.D. 1891. Description of the Collection of Post-Tertiary Mammals Gathered by the Novosibirsh Expedition. Apendix I. Zapiski Akademii Nauk 65: 706. St. Petersbourg. (Russian).

Danilov, I.D., A.I. Popov, and G.I. Smirnova. 1971. Geologo-geomorphological and Permafrost Structure in the Taimyr Station Area, in Tikhomirov, B.A., ed. Biogeocenoses of Taimyr Tundra and Their Productivity. Nauka, Leningrad, pp. 17-34. (Russian).

Davis, O.K., L. Agenbroad, P.S. Martin, and J. Mead. 1984. The Pleistocene Dung Blanket of Bechan Cave, Utah. Carnegie Museum of Natural History. Special Publication No. 8: 267-282. Pittsburgh.

Davitashvili, L.Sh. 1969. Causes of Extinction of Organisms. Nauka, Moscow, 440 pp. (Russian).

Davitaya, F.F. 1964. Prediction of Warmth Provision and Some Problems of Seasonal Development of Nature. Gidrometeoizdat, Moscow. 130 pp. (Russian).

Dylis, N.V. 1948. On Self-Pollination and Disperse of Pollen of Larch. Dokl. Akad. Nauk SSSR 60(8): 203-205. (Russian).

Egorov, A.D. 1960. Chemical Composition of Fodder Plants of Yakuta (Meadows and Pastures). Izd-vo Akad. Nauk SSSR, Moscow. 336 pp. (Russian).

Egorova, T.V. 1977. Carpological Analysis of Plant Remains of the Selerikhan Fossil Horse. Anthropogene Fauna and Flora of the Siberian North East. Nauka, Leningrad, pp. 218-221. (Russian).

Eiseley, L.C. 1942. Post-Glacial Climatic Amelioration and the Extinction of *bison taylori*. Science 95: 646-647.

Erdtman, G. 1960. The Acetolysis Method. A Revised Description. Sven. Bot. Tidskr 54: 561-564.

Evseev, V.P., I.A. Dubrovo, N.V. Rengarten, and A.Ya. Stremyakov. 1982. Location of the Yuribei Mammoth: Geology, Taphonomy, Paleo-geography, in Sokolov, V.E., ed. Yuribei Mammoth. Nauka, Moscow, pp. 5-19. (Russian).

Fedorova, R.V. 1952. Quantitative Consistent Patterns of Air Transportation of Pollen of Tree Species. Trudy Instituta Geografii 52: 91-103. (Material on Geomorphology and Paleogeography of the USSR 7). Izd-vo Akad. Nauk SSSR, Moscow. (Russian).

Fedorova, R.V. 1956. Air Transportation of Pollen of Some Herbaceous Plants. Izv. Akad. Nauk SSSR. Ser. georg. 1: 104-109. (Russian).

Ferrand, W.R. 1961. Frozen Mammoths and Modern Geology. Science 133: 729-735.

Flerov, K.K. 1931. Trunk of Mammoth (*Elephas primigenius* Blum.) Found in the Kolyma Distrit (Siberia). Izv. Akad. Nauk SSSR, Ser. 7: 863-870. (Russian).

Flerov, K.K. 1955. Main Features of Formation of Mammal Fauna of the Quaternary Period in the Northern Hemisphere, in Gromov, V.N., and Ivanova, I.K., eds. Trudy Komissii po izucheniyu Chetvertich-nogo Perioda 12: 121-126. (Russian).

Flerov, K.K. 1965. On the Origin of Canadian Fauna in Connection with the History of Bering Land-Bridge, in Gromov, V.I., and Ivanova, I.K., eds. Quaternary Period and its History. Nauka, Moscow. (Russian).

Flerov, K.K. 1979. On the Reorganization of Theriofauna of the Northern Hemisphere in the Pleistocene. Dokl. Akad. Nauk SSSR 246(4): 971-973. (Russian).

Flerov, K.K., B.A. Trofimov, and N.M. Yanovskaya. 1955. History of Mammal Fauna in the Quaternary. Izd-vo MGU, Moscow. 38 pp. (Russian).

Garutt, V.E. 1951. Change in the Structure of Proboscidea, Related to Habitate. Dokl. Akad. Nauk SSSR 77(3): 513. (Russian).

Garutt, V.E. 1965. Fossil Elephants of Siberia. Trudy NIIGA 143: 106-130. Leningrad. (Russian).

Garutt, V.E., E.P. Meteltseva, and B.A. Tikhomirov. 1970. New Data on the Food of Woolly Rhinoceros in Siberia, in Tolmachev, A.I., ed. The Arctic Ocean and its Coast in the Pleistocene. Gidrometeoizdat, Leningrad, pp. 113-125. (Russian).

Geobotanical Map of the USSR. 1954. Edited by Lavrenko, E.M. Nauka, Leningrad.

Gerasimov, I.P. 1961. State-of-the-art in the Teaching of Glacial Period and its Role in Study of the Quaternary Period (Anthropo-gene) Over the USSR Territory. Izv. Akad. Nauk SSSR, Ser. geogr. 4: 3-9. (Russian).

Gerasimov, I.P. and A.A. Velichko, 1984. Complex Paleogeographic Atlases-Monographs for the Anthropogene and Their Prediction Value, in Nikiforova, K.V. et al., eds. Reports Presented at the 27th International Geological Congress, Moscow, USSR. August 4-14, 1984. Nauka, Moscow, Vol. 3: 57-66. (Russian).

Gerz, O.F. 1902. Reports of the Head of the Berezovka Mammoth Unearthing

Expedition of the Imperial Academy of Sciences, in Izvestiya Imperator-skoi Akademii Nauk 16(4): 137-174. (Russian).

Giterman, R.E. 1963. Stages of Development of Quaternary Vegetation of Yakutia and Their Significance for Stratigraphy. Izd-vo Akad. Nauk SSSR. Moscow. 191 pp. (Russian).

Giterman, R.E. 1972. To the Palynological Characteristics of Karginsky Deposits in the Lower Kolyma River Basin, in Grichuk, V.P., ed. Palynology of the Pleistocene. Nauka, Moscow, pp. 119-132. (Russian).

Giterman, R.E. 1985. History of Vegetation of the North East of the USSR in the Pliocene and Pleistocene. Nauka, Moscow. 90 pp. (Russian).

Giterman, R.E. and L.V. Golubeva. 1965. History of Development of Vegeta-tion of Siberia in the Anthropogene, in Saks, V.N. et al., eds. Main Pro-blems of Study of the Quaternary Period. Nauka, Moscow, pp. 365-374. (Russian).

Giterman, R.E., L.V. Golubeva, E.D. Zaklinskaya, E.V. Koreneva, O.V. Matveeva, and L.A. Skiba. 1968. Main Stages of Development of Vegeta-tion of North Asia in the Anthropogene. Akad. Nauk SSR 177(4): 270. (Russian).

Glushkova, O.Yu. 1982. Glaciation of the North East Territory of the USSR at the End of Late Pleistocene, in Velichko, A.A., ed. Evolution of Nature of the USSR Territory in the Late Pleistocene and Holocene. Nauka, Moscow, pp. 78-83. (Russian).

Gorlova, R.N. 1982. Plant Macroremains from Gastrointestinal Tract of the Yuribei Mammoth, in Sokolov, V.E., ed. Yuribei Mammoth. Nauka, Moscow, pp. 37-43. (Russian).

Gorodkov, B.N. 1944. Tundras of the Ob-Yenisei Divide. Sovetskaya Botanika 3: 3-20. (Russian).

Grant, V. 1977. Organismic Evolution. San Francisco.

Grichuk, M.P. 1961. Main Features of Variation of Vegetational Cover of Siberia During the Quaternary, in Markov, K.K., ed. Paleogeography of the Quaternary Period of the USSR. Izd-vo MGU, Moscow, pp. 189-205. (Russian).

Grichuk, M.P. 1973. On the Pleistocene History of Flora in the Indigirka-Kolyma

Mountainous Region, in Grichuk, V.P., ed. Palynology of the Pleistocene and Pliocene. Nauka, Moscow, pp. 116-120. (Russian).

Grichuk, M.P. 1976. On Changes in Species Composition of Flora of the North East of Eurasia in the Late Cenozoic, in Kontrimavichus, V.L., ed. Beringia in the Cenozoic. Nauka, Vladivostok, pp. 145-155. (Russian).

Grichuk, V.P. 1940. Methods of Treatment of Sedimentary Rocks Poor in Organic Remains for the Purposes of Pollen Analysis. Problemy Fizicheskoi Geografii 8: 53-58. (Russian).

Grichuk, V.P. 1969. Experience of Reconstruction of Some Elements of Clilmate of the Northern Hemisphere in Atlantic Time of the Holocene, in Neishtadt, M.I., ed. The Holocene. Nauka., Moscow, pp. 41-57. (Russian).

Grichuk, V.P. 1982. Vegetation of Europe in the Late Pleistocene, in Gerasimov, I.P., and Velichko, A.A., eds. Paleogeography of Europe for the Last 100,000 Years. Atlas-Monography. Nauka, Moscow, pp. 92-100. (Russian).

Gromov, V.I. 1948. Paleontological and Archeological Evidence for Stratigraphy of Continental Quaternary Deposits in the USSR. Trudy Instituta Geologicheskikh Nauk 64 (17): 522. (Russian).

Gromov, V.I. 1950. Essay of the History of Quaternary Fauna in the USSR, in Materials on the Quaternary Period of the USSR 2: 37-49. Izd-vo Akad. Nauk SSSR, Moscow. (Russian).

Guenter, E.W. 1987. Zur Frage Des Aussterbens von Eiszeitlichen Grosssaugerrn Insbesondere der Mammute - In Mitteleuropa. Z. Cranium 4 (2): 67-75.

Guthrie, R.D. 1968. Paleoecology of the Large-Mammal Community in the Interior Alaska During the Late Pleistocene. The American Midland Naturalist 79 (2) 346-363.

Guthrie, R.D. 1976. Environmental Influences on Body Size, Social Organs, Population Parameters and Extinction of Beringian Mammals, in Kontrimavichus, V.L., ed. Beringia in the Cenozoic. Vladivostok, pp. 296-321. (Russian).

Guthrie 1982. Mammals of the Mammoth Steppe as Paleoenvironmental Indicators in D.M. Hopkins, et al. eds. Paleoecology of Beringia, 307-329. Academic Press, New York.

Hansen, R.M. 1980. Late Pleistocene Plant Fragments in the Dung of Herbivores, Cowboy Cave. University of Utah Anthropological Paper 104: 179-189.

Heinz, A.E. and V.E. Garutt. 1964. Absolute Age of the Mammoth and Woolly Rhinoceros Fossil Remains from Siberian Permafrost as Established by Radiocarbon Data. Dokl. Akad. Nauk SSSR, pp. 1367-1370. (Russian).

Ilyin, A.Ya. and I.N. Smirnov. Marxism-Leninism Philosophy and Theory of Evolution, in Bykhovsky, B.E. et al., eds. Philosophical Problems of Theory of Evolution (Material to Symposium). Part 1. Nauka, Moscow, pp. 9-29. (Russian).

Isaeva, L.L., N.V. Kind, S.M. Andreeva, G.V. Ivanenko, M.V. Nikolskaya, L.D. Sulerzhitsky, and E.L. Fisher. 1980. Geochronology and Paleogeography of the Late Pleistocene of the North Siberian Lowland from Radiocarbon Data, in Ivanova, I.K., and Kind, N.V., eds. Geochronology of the Quaternary Period. Nauka, Leningrad, pp. 191-197. (Russian).

Ivanova, E.I. 1981. Morphology of Fragments of Blood Vessels and Some Organs of Baby Mammoth's Neck, in Vereschagin, N.K., and Mikhelson, V.M., eds. Magadan Baby Mammoth. Nauka, Leningrad, pp. 128-154. (Russian).

Jennings, J.D. 1980. Cowboy Cave. University of Utah Anthropological Papers 104: 224. Salt Lake City.

Kahlke. H.D. 1976. Southern Boundary of the Late Pleistocene Eurosiberian Faunal Complex in East Asia, in Kontrimavichus, V.L., ed. Beringia in the Cenozoic. Vladivostok, pp. 263-271. (Russian).

Kaplina, T.N. 1979. Spore and Pollen Spectra of the Glacial Complex of Primorsky Lowlands of Yakutia. Izv. Akad. Nauk SSSR, Ser. georg. 2: 85-93. (Russian).

Kaplina, T.N., O.V. Lakhtina, and N.O. Rybakova. 1980. History of Development of Landscapes and Frozen Strata of the Kolyma Lowland from Radiocarbon, Cryolithological and Palynological Data (Exemplified by Stanichkovsky Yar Section on the Maly Anyui River). in Ivanova, I.K., and Kind, N.V., eds. Geochronology of the Quaternary Period. Nauka, Moscow, pp. 243-253. (Russian).

Karevskaya, I.A. 1972. History of Development of Vegetation of the Pleistocene of the Upper Kolyma River as Related to Problems of Paleogeomorphology. Thesis of Dissertation, M.V. Lomonosov Moscow University,

Moscow. 34 pp. (Russian).

Karevskaya, I.A. 1979. Characteristics of Recent Spore and Pollen Spectra of Deposits of Different Genesis in the Lower Kukhtui, Urak and Okhta Rivers, in Grichuk, V.P., ed. Palynological Studies in the North East of the USSR. Akad. Nauk SSSR, Far East Scientific Center, Vladivostok, pp. 87-89. (Russian).

Karpenko, A.S. 1958. Principal Regular Features in the Distribution of Vegetation in the Indigirka Lowland Area (North East Siberia). Botanical Journal 43 (1): 70-75. (Russian).

Khotinsky, N.A. 1977. The Holocene of Northern Eurasia. Nauka, Moscow. 197 pp. (Russian).

Khotinsky, N.A. 1981. Tracks of the Past Lead to the Future. Mysl, Moscow. 160 pp. (Russian).

Kind, N.V. 1973. Chronology of the Late Anthropogene According to Radiometric Data, in Gromov, V.I., ed. Stratigraphy, Paleontology 4: 5-49. (Russian).

Kind, N.V. 1974. Geochronology of the Late Anthropogene from Isotopic Data. Nauka, Moscow, vyp. 257: 254. (Russian).

Kind, N.V. and B.N. Leonov (eds). 1982. The Anthropogene of Taimyr Peninsula. Nauka, Moscow. 181 pp. (Russian).

Klimanov, V.A. 1976. To the Reconstruction of Quantitative Characteristics of Climate of the Past. Vestnik LGU, Ser. 5, Geografia 2: 92-96. (Russian).

Klimanov, V.A. and L.D. Nikilorova. 1982. Climatic Variations in the North East of Europe During the Last 2000 years. Dokl. Akad. Nauk SSSR 267 (1): 164-167. (Russian).

Komarov, V.L. 1926. Introduction to the Study of Vegetation of Yakutia. Trudy Komissii po izucheniyu Yakutskoi ASSR 1. Izd-vo Akad. Nauk SSSR, Leningrad.

Kondratieva, K.A., N.I. Trush, N.I. Chizhova, and N.O. Rybakova. 1976. On the Characteristics of Pleistocene Deposits in Mus-Khaya Outcrop on the Yana River, in Kudryavtsev, V.A., ed. Permafrost studies. Izd-vo MGU, Moscow, vyp. 15:60-93. (Russian).

Korobko, Yu.A. and K.M. Kirsanov. 1979. Female Genital System in Fossel American Bison, in Sokolov, V.E., ed. European Bison. Morphology, Systematics, Evolution, Ecology. Nauka, Moscow, pp. 425-433. (Russian).

Korzhuev, S.S. and R.V. Fedorova. 1962. Chekvurovka Mammoth and Its Habitats. Dokl. Akad. Nauk SSSR 143 (1): 181-813. (Russian).

Koshkina, T.V. 1961. New Data on the Nutrition of the Norwegian Lemming (Lemmus lemmus). Byulleten Moskovskogo obshchevstva ispytatelei prirody, Otelenie Biologii 66 (6): 15-32. (Russian).

Kozhevnikov, Yu.P. 1981. Ecology-floristical Studies on the Indigirka, Kolyma Rivers and in the North East of the Putorana Plateau. Part 1. VINITI, Leningrad. 239 pp. (Russian).

Kozhevnikov, Yu.P. and A.P. Khokhrakov. 1976. On the Flora of Koni Peninsula, in Kontrimavichus, V.L. ed. Flora and Vegetation of the Magadan Region. Vladivostok, pp. 53-64. (Russian).

Kozhevnikov, Yu.P. and V.V. Rapota. 1981. Botanical-Ecological Observations in the Eastern Part of Byrranga Mountains and Adjacent Rolling Plain (Taimyr). Botanical Journal 66 (7): 549-555. (Russian).

Kubiak, H. 1982. Morphological Characters of the Mammoth: an Adaptation to the Artic Steppe Environment, in Hopkins, D.M. et al., eds. Paleoecology of Beringia. Academic Press, New York - London, pp. 281-289.

Kultina, V.V., N.E. Lovelius, and V.V. Kostyukevich. 1974. Palynological and Geochronological Studies of Holocene Deposits in the Novaya River Basin on Taimyr. Botanical Journal 59 (9): 1310-1316. (Russian).

Kupriyanov, L.A. 1957. Pollen Analysis of Plant Remains from the Berezovka Mammoth Stomach (on the problem of Vegetation Pattern for Berezovka Mammoth Epoch, in Sbornik pamyati Afrikana Nikolaevicha Kristofovicha. Mosco-Leningrad, pp. 334-358. (Russian).

Kuzmina, I.E. 1977. On the Origin and History of Theriofauna in the Siberian Artic, in Skarlato, O.A., ed. Anthropogene Fauna and Flora of the Siberian North East. Nauka, Leningrad, pp. 18-55. (Russian).

Larin, I.V. 1958. Food Value of Sedges. Doklady VASKHNIL, Moscow 8: 15-22. (Russian).

Lavrenko, E.M. 1981. On the Vegetation of the Pleistocene Periglacial Steppes of the USSR. Votanical Journal 66 (3): 313-327. (Russian).

Lavrushin, Yu.A., A.L. Devirts, R.E. Giterman, and N.G. Markova. 1963. First Data on the Absolute Chronology of Main Events of the Holocene of the North East of the USSR, in Gromov, V.N., and Ivanova, I.K., eds. Bulletin of the Commission Study of the Quaternary Period No. 28. Izd-vo Akad. Nauk SSR, Moscow, pp. 112-126. (Russian).

Lazarev, P.A. 1982. Fauna of Mammals and Biostratigraphy of the Cenozoic of Yukutia. XI Congress INQA, Moscow, August 1982. Abstracts of Reports 2: 138. (Russian).

Lazukov, G.I. 1970. Anthropogene of the Northern Part of Western Siberia (Stratigraphy). Izd-vo MGU, Moscow, pp. 151-174. (Russian).

Lazukov, G.I. 1972., Anthropogene of the Northern Part of Western Siberia (Paleogeography). Izd-vo MGU, Moscow. 125 pp. (Russian).

Lazukov, G.I. 1973. Beringia and Problems of Formation of Cold-Enduring (Arctic) Flora and Fauna, in Giterman, R.E. et al., eds. Beringian Land and Its Significance for the Development of Holarctic Floras and Faunas in the Cenozoic, in Giterman, R.E. et al., eds. All-Union Symposium. Abstracts of Reports. Khabarovsk, pp. 72-76. (Russian).

Levkovskaya, G.M. 1977. History of Holocene Afforestation of the Artic in the Light of Radiocarbon Dates, in Yakhimovich, V.L, ed. Results of Biostratigraphical, Lithological and Physical Studies of the Pliocene and Pleistocene of the Volga-Ural Region. Ufa, pp. 15-35. (Russian).

Lovelius, N.V. 1979. Variability of Tree Grows. Dendroindicators of Natural Processes and Anaternary Influence. Nauka, Leningrad. 229 pp. (Russian).

Lozhkin, A.V. 1975. New Data on the Age of Quaternary Deposits in the Lower Indigirka River. Dokl. Akad. Nauk SSSR 275 (5): 1143-1146. (Russian).

Lozhkin, A.V. 1977. Living Conditions of Berelekh Mammoth Population, in Skarlato, O.A., ed. Mammoth Fauna of the Russian Plain and East Siberia. Nauka, Leningrad, pp. 67-68. (Russian).

Makeev, V.M. and D.P. Ponomareva. 1988. Paleogeography of Kotelny Island in the Holocene. International Conference "Problems of the Holocene", Tbilisi, October 17-22, 1988. Abstracts of Reports. Tbilisi, p. 69. (Russian).

Makeev, V.M., Kh.A. Arslanov, and V.E. Garutt. 1979. The Age of Mammoths of Severnaya Zemlya and some Problems of Paleogeography of the Late Pleistocene. Dokl. Akad. Nauk SSSR 245 (2): 421-424. (Russian).

Markov, K.K., G.I. Lazukov, and M.P. Grichuk, 1961. Main Regular Features of the Development of Nature of the USSR Territory During the Quaternary (Glacial Epoch-Anthropogene). Izv. Akad. Nauk SSSR, Ser. geogr 4: 10-13. (Russian).

Markov, K.K., A.A. Velichko, G.I. Lazukov, and V.A. Nikolaev. 1968. Pleistocene. Vysshaya Shkola, Moscow. 303 pp. (Russian).

Martin, P.S. 1982. The Pattern and Meaning of the Holarctic Mammoth Extinction, in Hopkins, D.M. et al., eds. Paleoecology of Beringia. Academic Press, Inc., pp. 339-409.

Martin, P.S., B.E. Sabels, and D. Shutler. 1961. Rampart Cave Coprolite and Ecology of the Shasta Ground Sloth. American Journal of Science. 259: 102-27.

Matthews, J.V., Jr., T.W. Anderson, M. Boyko-Diakonow, R.M. Mathewes, J.H. McAndrews, R.J. Mott, P.T.H. Richard, J.C. Ritchie, and C.E. Schweger. 1989. Quaternary Environments in Canada as Documented by Paleobotanical Case Histories; Chapter 7, in Fulton, R.J., ed. Quaternary Geology of Canada and Greenland. Geological Survey of Canada, Geology of Canada, No. 1 (also Geological Society of America, The Geology of North America, v. K-1), pp. 481-539.

Matveeva, N.V. 1979a. Flora and Vegetation of Maria Pronchishcheva Bay, in Aleksadrova, V.D., ed. Arctic Tundras and Polar Deserts of Taimyr. Nauka, Leningrad, pp. 78-109. (Russian).

Mead, J.I., L.D. Agenbroad; O.K. Davis, and P.S. Martin. 1986. Dung of Mammuthus in the arid Southwest, North America. Quaternary Research 25:121-127.

Medvedev, L.N. and N.V. Voronova, 1977. Anaylsis of Geological Sections of Mammoth Cemetery in Northern Yakutia, in Skarlato, O.A., ed. Mammoth Fauna of the Rusian Plain and East Siberia. Nauka, Leningrad, pp. 72-77. (Russian).

Middendorf, A.F. 1860. On Siberian Mammoths. Vestnik Estestvennykh Nauk 26027: 843-847. (Russian).

Mironenko, O.N. and L.N. Savina. 1975. To the History of Forest Vegetation of Middle Siberia on its Northern Limit, in Savina, L.N., ed. History of Forests of Siberia in the Holocene. ILiD, Krasnoyarsk, pp. 37-59. (Russian).

Miroshnikov, L.D. 1958. Remains of Ancient Forest Vegetation on Taimyr Peninsula. Priroda 2: 106-107. (Russian).

Neishtadt, M.I. 1957. History of Forests and Paleogeography of the USSR in the Holocene. Izd-vo Akad. Nauk SSSR, Moscow. 402 pp. (Russian).

Neishtadt, M.I. and N.A. Steklov. 1982. On So Terms of the Holocene and Its Units, in XI Congress INQA, Moscow, August 1982. Abstracts of Reports 3: 232-233. (Russian).

Nikitin, V.P. 1981. Plant Remains from Baby Mammoth's Intestine, in Vereschagin, N.K., and Mikhelson, V.M., eds. Magadan Baby Mammoth. Nauka, Leningrad, pp. 242-246. (Russian).

Nikolskaya, M.V. 1980. Paleobotanical Characteristics of Upper Pleistocene and Holocene Deposits of Taimyr, in Saks, V.N. et al., eds. Paleopalynology of Siberia. Nauka, Moscow, pp. 97-111. (Russian).

Nikolskaya, M.V., N.V. Kind, L.D. Sulerzhitsky and M.N. Cherkasova. 1980. Geochronology and Paleophytological Characteristics of the Holocene of Taimyr, in Ivanova, I.K., and Kind, N.V., eds. Geochronology of the Quaternary Period. Nauka, Moscow, pp. 176-183. (Russian).

Norin, B.N. (Ed.). 1978a. Ary-Mas. Natural Environments, Flora, Vegetation. Nauka, Leningrad. 187 pp. (Russian).

Norin, B.N. 1978b. Vegetational Cover of Ary-Mas Urochishche, in Norin, B.N., ed. Ary-Mas. Natural Environments, Flora, Vegetation. Nauka, Leningrad, pp. 124-162. (Russian).

Ovander, M.G., D.K. Bashlavin, and S.N. Zhigulevtseva. 1982. Late Cenozoic Deposits of the Yana-Indigirka Lowland, in XI Congress INQA, Moscow, August 1982. Abstracts of Reports 2: 209-210. (Russian).

Pavlova, M.V. 1906. Description of Fossil Mammals Collected by the Russian Polar Expedition in 1900-1903. Part I. Zapiski Imperatorskoi Akademii Nauk, Ser. VIII Fiziko-Math. Otd. 21(1): 1-40. St. Petersbourg. (Russian).

Pavlova, M.V. 1910. Description of the Fossil Remains of Mammals in the

Troitskosavsku-Kyakhta Museum. Trudy Troitskosavsk-Kyakhta Ord. Priamyrskogo Otdela Russkog Geograficheskogo Obshchestva 13(1): 21-59. (Russian).

Pavlova, M.V. 1924. Causes of Extinction of Animals in the Past Geological Epochs, in Present Problems of Natural Science. Gosizdat, Moscow-Petrograd, pp. 5-88. (Russian).

Piavchenko, N.I. 1968. Reflection of the Present-Day Composition of Forests in Recent Pollen Spectra. Botanical Journal 53(2): 174-189. (Russian).

Piavchenko, N.I. 1982. Dynamics of Forest Vegetation of West Siberia in the Holocene, in XI Congress INQA, Moscow, August 1982. Abstracts of Reports 2: 236-237. (Russian).

Popov, A.I. 1950. Taimyr Mammoth. Voprosy Geografii 23: 296-305. (Russian).

Popov, A.I. 1959. A Mammoth from Taimyr and Problem of the Mammoth Fauna Remnants in the Siberian Quaternary Deposits, in Markov, K.K., and Popov, A.I., eds. Ice Age in Europe - A Section of the USSR, and in Siberia. MGU, Moscow, pp. 259-275. (Russian).

Popov, A.I. 1982a. Genesis and Paleogeographic Conditions of Formation of the Sedimentary-Cryogenic (Edoma) Complex on the Primorsky Plains of Subarctic, in XI Congress INQA, Moscow, August 1982. Abstracts of Reports 2: 259-260. (Russian).

Popov, A.I. 1982b. On the Genesis and Conditions of Formation of the Sedimentary-Cryogenic (Edoma) Complex on the Primorsky Plains of Subarctic. Vestnik MGU, Ser. Georg. 6: 60-65. (Russian).

Popov, Yu.N. 1948. Find of Fossil Mammal Carcasses in the Pleistocene Permafrost Layers of the Siberian Northeast. Byulleten Komissii po Izucheniyu Chetvertichnogo Perioda 13: 74-75. (Russian).

Rapota, V.V. 1981. Vascular Plants of the Bikada River (Eastern Taimyr) and their Feeding Value for Musk Oxen, in Solomakha, A.I. et al., eds. Ecology and Economic Use of Terrestrial Fauna of the Yenisei North. Siberian Branch of VASKhNIL, Novosibirsk, pp. 73-92. (Russian).

Reference Book of the Climate of the USSR. 1966. Gidrometeoizdat, Leningrad. Vyp. 25, ch. 2, 312 pp. Vyp. 33, ch. 2, 288 pp; 1967. Gidrometeoizdat, Leningrad. Vyp. 21, ch. 2, 504 pp.; 1969. Gidrometeoizdat, Leningrad. Vyp. 25, ch. 4, 275 pp.; 1969. Krasnoyarsk. Vyp. 21, ch. 2, 649 pp.; 1970.

Krasnoyarsk. Vyp. 21, ch. 1, 1000 pp.; 1970. Magadan, Vyp. 33, ch. 2, 331 pp.; 1972. Yakutsk. Vyp. 24, ch. 1, 472 pp.; 1973. Yakutsk. Vyp. 24, ch. 2, 572 pp. (Russian).

Ritchie, J.C. 1984. Holocene Pollen Record of Boreal Forest History from the Travaillant Lake Area, Lower Mackenzie River Basin Canad. J. Bot. 62 (7): 1385-1392.

Ritchie, J.C., L.C. Cwynar, and R.W. Spear. 1983. Evidence from Northwest Canada for an Early Holocene Milankovitch Thermal Maximum. Nature 305 (5930): 126-128.

Romanova, E.M. 1978. Micro- and Mezoclimates of Taimyr, in Tomilin, B.A., ed. Structure and Functions of Biogeocenoses of Taimyr Tundra. Nauka, Leningrad, pp. 5-29. (Russian).

Rybakova, N.O. 1979. Spore and Pollen Spectra of Upper Quaternary Deposits of the Yana-Indigirka Lowland, in Grichuk, V.P., ed. Palynological Studies in the North East of the USSR. Vladivostok, pp. 62-66. (Russian).

Rybakova, N.O. and M.B. Chernysheva. 1979. Data on Holocene Flora of Yakutia. Byulleten Moskovskogo Obshchestva Ispytatelei Prirody. Otdel Geol. 54 (4): 125-131. (Russian).

Rybakova, N.O. and L.G. Pirumnova. 1983. Micropaleobotanical Studies of Middle-Upper Quaternary Deposits of the Khroma River Basin (Yana-Indigirka Lowland). Vestnik MGU, Ser. 4, Geologia 5: 28-33. (Russian)

Sher, A.V. 1970. Pleistocene Mammal Fauna of Plain Coasts of the East Siberian Sea and the Problem of Beringia Land, in Tolmachev, A.I., ed. The Arctic Ocean and its Coast in the Cenozoic. Gidrometeoizdat, Leningrad, pp. 516-524. (Russian).

Sher, A.V. 1971. Mammals and Stratigraphy of the Pleistocene in the Far East of the USSR and North America. Nauka, Moscow, 310 pp. (Russian).

Sher, A.V., T.N. Kaplina, R.E. Giterman, A.V. Lozhkin, A.A. Arkhangelov, E.N. Virina, V.S. Zazhigan, S.V. Kisilev, and Yu.V. Kuznetsov. 1979. Guidebook of Excursion XI, in Sher, A.V., and Kaplina, T.N. eds. Late Cenozoic Deposits of the Kolyma Lowland. XIV Pacific Scientific Congress. Akademia Nauk, Moscow. 112 pp. (Russian).

Shilo, N.A. and A.V. Lozhkin. 1981. Radiocarbon Dating of the Baby Mammoth, in Vereshchagin, N.K., and Mikhelson, V.M., eds. Magadan Baby

Mammoth. Nauka, Leningrad, p. 48. (Russian).

Shilo, N.A., A.V. Lozhkin, E.E. Titov, and Yu.V. Shumilov. 1983. Kirgilyakh Mammoth. Nauka, Moscow. 209 pp. (Russian).

Shmalgauzen, I.I. 1968. Factors of Evolution. Theory of Stabilizing Selection. Nauka, Moscow. 451 pp. (Russian)

Shumilova, L.V. 1962. Botanical Geography of Siberia. Izd-vo Tomskogo Universiteta, Tomsk. 433 pp. (Russian).

Shvedchikov, G.V. 1975. Arctophila fulva (Trin.) as an Edificator of Hygrophytons in the North of Yakutia. Thesis of Dissertation, Botanical Institute, Academy of Sciences of the USSR, Leningrad, 20 pp. (Russian).

Simpson, G.G. 1944. Tempo and Mode in Evolution.

Sladkov, A.N. 1967. Introduction into Spore and Pollen Analysis. Nauka, Moscow. 270 pp. (Russian).

Sokolova, M.V. 1977. Vascular Plants of the Fossil Horse Burial Site, in Skarlato, O.A., ed. Anthropogene Fauna and Flora of the Siberian North East. Nauka, Leningrad, pp. 254-264., (Russian).

Solonevich, N.G. and V.V. Vikhireva-Vasilkova. 1977. Preliminary Results of a Study of Plant Remains from the Gastrointestinal Tract of the Shandrin Mammoth (Yakutia), in Skarlato, O.A., ed. Anthropogene Fauna and Flora of the Siberian North East. Nauka, Leningrad. pp. 208-221. (Russian).

Solonevich, N.G., B.A. Tikhomirov, and V.V. Ukraintseva. 1977. Prelimianry Results of the Investigation into the Remains from Shandrin Mammoth Gastrointestinal Contents (Yakutia), in Skarlato, O.A., ed. Anthropogene Fauna and Flora of the Siberian North East. Nauka, Moscow, pp. 277-280. (Russian).

Soper, J.D. 1941. History, Range, and Home Life of the Northern Bison. Ecological Monographs 11 (4): 349-442.

Spaulding, W.G. and P.S. Martin. 1979. Ground Sloth Dung of the Guadalupe Mountains, in Genoway, H.H., and Baker, R.J., eds. Biological Investigation in the Guadalupe Mountains National Park, Texas. National Park Service Proc. and Trans., Washington, D.C. Ser. 4: 259-269.

Stanishcheva, O.N. 1982. Natural Environment of the Area of Yuribei Mammoth's Death from Paleocarpological Data, in Sokolov, V.E., ed. Yuribei Mammoth. Nauka, Moscow, pp. 30-35. (Russian).

Sukachev, V.N. 1914. Research into the Plant Remains from the Food Contents of the Berezovka Mammoth Find, in Scientific Results of the Expedition, Established by the Academy of Sciences for Unearthing the Mammoth on the Berezovka River in 1901. St. Petersbourg, pp. 1-17. (Russian).

Sulerzhitsky, L.D. 1976. Radiocarbon Method and Dynamics of Distribution of Forests in the Tundra Zone, in Denisman, L.G., ed. History of Biogeocenoses of the USSR in the Holocene. Nauka, Moscow, pp. 146-149. (Russian).

Szafer, W. 1946. Flora Pliocenska z Kroscienka n/Dunajcem. Rozprawy wydzialu Matematyczno-pryrodniczego, tom 72. Dzial B (seria III), n. 1-2. Krakow, 213 S.

Tikhomirov, B.A. 1941. On the Forest Phase in Post-Glacial History of Vegetation in the North of Siberia and its Relics in Present-Day Tundra, in Materials on the History of Flora and Vegetation of the USSR. Izd-vo Akad. Nauk SSSR, vyp. 1: 315-374. (Russia).

Tikhomirov, B.A. 1950a. To the Characteristics of Vegetational Cover of the Mammoth Epoch on Taimyr. Botanical Journal 35 (5): 482-497. (Russian).

Tikhomirov, B.A. 1950b. Data on Transportation of Pollen of Free Species North of the Forest Boundary. Dokl. Akad. Nauk SSSR 71 (4): 755. (Russian).

Tikhomirov, B.A. 1958. On North Siberian Environmental Conditions and Vegetation of Mammoth Epoch, in Tikhomirov, B.A., ed. Problems of the North, 1. Nauka, Leningrad, pp. 168-188. (Russian).

Tikhomirov, V.A. 1962. Main Stages of Development of Vegetation of the North of USSR Related to Climatic Variations and Human Activity. Byulleten MOIP 67 (1): 34-57. (Russian).

Tikhomirov, B.A. and L.A. Kupriyanova. 1954. Pollen Anaylsis of Plant Remains from the Forage Content of the Berezovka Mammoth. Dokl. Akad. Nauk SSSR 95 (6): 1313-1315. (Russian).

Tikhomirov, B.A. and V.V. Kultina (Ukraintseva). 1973. Pollen and Spore Analysis from the Selerikhan Horse Stomach. Dokl. Akad. Nauk SSSR 209 (6): 1464-1466. (Russian).

Titov, E.E. 1982. Burial Conditions and Successive Displacements of the Kirgilyakh Baby Mammoth, in XI Congress INQA, Moscow, August 1982. Abstracts of Reports 1: 254-255. (Russian).

Tomirdiaro, S.V. 1972. Permafrost and Human Activity in Highlands and Lowlands. Magadanskoe Knizhnoe Izdatelstvo, Magada. 174 pp. (Russian).

Tomirdiaro, S.V. 1980. Loess-Ice-Laid Deposits of East Siberia in the Late Pleistocene and Holocene. Nauka, Moscow. 183 pp. (Russian).

Tomskaya, A.I. 1981. Palynology of the Cenozoic of Yakutia. Nauka, Siberian Branch, Novosibirsk. 221 pp. (Russian).

Trofimov, B.A. 1955. To the Problem of Origin and Development of Mammal Fauna of the Quaternary Period of Temperate and Northern Zones, in Gromov, V.I., and Ivanova, I.K., eds. Transactions of the Commission on the Study of the Quaternary Period 12: 106-120. (Russian).

Tuhkanen, Sakari. 1980. Climatic Paramenters and Indices in Plant Geography. Acta Phytogeogr. Suec. 67. Upsala. 109 pp.

Tyulina, L.N. 1937. Forest Vegetation of the Khatanga Area at its Northern Limit. Leningrad, Trudy ANIL 63: 83-180. (Russian).

Ukraintseva (Kultina), V.V. 1977. Flora and Vegetation Reconstruction for Selerikhan Fossil Horse Life-Time Based on Palynological Data, in Skarlato, O.A. ed. Anthropogene Fauna and Flora of the Siberian North East. Nauka, Leningrad, pp. 223-253. (Russian).

Ukraintseva, V.V. 1979. Vegetation of Warm Intervals of Late Pleistocene and Extinction of Some Large Herbivorous Mammals. Botanical Journal 64 (3): 318-330. (Russian).

Ukraintseva, V.V. 1981a. Environment and the Causes of the Death of Baby Mammoth, in Vereschagin, N.K., and Mikhelson, V.M., eds. Magadan Baby Mammoth. Nauka, Leningrad, pp. 254-261. (Russian).

Ukraintseva, V.V. 1981. Vegetation of Warm Late Pleistocene Intervals and the Extinction of Some Large Herbivorous Mammals. Polar Geography and Geology. October-December. Y.H. Winston and Sons, pp. 189-203.

Ukraintseva, V.V. 1982a. Environment and the Causes of Mammoth Death in the Middle Reaches of the Yuribei River (Gydan Peninsula), in Sokolov,

V.E., ed. Yuribei Mammoth. Nauka, Moscow, pp. 19-29. (Russian).

Ukraintseva, V.V. 1982b. Forage of the Big Herbivouous Mammals of the Epoch of Mammoths. Third International Theriological Congress, Helsinki, August 15-20, 1982. Abstracts of Papers, p. 247.

Ukraintseva, V.V. 1985. Forage of the Large Herbivorous Mammals of the Epoch of Mammoth. Acta Zoologica Fennica 190: 215-220.

Ukraintseva, V.V. 1986a. On the Composition of the Forage of the Large Herbivorous Mammals of the Mammoth Epoch: Significance for Palaeobiological and Palaeogeographical Reconstruction. Quartarpalaontologie. 6: 231-238. Akademie-Verlag, Berlin.

Ukraintseva, V.V. 1986b. Larch Forests in the Early Holocene of Arctic. Priroda 8: 121. (Russian).

Ukraintseva, V.V. 1988. Paleobotanical Evidence of the Early Holocene Climate Warming at High Arctic Latitudes. International Conference "Problems of Holocene", October 18-22, 1988. Tbilisi. pp. 106-107.

Ukraintseva, V.V. and Yu.P. Kozhevnikov. 1979. The Vegetation of Kirgilyakh Mammoth Site (Upper Kolyma River Basin). Botanical Jounal 64 (8): 1091-1098. (Russian).

Ukraintseva, V.V. and Yu.P. Kozhevnikov. 1981. The Vegetation Cover in the Area of Taimyr Mammoth Find (Bolshaya Lesnaya Rasokha River, Southwestern Taimyr). Botanical Journal 66 (7): 978-992. (Russian).

Ukraintseva, V.V., C.C. Flerov, and N.G. Solonevich. 1978. Botanical Anaysis of Plant Remains from Gastrointestinal Tract of the Yukutina Bison. Botanical Journal 63 (7): 1001-1004. (Russian).

Ukraintseva, V.V., Kh.A. Arslanov, Zh.M. Belorusova, and M.S. Boch. 1981. Vegetation and Natural Conditions of the Bolshaya Lesnaya Rassokha River Basin in the Upper Pleistocene (in Connection with Mammoth Find). Botanical Journal 66 (10): 1444-1453. (Russian).

Ukraintseva, V.V., Kh.A. Arslonov, Zh.M. Belorusova, and V.N. Ustinov. 1989. First Data on Early Holocene Flora and Vegetation of Bolshoi Lyakhovsky Island (Novosibirski Islands). Botanical Journal 74 (6): 782-793. (Russian).

Van Campo, M. 1984. Relations entre la vegetation de l'Europe et les

Temperatures de surface oceaniques apres le dernier maximum glaciaire. Pollen at Spores 24 (3-4): 497-518.

Vangengeim, E.A. 1977. Paleontological Substantiation of Stratigraphy of the Anthropogene of North Eurasia (for Mammals). Nauka, Moscow, 172 pp. (Russian).

Vartanyan, S.L., V.E. Garutt, and A.V. Sher, 1993. Holocene dwarf mammoths from Wrangel Island in the Siberian arctic. Nature 362:337-340.

Vaskovsky, A.P. 1957. Spore and Pollen Spectra of Present-Day Vegetational Communities of the Far North East of USSR and their Significance for Reconstructions of Quaternary Vegetation, in Evangulov, B.B. ed. Materials Concerning the Geology and Mineral Resources of the North East of the USSR. Magadanskoe Knizhnoe Izdatelstvo, Magadan, vyp. 2: 130-178. (Russian).

Vaskovsky, A.P. 1958. New Data on the Limits of Distribution of Trees and Bushes Ofedificators in the Far North East of the USSR, in Evangulov, B.B., ed. Materials Concerning the Geology and Mineral Resources of the North East of the USSR. Magadanskoe Knizhnoe Izdatelstvo, Magadan, vyp. 13: 187-204. (Russian).

Vaskovsky, A.P. 1959. Essay of the Vegetation, Climate and Chronology of the Quaternary Period in the Upper Kolyma, Indigirka River Basins and on the Northern Coast of the Sea of Okhotsk, in Markov, K.K, and Popov, A.I., eds. Ice Age in the European Section of the USSR and in Siberia. MGU, Moscow, pp. 259-275. (Russian.)

Veinbergs, I., M. Voschilko, A. Shpetalenko, V. Stelle, and B.K. Rose. 1976a. Spore/Pollen Characteristics of New Sections of Late Quaternary Deposits in Nearshore Land Zone and Shelf of the SW Sea of Okhotsk, in Bartosh, T.D. et al., eds. Palynology in Continental and Marine Geologic Investigations. Zinatne, Riga, pp. 133-147. (Russian).

Veinbergs, I., M. Voschilko, V. Stelle, A. Savvaitov, M. Rosenblats, F. Kovalenko, and V. Veksler. 1976b. Spore/Pollen Spectra of Costal and Marine Deposits and Climatic-Floral Changes in Chaun Bay Area, in Bartosh, T.D. et al., eds. Palynology in Continental and Marine Geologic Investigations. Zinatne, Riga, pp. 119-131. (Russian).

Velichko, A.A. 1973. The Natural Process in the Pleistocene. Nauka, Moscow. 256 pp. (Russian).

Velichko, A.A. 1981. To the Problem of the Succession and Principal Structure of Main Climatic Pleistocene Rhythms, in Velichko, A.A., Problems of Pleistocene Palaeogeography in Glacial and Periaglacial Regions. Nauka, Moscow, pp. 220-246. (Russian).

Velichko, A.A. and M.A. Faustova. 1982. Problems of Substantiation of Maximal Limit of Late Pleistocene Glaciation in the North of Eurasia, in Velichko, A.A., ed. Evolution of Nature of the USSR Territory in the Late Pleistocene and Holocene. Nauka, Moscow, pp. 7-16. (Russian).

Vereshchagin, N.K. 1971. Excavations of the Mammoth Cemetery on the Berelekh River. Izd-vo Akad. Nauk SSSR 8: 85-93. (Russian).

Vereshchagin, N.K. 1975. On the Mammoth from the Shandrin River. Vestnik Zoologii 2: 81-84. (Russian).

Vereshchagin, N.K. 1977. Berelekh Mammoth Cemetery, in Skarlato, O.A., ed. Mammoth Fauna of the Russian Plain and East Siberia. Nauka, Leningrad, pp. 5-50. (Russian).

Vereshchagin, N.K. 1979. Why Did Mammoths Die Out? Nauka, Leningrad. 194 pp. (Russian).

Vereshchagin, N.K. 1981. Morphological Description of the Baby Mammoth, in Vereshchagin, N.K., and Mikhelson, V.M., eds. Magadan Baby Mammoth. Nauka, Leningrad, pp. 52-80. (Russian).

Vereshchagin, N.K. and P.A. Lazarev. 1977. Description of Parts of the Corpse and Skeletal Remains of the Selerikhan Horse, in Vereshchagin, N.K., and Mikhelson, V.M., eds. Anthropogene Fauna and Flora of the Siberian North East. Nauka, Leningrad, p. 85. (Russian).

Vereshchagin, N.K. and G.F. Baryshnikov. 1980. Palaeoecology of Late Mammoth Fauna in the Arctic Zone of Eurasia. Byulleten Moskovskogo Obshchestba Ispytatelei Prirody 2: 5-19. (Russian).

Vereshchagin, N.K. and A.I. Nikolaev. 1981. Remains of Other Mammals from the Baby Mammoth Burial Site, in Vereshchagin, N.K. and Mikhelson, V.M., eds. Magadan Baby Mammoth. Nauka, Leningrad, pp. 238-241. (Russian).

Vereshchagin, N.K. and A.I. Nikolaev. 1982. Excavations of the Khatanga Mammoth. Trudy Zoologicheskogo Instituda Akad. Nauk SSSR 3: 3-17. (Russian).

Vereshchagin, N.K. and G.F. Baryshnikov. 1985. Extinction of Mammals During the Quaternary in Northern Eurasia. Trudy Zoologicheskogo Instituta Akad. Nauk SSSR 131: 3-35. (Russian).

Vereshchagin, N.K. and V.V. Ukraintseva. 1985. The Origin and Stratigraphy of Berelekh Mammoth Burial Site. Trudy Zoologicheskogo Instituta Akad. Nauk SSSR 131: 104-112. (Russian).

Volkova, V.S. 1977. Stratigraphy and History of the Development of Vegetation of Western Siberia in the Late Cenozoic. Nauka, Moscow. 236 pp. (Russian).

Volkova, V.S. and T.P. Levina. 1982. Vegetation of the Holocene of Western Siberia from Palynological Data, in Velichko, A.A., ed. Evolution of Nature of the USSR Territory in the Late Pleistocene and Holocene. Nauka, Moscow, pp. 186-192. (Russian).

Volkova, V.S., M.P. Votakh, and V.A. Belova. 1984. Main Stages of Climatic Changes of Siberia in Quaternary Time, in Khlonova, A.F. et al., eds. Problems of Modern Palynology. Nauka, Siberian Branch, Novosibirsk, pp. 147-152. (Russian).

Young, S.B. 1971. The Vascular Flora of St. Lawrence Island With Special Reference to Floristic Zonation in the Arctic Regions. In Contributions from the Gray Herbaruim. No. 201. ed. R.C. Rollins and K. Raby. Howard University, Cambridge.

Yudichev, Yu.F. and A.I. Averikhin. 1982. On the Macro-Micromorphology of Abdominal Cavity of the Shandrin Mammoth and Causes of Its Death. Trudy Zoologicheskogo Instituta Akad. Nauk SSSR 111:35-37. (Russian).

Yurtsev, B.A., A.I. Tolmachev, and O.V. Rebristaya. 1978. Floristic Limits of the Arctic, in Tolmachev, A.I., and Yurtsev, B.A., eds. Arctic Floristic Region. Nauka, Leningrad, pp. 9-104. (Russian).

Zagwijn, W.H. 1963. Pleistocene Stratigraphy in the Netherland, Based on Changes in Vegetation and Climate. Verhandelingen van het kohinklijk Nederlands Geologisch mijnbouwkundig genootschap. Geol. Serie, deel 21 (2): 173-196.

Zaklinskaya, E.D. 1954. On the Problem of Vegetation Cover During Taimyr Mammoth Lifetime. Dokl. Akad. Nauk SSSR 98 (3): 171-174. (Russian).

Zavadsky, K.M. and E.I. Kolchinsky. 1977. Evolution of Evolution. Historico-

Critical Essays, Problems. Nauka, Leningrad. 236 pp. (Russian).

Zenkova, E.A. 1954. Liverworts from the Taimyr Mammoth Site. Botanical Journal 29 (6): 915-917. (Russian).

Zherekhova, I.E. 1977. Description and Measurements of Teeth of Berelekh Mammoths. Trudy Zoologicheskogo Instituta Akad. Nauk SSSR 72: 50-58. (Russian).

Zubakov, V.A. and I.I. Borzenkova. 1983. Paleoclimates of the Upper Cenozoic. Gidrometeoizdat, Leningrad. 213 pp. (Russian).

Zubkov, A.I. 1931. To the Problem of Climatic Variations in the North of Siberia in Post-Glacial Time, in Tolmachev, A.I., ed. Trudy Polyarnoi Komissii Akad. Nauk SSSR 5: 31-36. (Russian).

Pollen from the forage of the
Selerikhan Horse

PLATE I.   1.) *Picea obovata* Ledeb.; 2.) *Picea*
cf. *ajanensis* Fisch.; 3, 9.) *Pinus
pumila* Regel.; 4, 5.) *P. sylvestris*
L.; 6.) *Larix* cf. *gmelinii* (Rupr.)
Rupr.; 7, 8) *Juniperus* sp.; 10, 11.)
Populus suaveolens Fisch.; 1000[x].

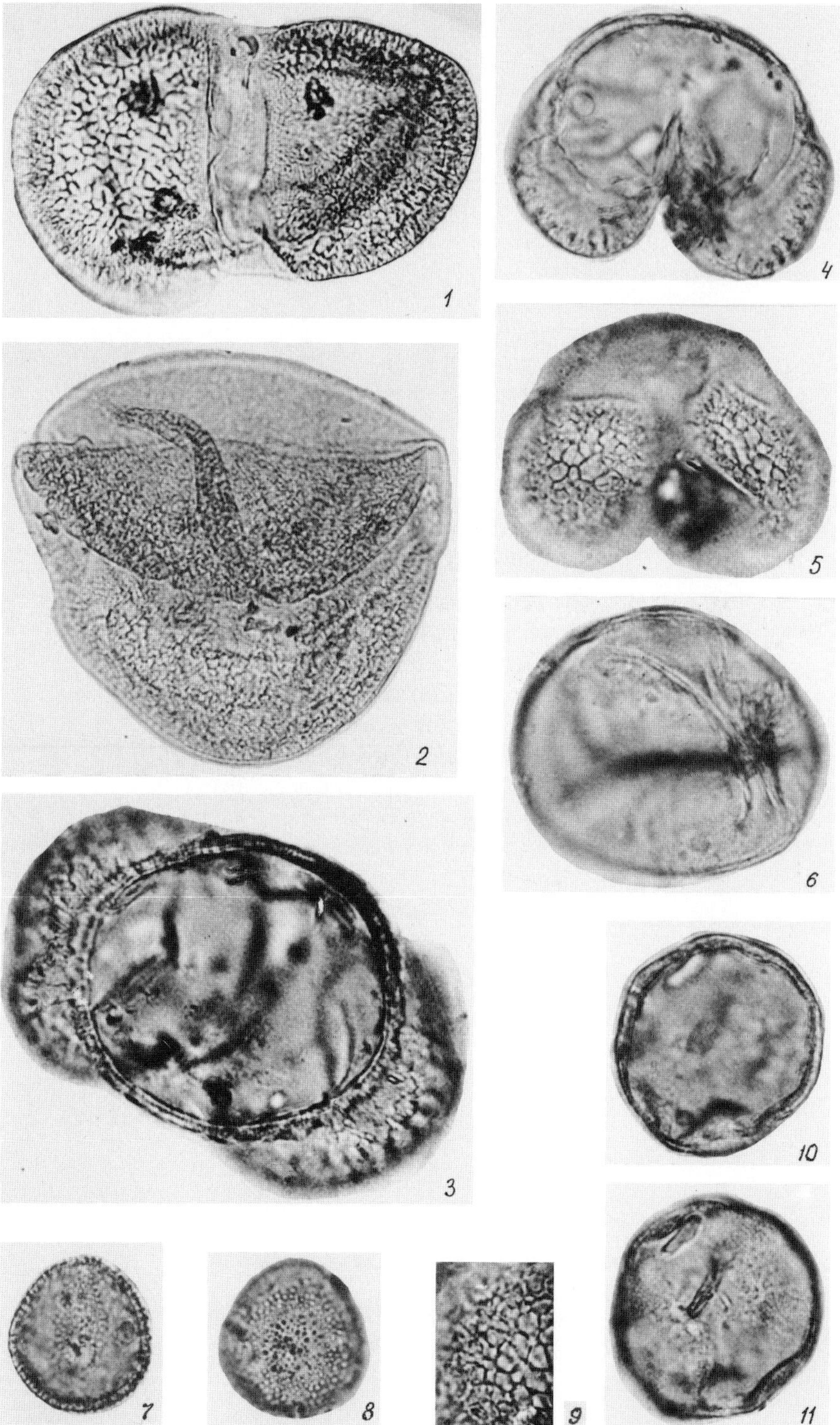

269

Pollen from the forage of the
Selerikhan Horse

PLATE II.   1-5.) *Salix* cf. *glauca* L.; 6-10.) *Salix*
cf. *caprea* L.; 11-15.) *Salix polaris*
Wehlenb.; 16, 17.) *Corylus* cf.
cornuta Marsh.; 18.) Alnus hirsuta
(Spach) Turcz. ex Rupr.; 19, 20, 22,
23.) *Betula platyphylla* Sukacz.; 24,
25.) *Betula exilis* Sukacz.; 26)
Betula sp.; 1000$^{\times}$.

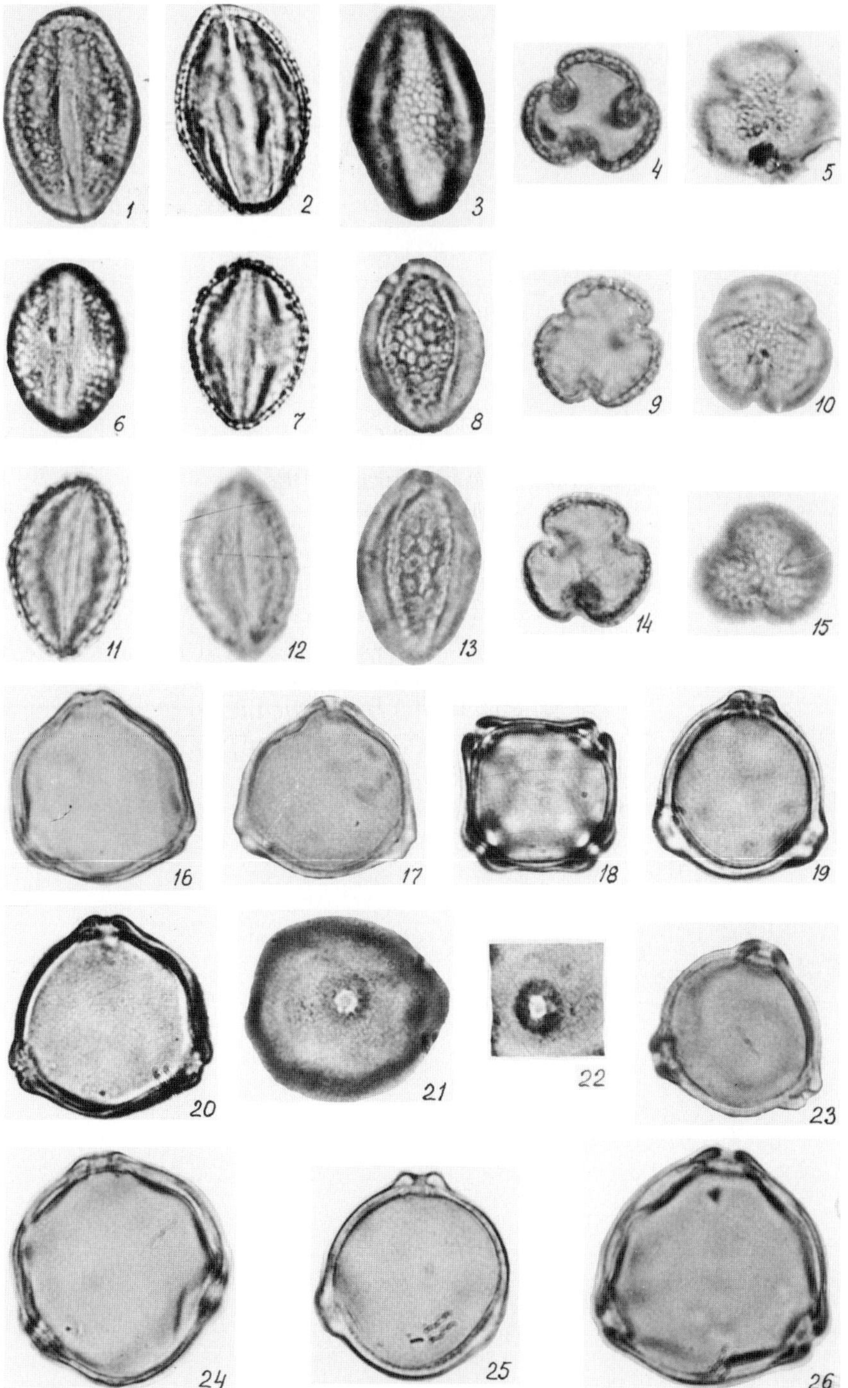

Pollen from the forage of the
Selerihkah Horse

PLATE III.   1, 2.) *Ulmus pumila* L.; 3-6, 8.)
Ulmus cf. *japonica* (Rehd.) Sarg.
(*Ulmus* cf. *propinqua* Koidz.); 7.)
Allium shoenoprasum L.; 10, 11.) *A.
strictum* Schrad.; 9) Poaceae gen.
et. sp.; 12-14.) *Poa arctica* R. Br.;
15.) *Carex* sp.; 16.) Poaceae gen. et.
sp.; 1000^{x}.

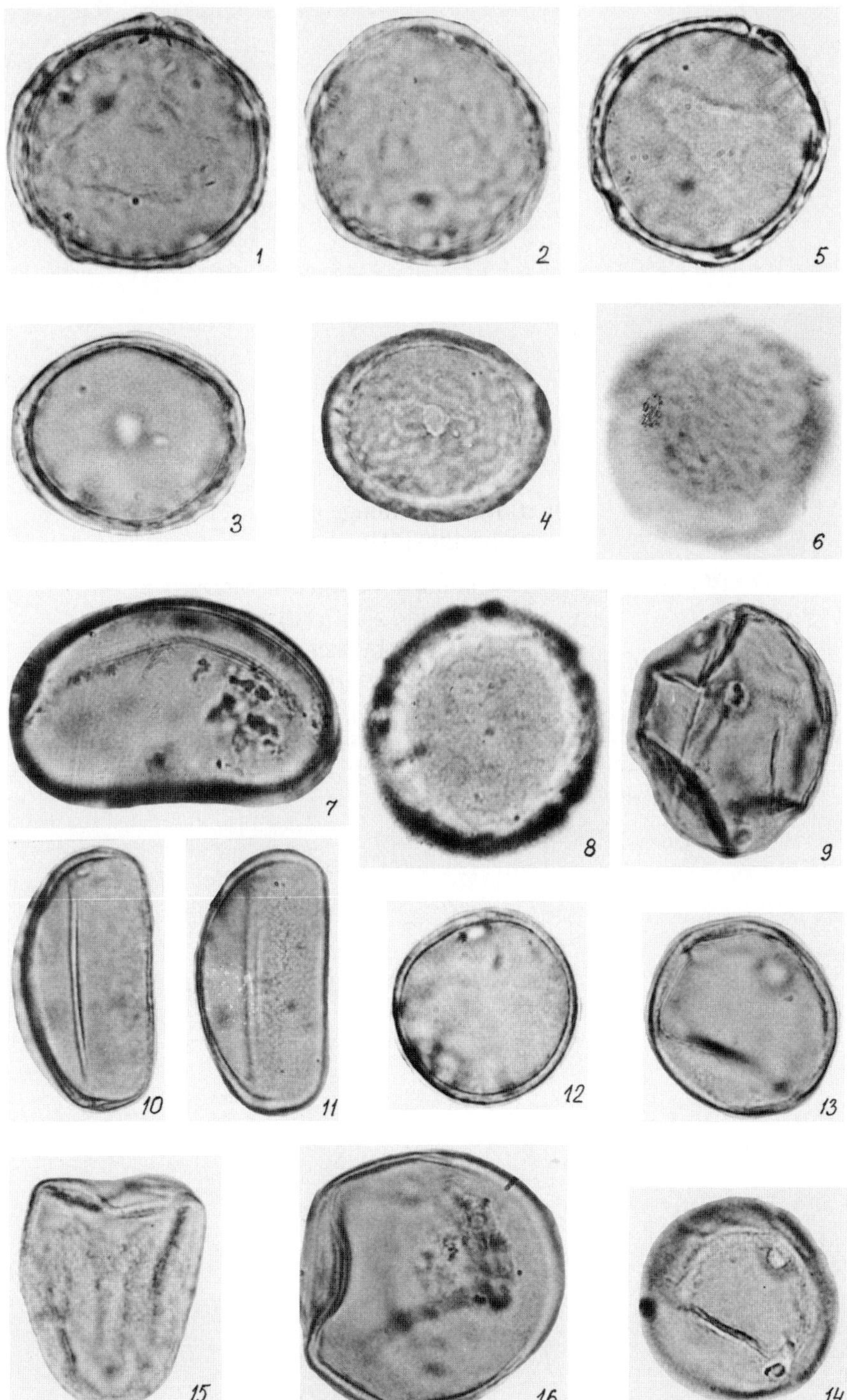

Pollen from the forage of the
Selerikhan Horse

<table>
<tr><td>PLATE IV.</td><td>1-4.) *Stellaria jucutica* Schischk.; 5, 6.) *Minuartia* sp.; 7, 8.) *Silene* Sp.; 9, 10.) Caryophyllaceae gen. et sp.; 11, 12.) *Lychnis sibirica* L.s.l.; 13, 14, 17, 18.) *Lychnis* sp.; 15, 16.) *Minauartia arctica* (Stev.) Aschers. et Graebn.; 1000^{x}.</td></tr>
</table>

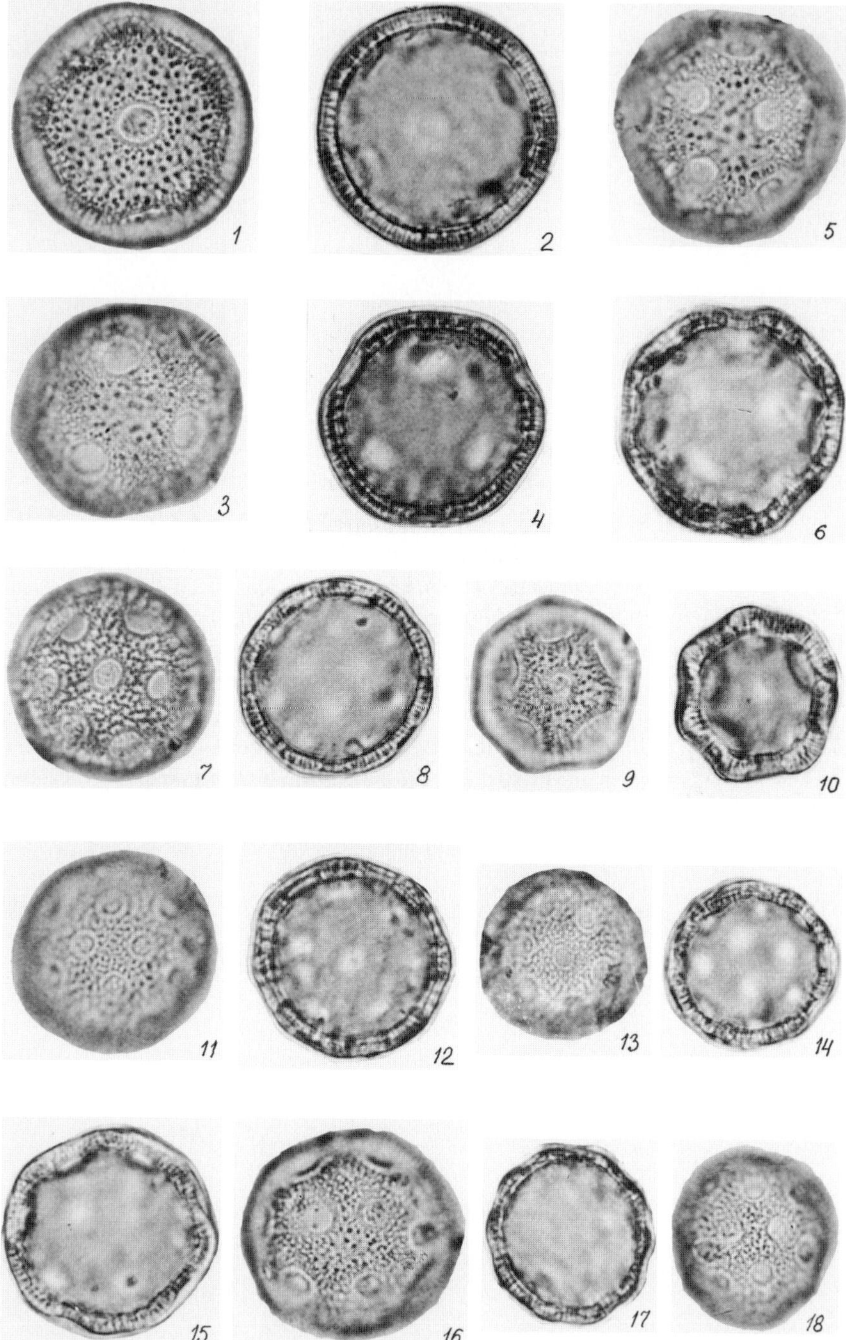

Pollen from the forage of the
Selerikhan Horse

PLATE V.   1-3.) *Polygonum aviculare* L.; 4-6.)
Ranunculus sp.; 7-8.) *Cicuta* sp.;
9-11.) *Castilleja pallida* Kunth.; 12.)
Thalictrum sp.; 13-15.) *Caltha* cf.
arctica R. Br.; 16.) *Nuphar pumila*
(Timm) DC.; 17.) *Sanquisorba officinalis* L.; 18, 19.) *Ranunculus af-
finis* R. Br.; 10, 21.) Rosaceae gen.
et sp.; 22, 24.) *Potentilla stipularis*
L.; 1000^{x}.

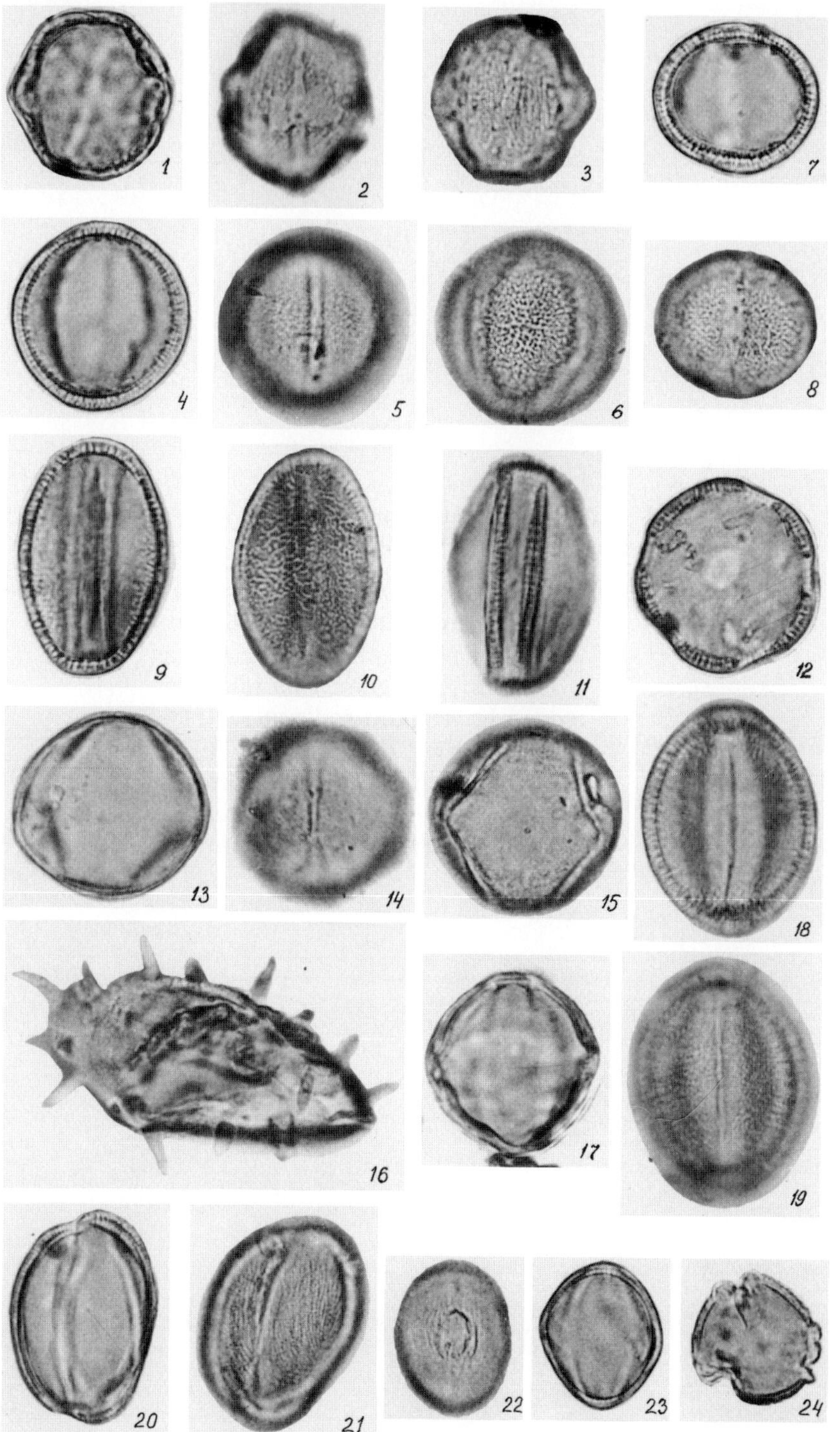

Pollen from the forage of the
Selerikhan Horse

PLATE VI.   1-4, 9, 10.) *Comarum palustre* L.; 5,
6.) *Pachyplerum alpinum* (Ledeb.)
R. Pol.; 7, 8.) *Potentilla* cf. *chinensis* Ser.; 11.) *Heracleum sibiricum*
L.; 12, 13.) *Oxytropis* sp.; 14.)
Epilobium palustre L.; 15, 16.)
Phlox sibirica L.; 17, 18.) *Artemisia*
sp.; 19, 20.) *Valeriana capitata* L.;
21, 22.) *Artemisia vulgaris* L.;
1000^{x}.

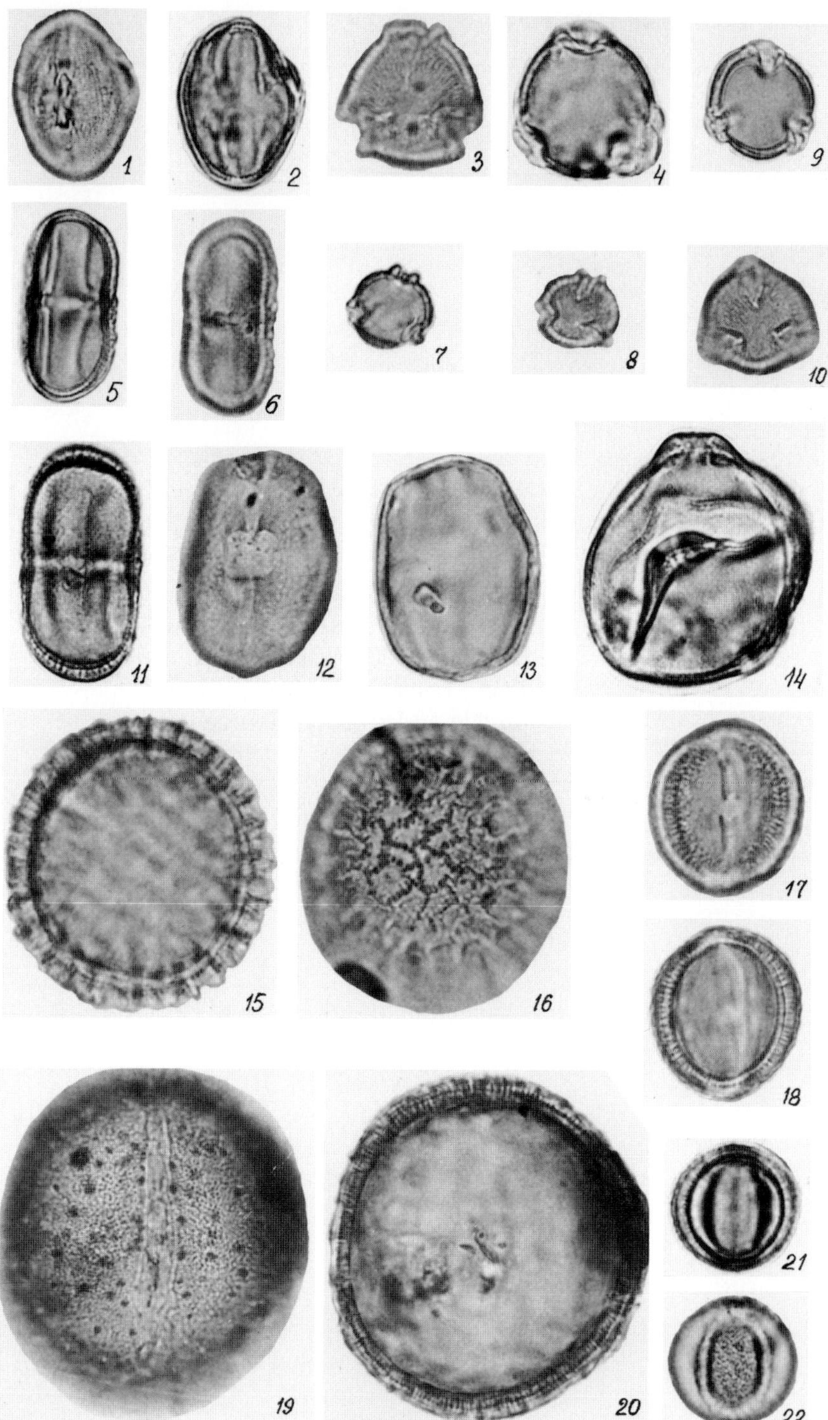

Pollen from the forage of the
Selerikhan Horse

PLATE VII. 1-4.) *Saussurea tilesii* (Ledeb.)
Ledeb.; 5-8.) *Senecio congestus* (R.
Br.) DC.; 9-10.) Saussurea sp.;
11-13.) *Lactuca sibirica* (L.) Maxim.; 14.) Monocotyledoneae inter.;
1000^{x}.

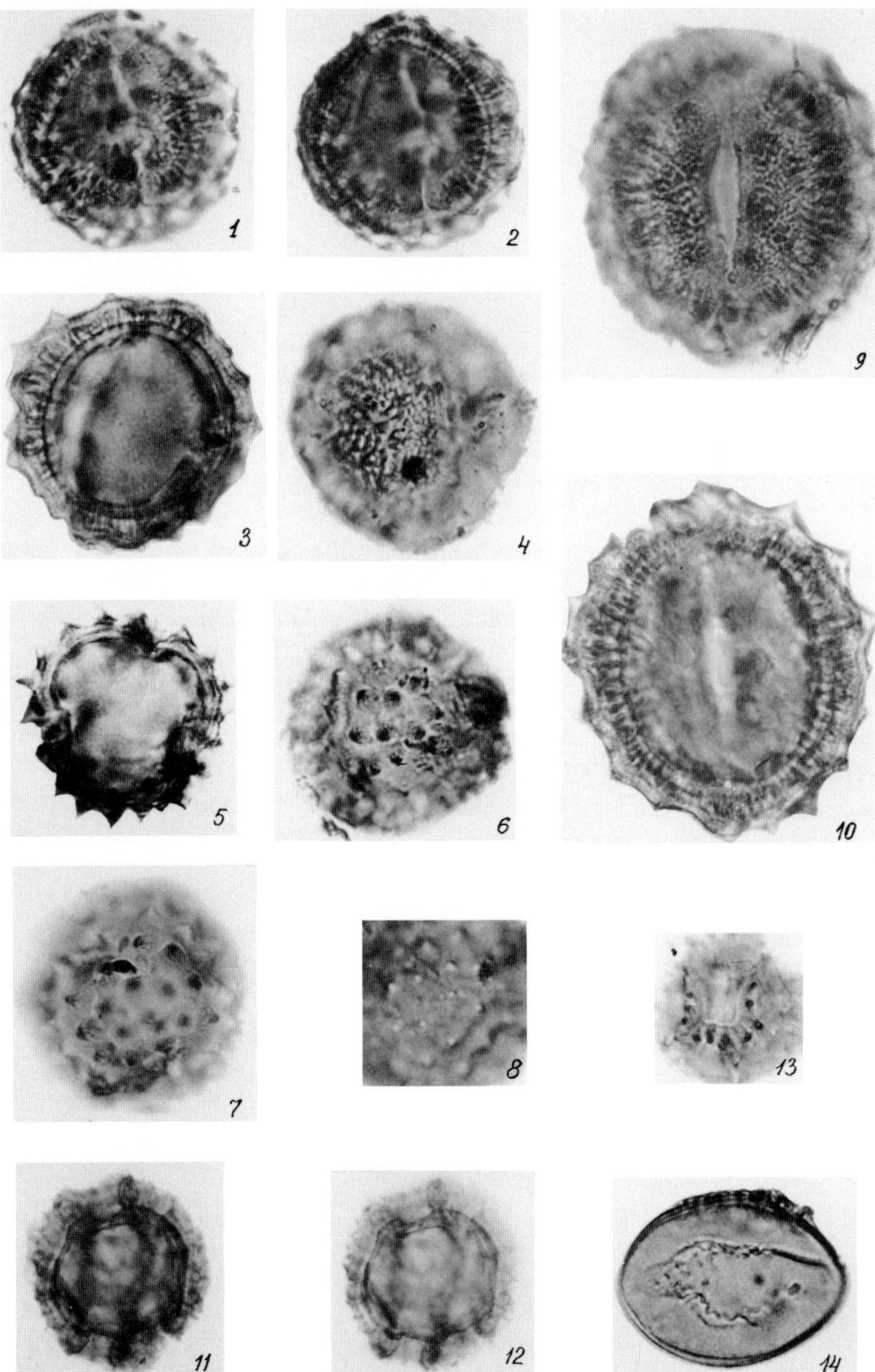

Pollen from the forage of the
Selerikhan Horse

PLATE VIII.   1, 2.) *Selaginella pleistocenica*
Ukraints.; 3, 4.) *S. rupestris* (L.)
Spring.; 5, 6.) *Lycopodium alpinum*
L.; 7, 8.) *Dicranum* sp.; 9.)
Polypodiaceae gen. et. sp.; 10, 11.)
Nymphaea tetragona Georgi.;
1000^{x}.

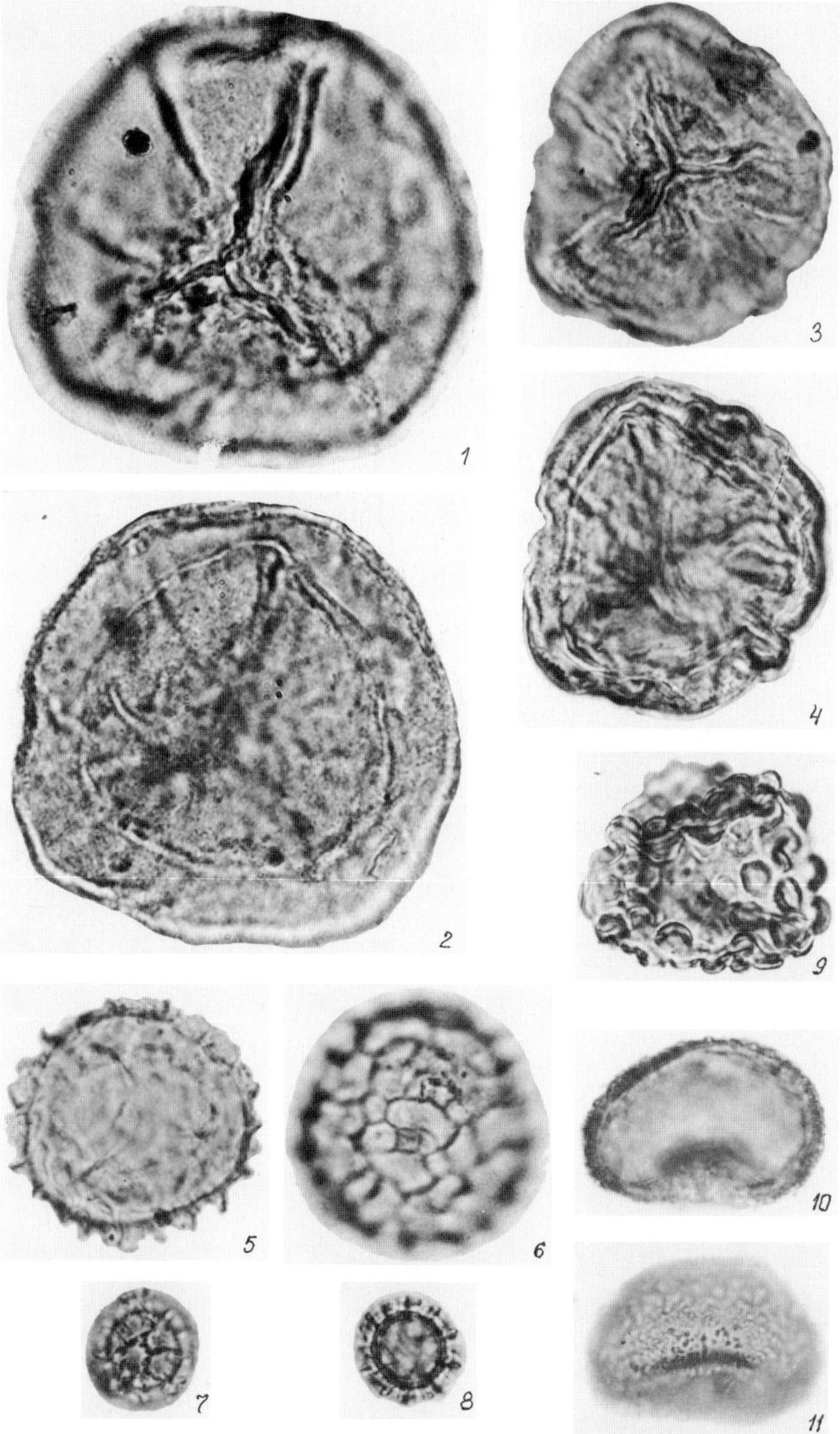

Pollen and spores of plants from the forage of
Mylakhchin Bison

PLATE IX. 1.) *Betula platyphilla* Sukacz.; 2.)
Alnus cf. *hirsuta* (Sphach) Turcz.
ex. Rupr.; 3.) *Ulmus* sp.; 4-6.) *Salix*
sp.; 7.) *Potentilla* sp.; 8, 9.) *Arctophila fulva* (Trin.) Anderss.; 10,
11.) Poa sp.; 14, 15.) Poaceae gen.
et. sp.; 12, 13.) *Typha latifolia* L.;
16.) Cyperaceae gen. et sp.; 17, 18.)
Carex sp.; 1000$^{\times}$.

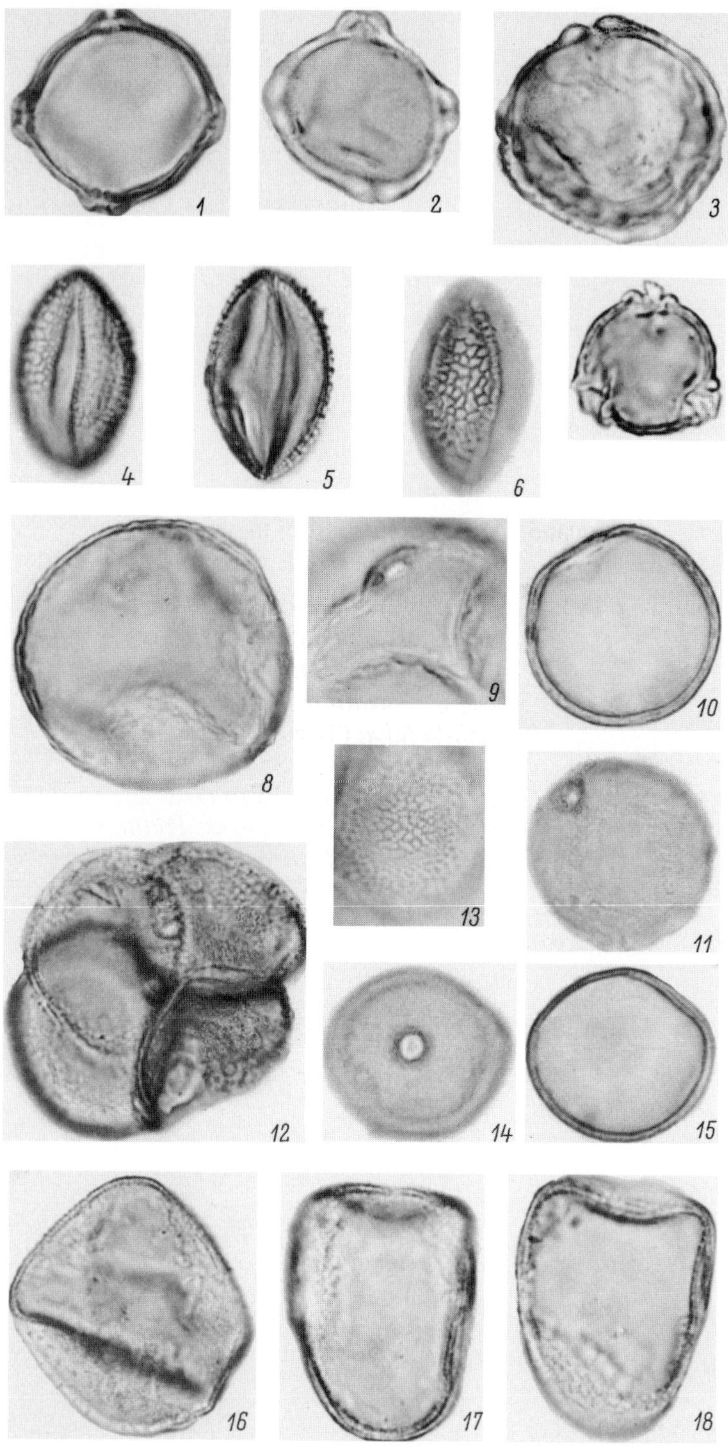

Pollen and spores of plants from the forage of
Mylakhchin Bison

PLATE X.   1, 2.) *Centiana tenella* Rottb.; 3, 4.)
Stellaria jacutica Schischk.; 5.)
Pelemonium acutiflorum Willd. ex
Roem. et Schult.; 6.) *Nardosmia*
sp.; 7.) *Lathyrus pilosus* Cham.; 8.)
Chenopodiaceae gen. et sp.; 9.)
Artemisia sp.; 10, 11.) Bryales sp.;
12.) *Sphagnum* sp.; 13.) cf. *Pottia*
sp.; 14.) *Polytrichum* sp.; 15.)
Bryales sp.; 16.) *Polypodium* sp.;
1000^{x}.

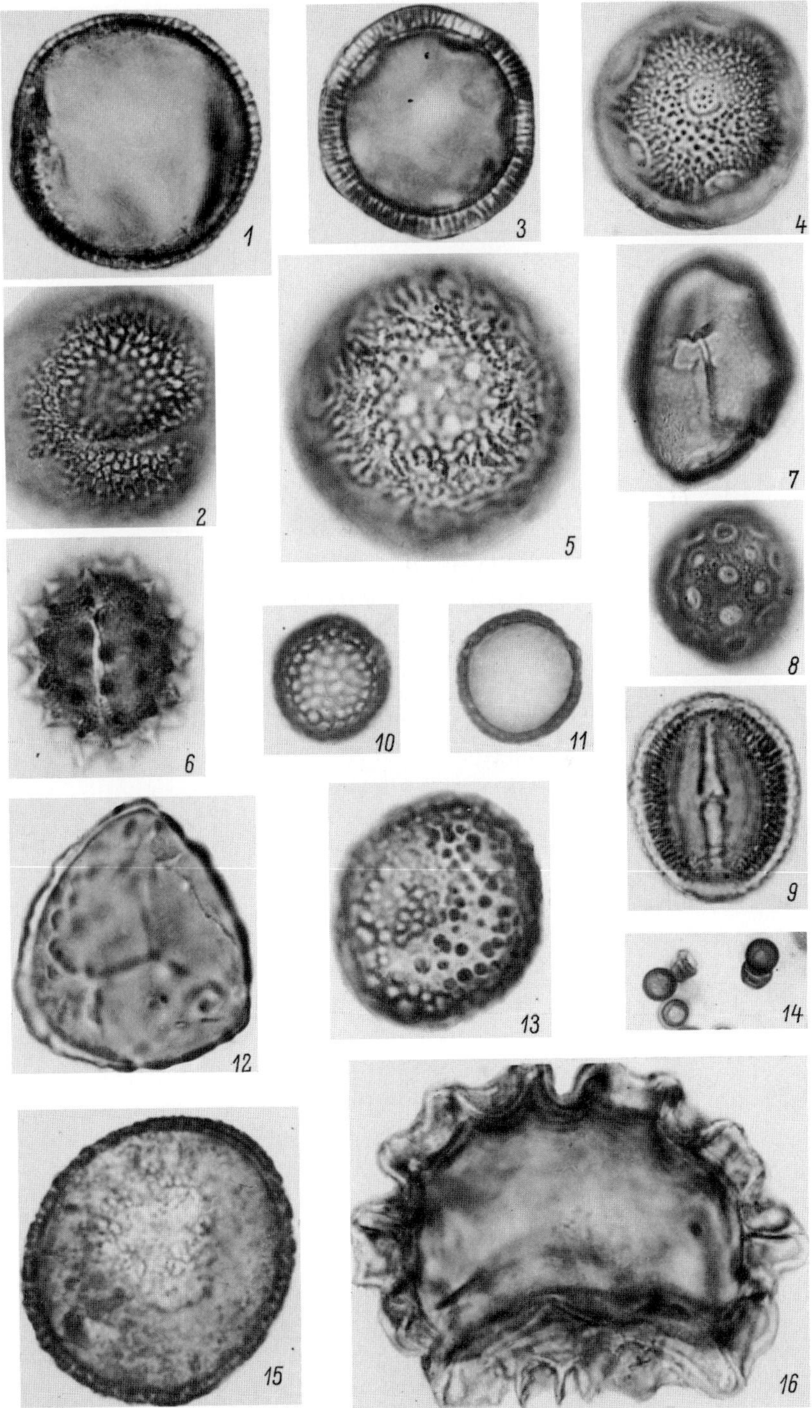

Pollen and spores of plants from the
gastrointestinal contents of Yuribei Mammoth

PLATE XI.   1.) *Larix* sp.; 2.) *Pinus sibirica* Du Tour.; 3, 4.) *Betula* sp. (sect. *Betula*); 5.) *Betula exilis* Sukacz.; 6-8.) *Betula* sp. (sect. *Nanae*); 9, 10.) *Salix* sp.; 11, 12.) *Salix glauca* L.; 13, 14.) *Alnus fruticosa* Rupr.; 1000^{x}.

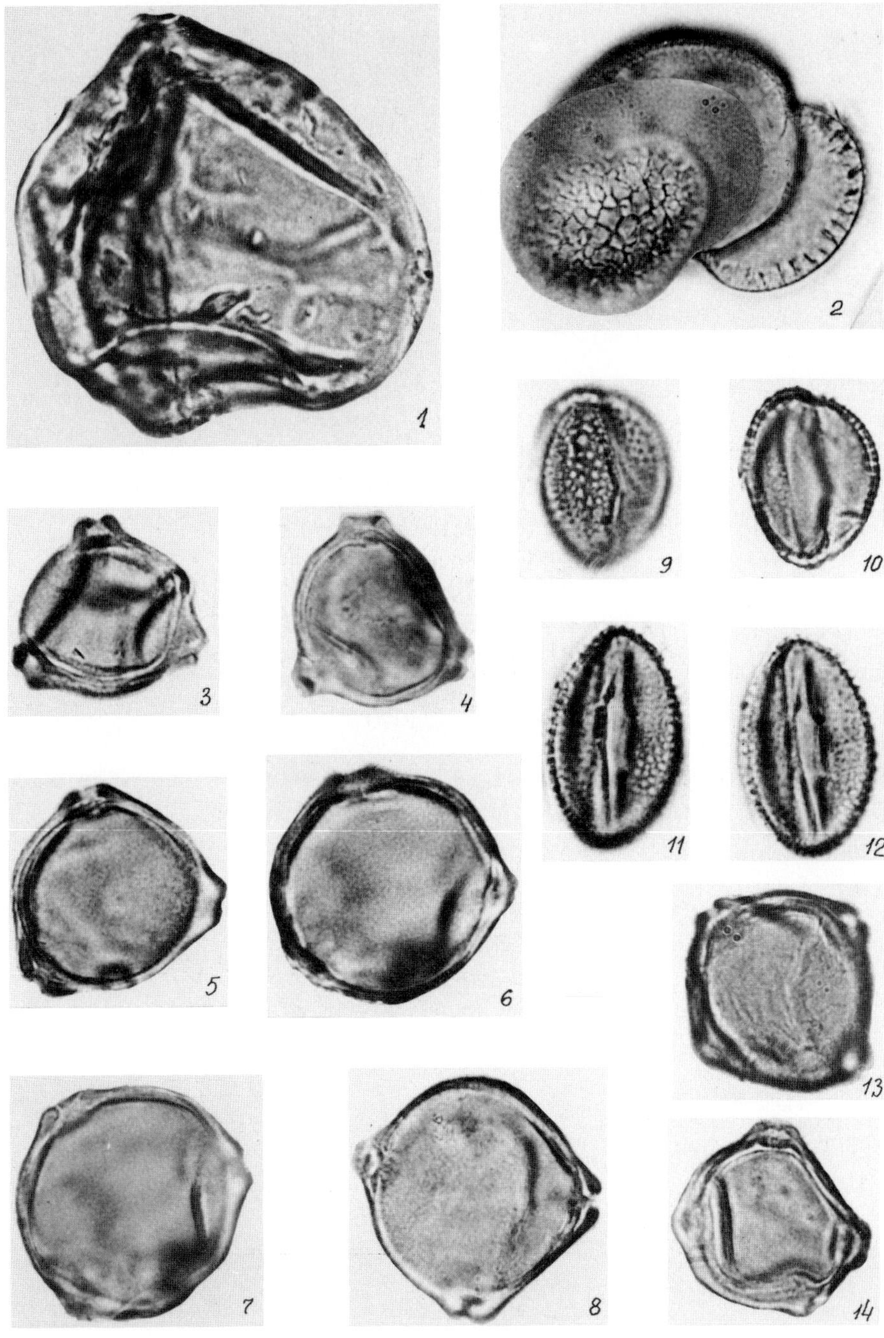

Pollen and spores of plants from the
gastrointestinal contents of Yuribei Mammoth

PLATE XII.   1.) *Carex* cf. *chordorrhisa* Ehrh.;
2.) Cyperaceae gen. et sp.; 3.)
Eriophorum cf. *polystachyon* L.; 4.)
Cyperaceae gen. et sp.; 5, 6, 8.)
Arctophila fulva (Trin.) Anderss.; 7,
9.) Poaceae gen. et sp.; 10, 11.)
Stellaria sp.; 1000$^{\times}$.

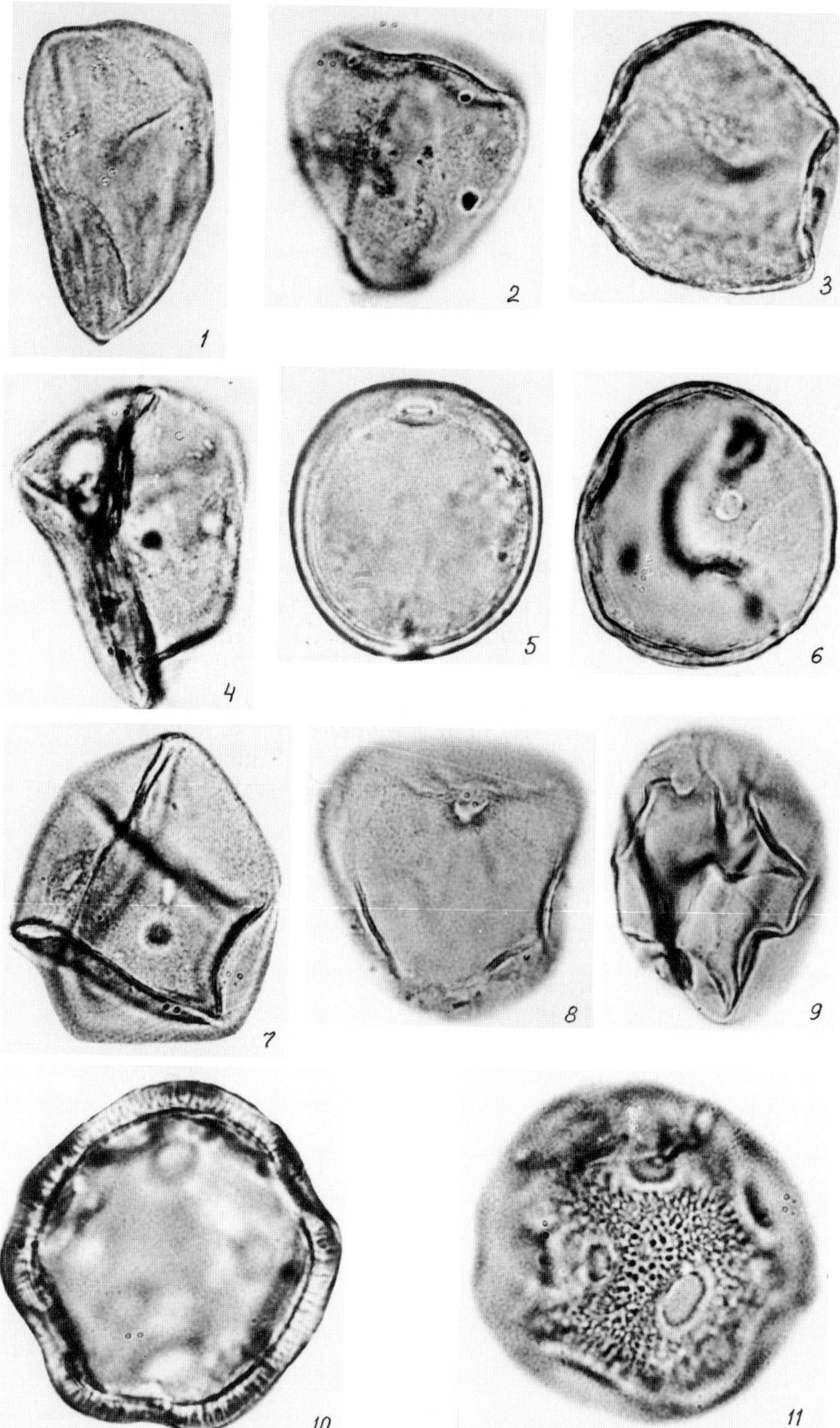

1
2
3
4
5
6
7
8
9
10
11

Pollen and spores of plants from the
gastrointestinal contents of Yuribei Mammoth

PLATE XIII.   1.) *Poa* sp.; 2, 3.) *Dryas* sp.; 4.)
Ranunculus sp.; 5, 6.) *Dryas* sp.;
7-9.) *Artemisia* sp. 1; 10-12.)
Artemisia sp. 1; 13.) *Equisetum* sp.;
14, 15.) *Lycopodium* sp.; 16, 17.)
Thalictrum alpinum L.; 1000$^{\times}$.

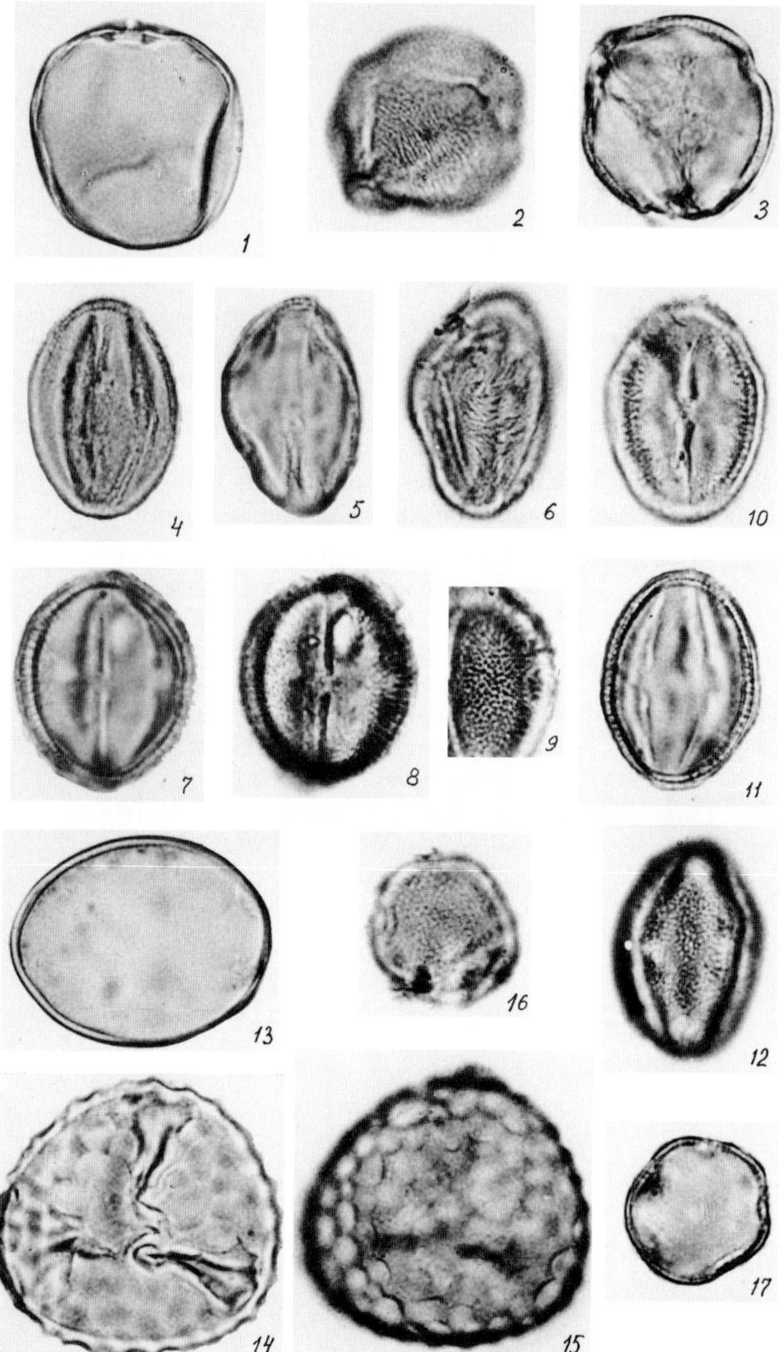

Pollen and spores of plants from the
gastrointestinal contents of Yuribei Mammoth

PLATE XIV.   1.) *Equisetum* sp.; 2, 3.) *Bryales* sp.
1; 4.) *Bryales* sp. 2; 5, 6.) *Bryales*
sp. 3; 7, 8.) *Bryales* sp. 4; 9, 10, 11,
12.) *Bryales* sp. 5; 18.) cf.
Drepanocladus sp.; 13.) *Equisetum*
sp.; 14, 15.) cf. *Calliergon* sp.; 16,
17.) *Bryales* sp. 6; 10.) *Bryales* sp. 7;
20.) *Bryales* sp. 8; 21.) *Bryales* sp.
9; 22, 23.) *Aulacomnium* cf.
turgidum (Wahl.) Schwaegr.; 24,
25.) *Sporites* indctcr.; 26, 27.)
Sporites inteter.; 1000^{x}.

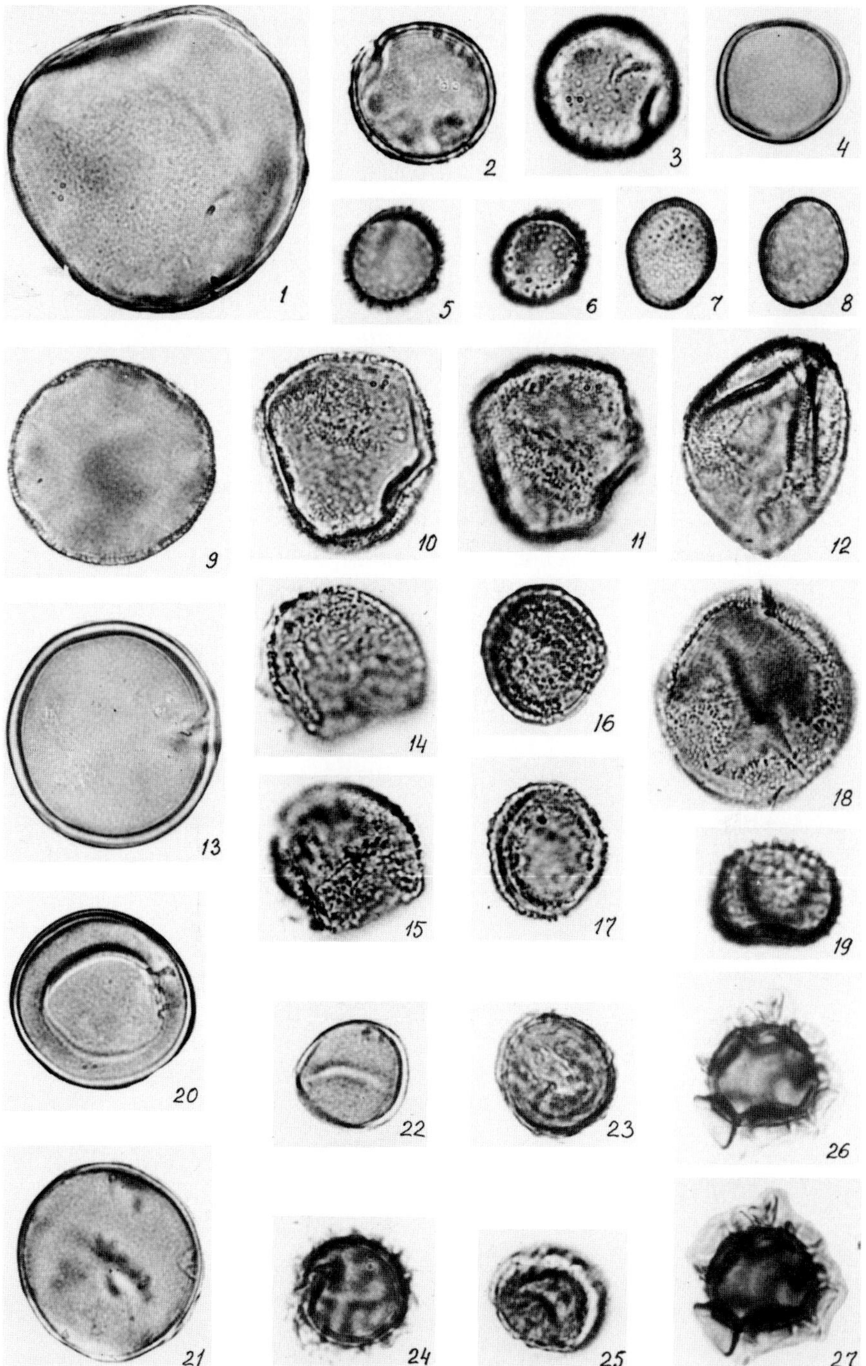

1
2
3
4
5
6
7
8
9
10
11
12
13
14
15
16
17
18
19
20
21
22
23
24
25
26
27

Pollen and spores of plants from the forage
of Mylakhchin Bison

PLATE XV.   1.) *Larix gmelinii* (Rupr.) Rupr.; 2,
3.) *Pinus sibirica* (Rupr.) Mayr.; 4.)
Betula sp. (sect. *Betula*); 5.) *Arctophila fulva* (Trin.) Anderss.; 6.)
Polypodium sp.; 1250^{x}.

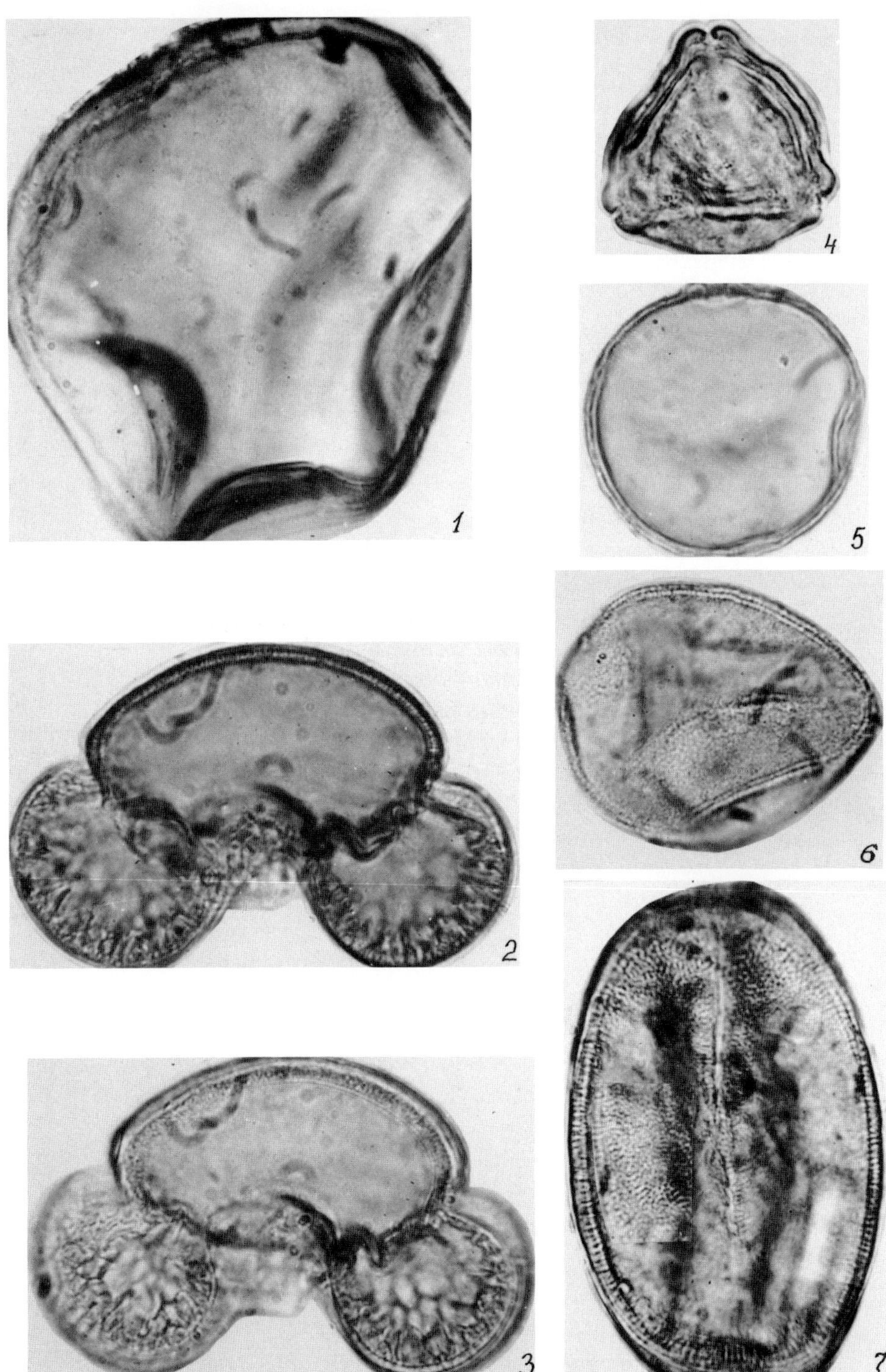

1
2
3
4
5
6
7

Pollen and spores of plants from the forage
of Mylakhchin Bison

PLATE XVI.   1, 2.) *Minuartia macrocarpa* (Pursh.) Ostenf.; 3-5.) *Stellaria jacutica* Schischk.; 6-9.) *Polemonium* sp.; 10, 11.) *Hedysarum hedysaroides* (L.) Schinz. et Thell.; 12-14.) *Draba* sp.; 15, 16.) *Ranunculus* sp.; 17.) *Sanguisorba officenalis* L.; 1250[x].

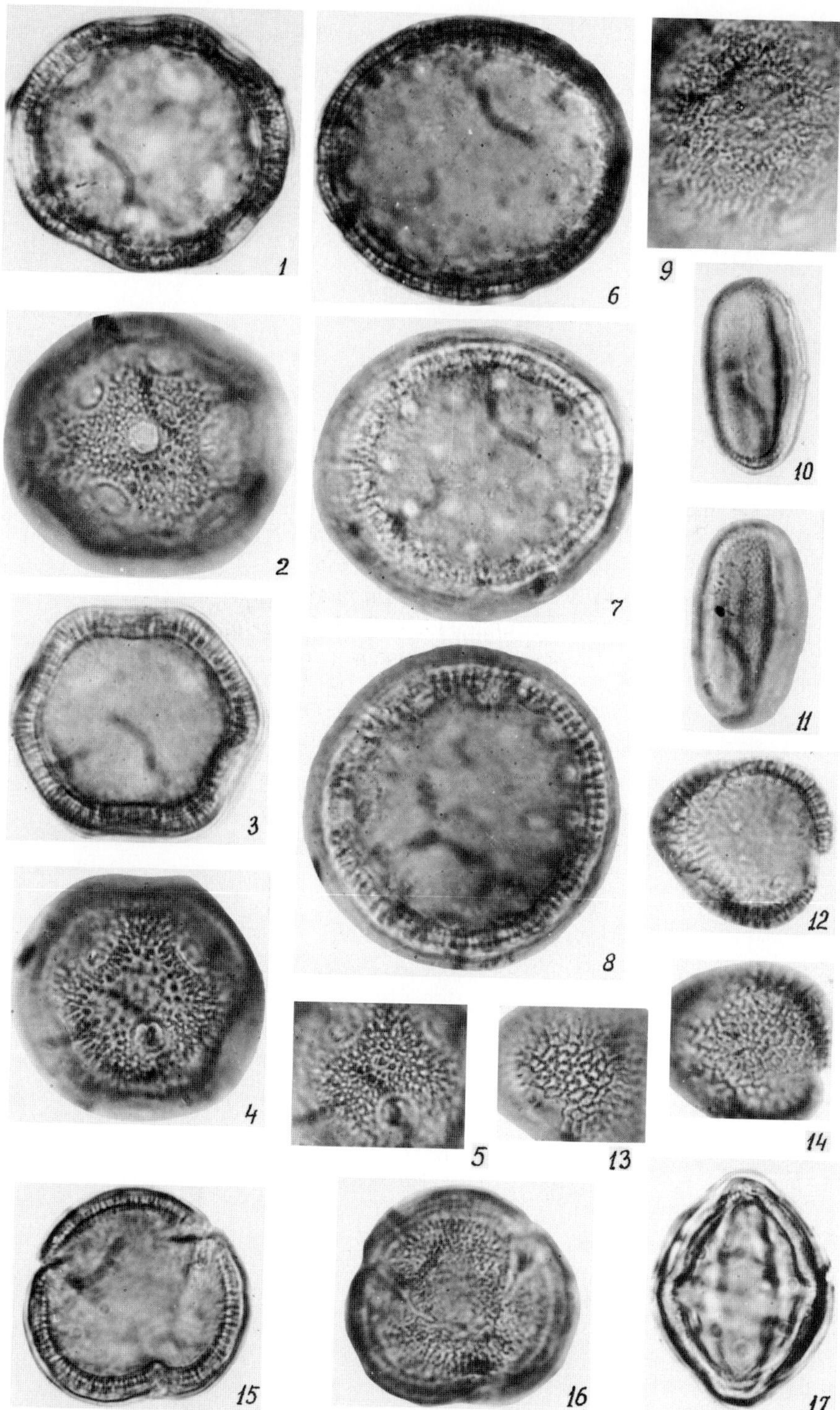

Pollen and spores of plants from the forage
of Mylakhchin Bison

PLATE XVII.   1-4.) *Centiana* sp.; 5-8.) *Valeriana*
capitata L.; 9, 10.) *Valeriana* sp.;
1250^x.

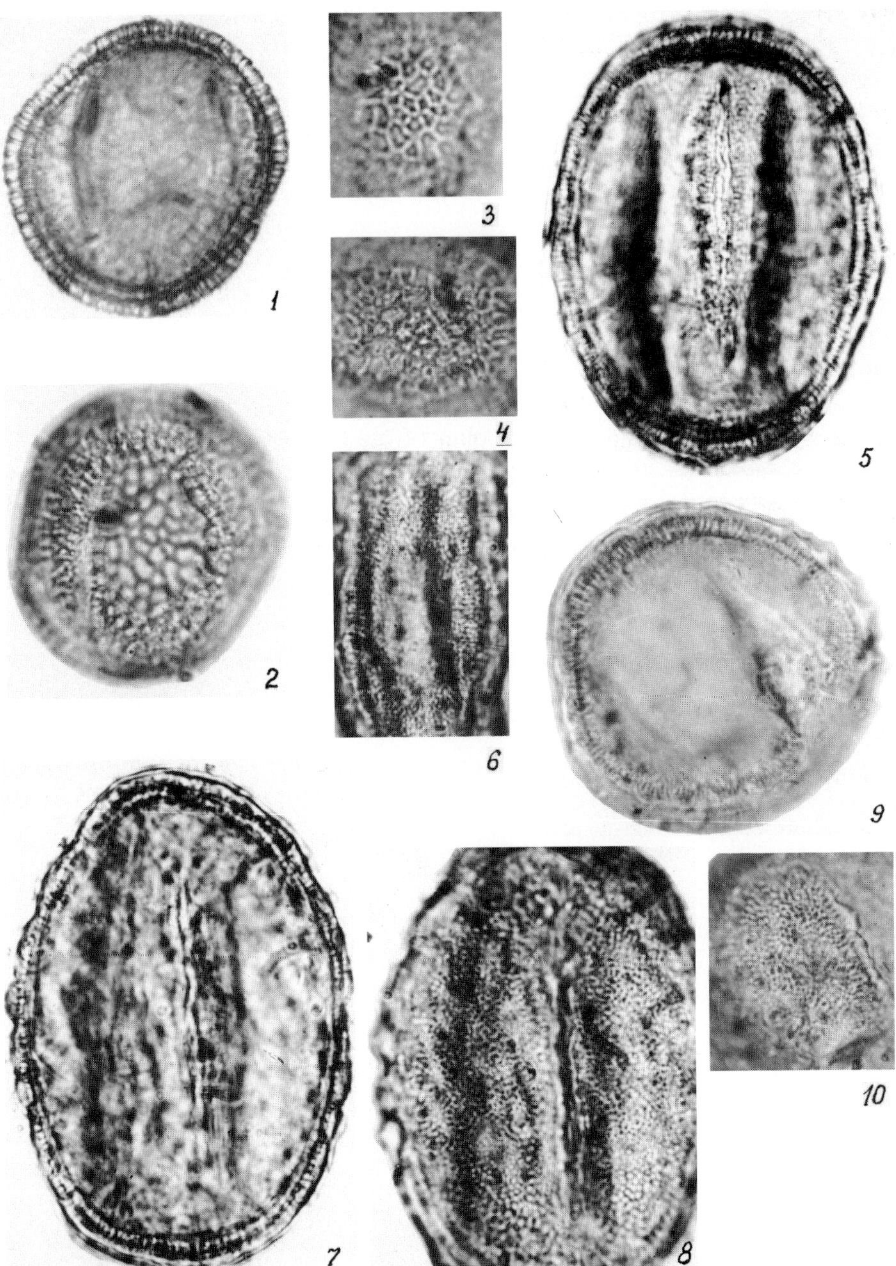

1
2
3
4
5
6
7
8
9
10

Pollen and spores of plants from the forage
of Mylakhchin Bison

PLATE XVIII.   1-4.) *Lactuca sibirica* (L.) Maxim.;
5, 6.) *Lactuca* sp.; 7, 8.) *Senecio
congestus* (R. Br.) DC.; 9, 10.)
Artemisia sp.; 11, 12.) *Artemisia* sp.;
1000^x.

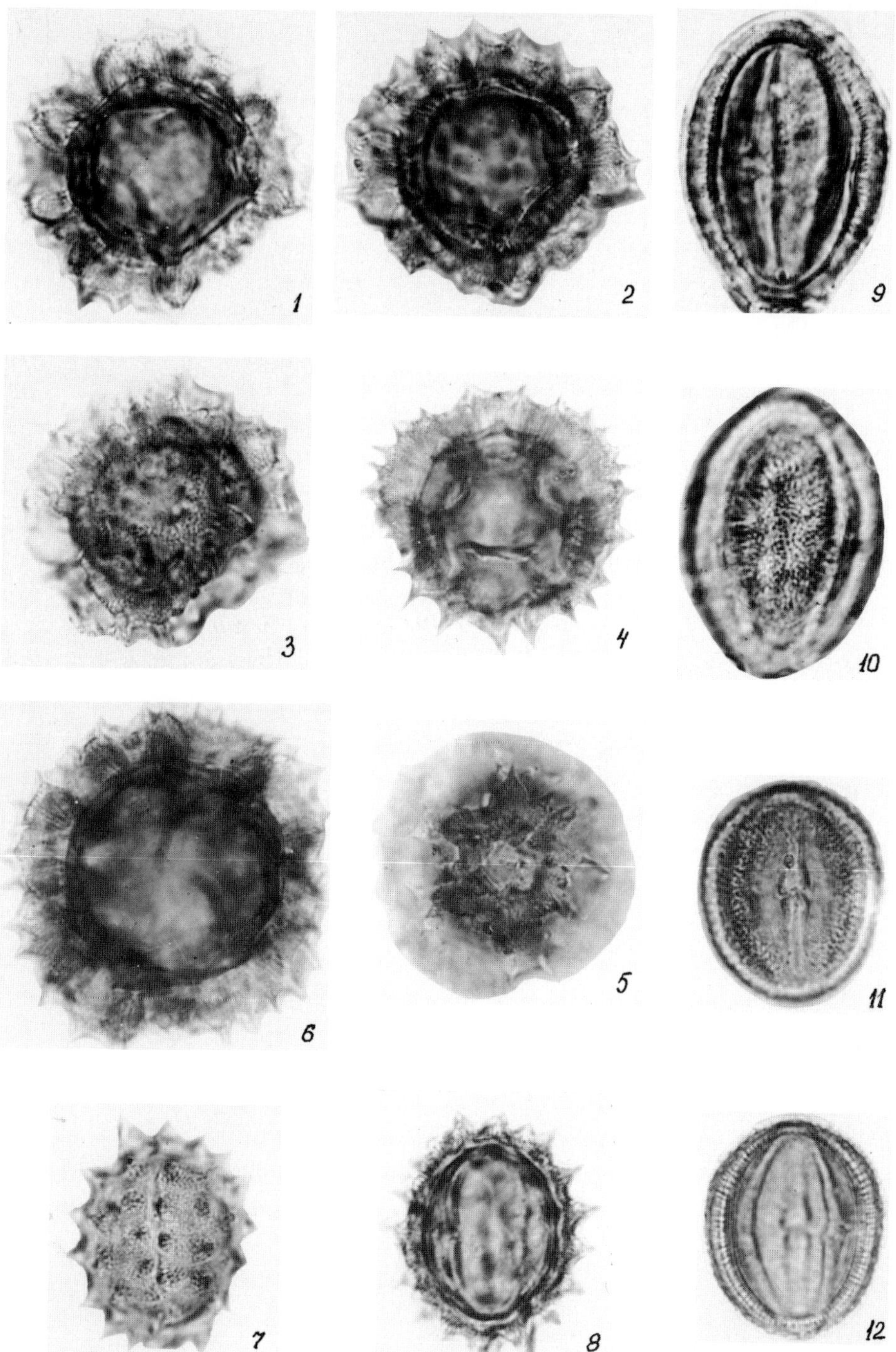

Pollen and spores of plants and forage
of Mylakhchin Bison

PLATE XIX.   1, 2.) *Selaginella selaginoides* (L.)
Link; 3, 4.) Polypodiaceae gen. et
sp.; 5, 6.) *Pottia* sp.; 7, 8.) *Bryales*
sp.; 9, 10.) *Dicranum* sp.; 11.)
Equisetum sp.; 1250[x].

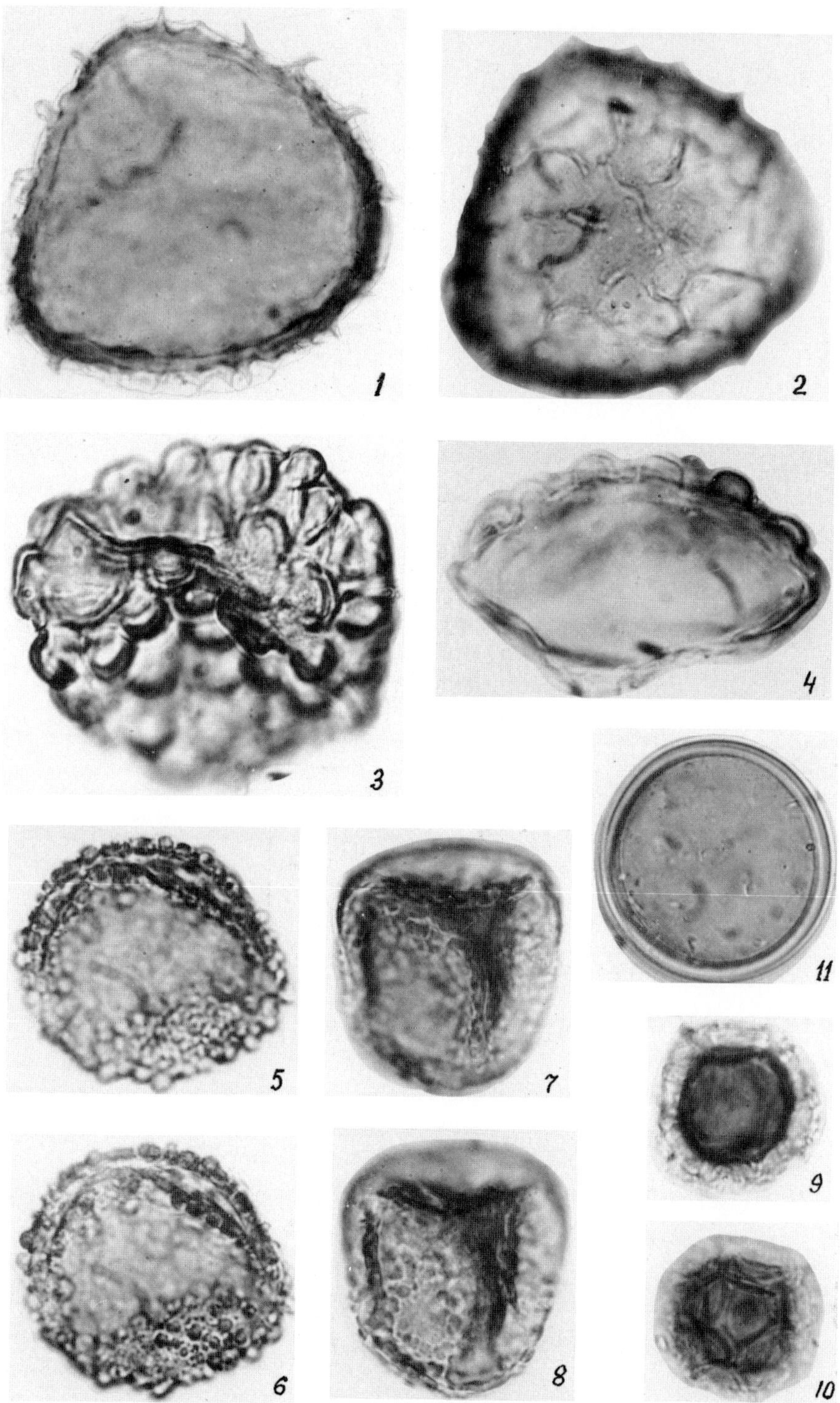

1
2
3
4
5
6
7
8
9
10
11

Pollen and spores of plants from the forage
of Mylakhchin Bison

PLATE XX. 1-4.) *Lactuca sibirica* (L.) Maxim.; 5-6.) *Lactuca* sp.; 7, 8.) *Senecio congestus* (R. Br.) DC.; 9, 10.) *Artemisia* sp.; 11-12.) *Artemisia* sp.; 1250^x.

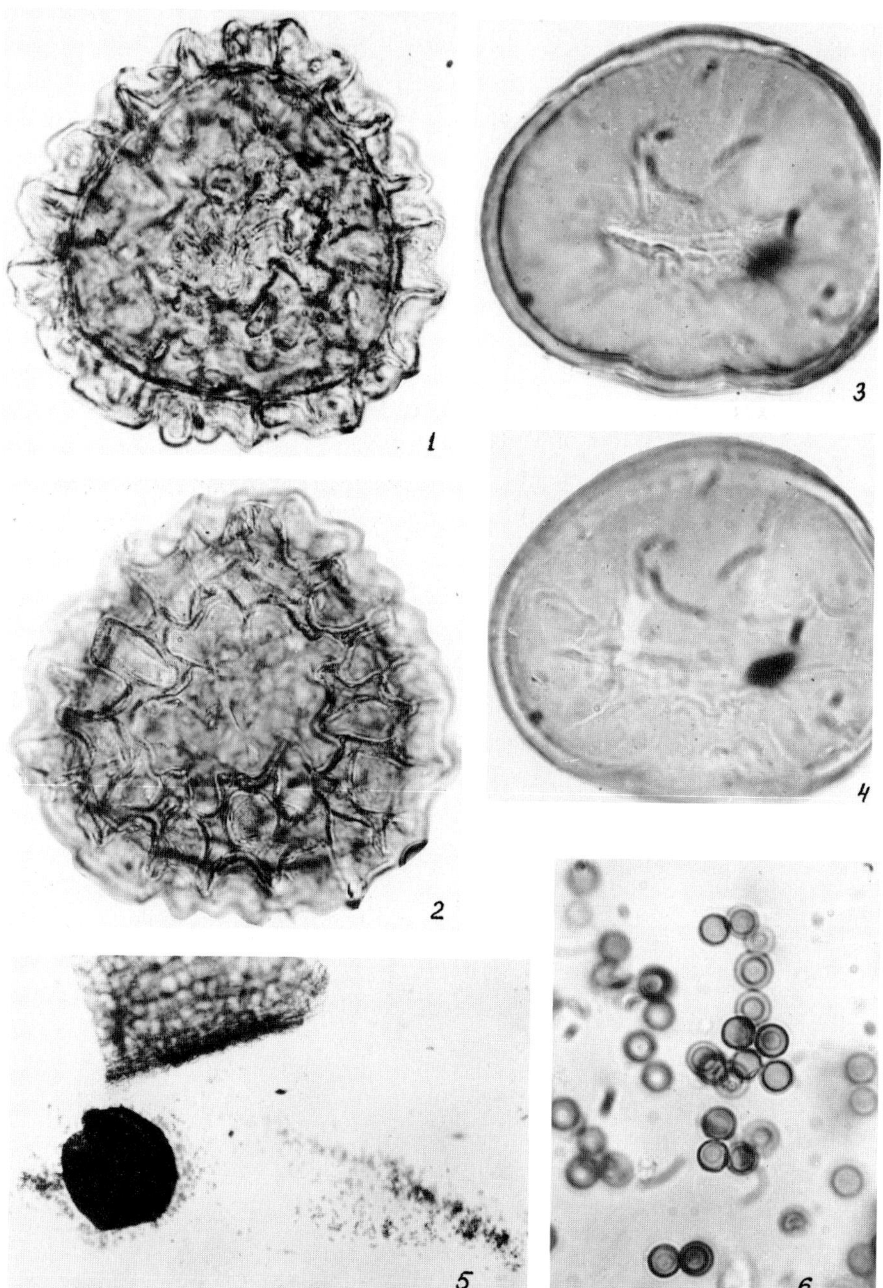

Index

The concept of any index is to facilitate the reader in finding certain words that lay hidden within the voluminous pages of text (in this case, 241 pages). This book is about both plants and animals, but by far, it is about the details concerning the plant communities in which certain selected large herbivores lived. More often than not, plants are discussed using their Latin name. In contrast, the animals are rarely referred to by their Latin name. We needed to reach a compromise. The author has organized the discussion of plants and plant communities by the animal locality; therefore, we felt that the reader wishing to find more information about a certain taxon of herbivore, to use the table of contents. The index below does not contain common or Latin names for animals. We felt that to list all the genera and species of plants discussed by Dr. Ukraintseva would have been overly cumbersome. What we have done is select certain tree and shrub species that most readers may want to be able to locate. This we have done only to the generic level. Although we have omitted indexing numerous herbaceous species and all of the graminoids, we hope that the reader will find the abbreviated index of use.